Chemical and Biological Microsensors

Chemical and Biological Microsensors

Applications in Liquid Media

Edited by
Jacques Fouletier
Pierre Fabry

First published 2003 in France by Hermes Science/Lavoisier entitled: *Microcapteurs chimiques et biologiques : applications en milieu liquide* © LAVOISIER 2003
First published 2010 in Great Britain and the United States by ISTE Ltd and John Wiley & Sons, Inc.

Apart from any fair dealing for the purposes of research or private study, or criticism or review, as permitted under the Copyright, Designs and Patents Act 1988, this publication may only be reproduced, stored or transmitted, in any form or by any means, with the prior permission in writing of the publishers, or in the case of reprographic reproduction in accordance with the terms and licenses issued by the CLA. Enquiries concerning reproduction outside these terms should be sent to the publishers at the undermentioned address:

ISTE Ltd
27-37 St George's Road
London SW19 4EU
UK

www.iste.co.uk

John Wiley & Sons, Inc.
111 River Street
Hoboken, NJ 07030
USA

www.wiley.com

© ISTE Ltd 2010

The rights of Jacques Fouletier and Pierre Fabry to be identified as the authors of this work have been asserted by them in accordance with the Copyright, Designs and Patents Act 1988.

Library of Congress Cataloging-in-Publication Data

Chemical and biological microsensors : applications in fluid media / edited by Jacques Fouletier, Pierre Fabry.
   p. cm.
  "First published in France in 2003 by Hermes Science/Lavoisier entitled: Microcapteurs chimiques et biologiques : applications en milieu liquide."
  Includes bibliographical references and index.
  ISBN 978-1-84821-142-1
  1. Chemical detectors. 2. Biosensors. 3. Microfluidics. I. Fouletier, Jacques. II. Fabry, Pierre.
  TP159.C46C387 2009
  681'.2--dc22

2009039567

British Library Cataloguing-in-Publication Data
A CIP record for this book is available from the British Library
ISBN 978-1-84821-142-1

Printed and bound in Great Britain by CPI Antony Rowe, Chippenham and Eastbourne

Table of Contents

Foreword. xi

Chapter 1. General Features 1
Bernard MICHAUX

    1.1. Definitions 1
        1.1.1. Sensors 1
        1.1.2. Qualities of measuring sensors 2
        1.1.3. Chemical and biochemical sensors. 5
    1.2. Classification. 6
        1.2.1. Mass variation. 6
        1.2.2. Optics. 7
        1.2.3. Electrical parameters. 8
    1.3. Specific problems of chemical sensors. 14
        1.3.1. pH measurement: Nernst equation for the glass electrode 14
        1.3.2. Ionometry 18
        1.3.3. Conclusion concerning chemical sensors 20
    1.4. Advantages and drawbacks of chemical microsensors 21
    1.5. Perspectives 22
    1.6. Bibliography 23

Chapter 2. Chemical Sensors: Development and Industrial
Requirements 25
Jacques FOULETIER and Pierre FABRY, based on discussions with
Jacques FOMBON

    2.1. Introduction. 25
        2.1.1. Overview of the various types of chemical sensors 25
        2.1.2. General trends. 26

2.2. Modern research and development (R&D) management methods
applied to sensors.................................... 26
  2.2.1. Preliminary phases in the industrialization of a new sensor..... 26
  2.2.2. Development and industrialization.................... 31
  2.2.3. Information and establishment of the commercial network..... 32
2.3. Applications and inventory of the needs.................... 33
  2.3.1. pH sensors.................................. 33
  2.3.2. Potentiometric selective electrodes................... 33
  2.3.3. Amperometric specific electrodes..................... 36
  2.3.4. Conductimetric sensors........................... 36
  2.3.5. Biosensors................................... 36
  2.3.6. Biomedical applications.......................... 37
2.4. New needs and industrial applications..................... 37
  2.4.1. Creation of new sensors.......................... 38
  2.4.2. Reliability and absence of maintenance................. 38
  2.4.3. Miniaturization and robustness...................... 38
  2.4.4. Lowering the production and use costs.................. 39
2.5. The sensor in the measuring chain....................... 39
  2.5.1. General features............................... 39
  2.5.2. Quality and metrology of the standards and of the instruments... 40
2.6. Conclusions and prospects............................ 43
2.7. Bibliography..................................... 43

Chapter 3. Sensitivity and Selectivity of Electrochemical Sensors...... 45
Pierre FABRY and Jean-Claude MOUTET, and translated by J.C. POIGNET
and Pierre FABRY

3.1. General concepts.................................. 45
  3.1.1. Various kinds of electrochemical sensors................. 46
  3.1.2. Interference and selectivity........................ 49
  3.1.3. Nature and shape of materials....................... 50
3.2. Models for the sensitivity and selectivity of potentiometric
sensors.. 51
  3.2.1. Basic concepts................................. 51
  3.2.2. Ionic conducting membranes of the first kind............. 55
  3.2.3. Ionic conducting membranes of the second kind............ 59
  3.2.4. Liquid and organic membranes....................... 63
3.3. Case of amperometric sensors.......................... 64
  3.3.1. Principle of sensitivity........................... 64
  3.3.2. Selectivity model............................... 65
3.4. Molecular recognition and sensors....................... 68
3.5. Characterization methods............................. 70
  3.5.1. Definition and determination of the detection limit.......... 70

3.5.2. Determination methods of selectivity coefficients	73
3.6. Bibliography	77

Chapter 4. Potentiometric Sensors (Ions and Dissolved Gases) — 81
Annie PRADEL and Eric SAINT-AMAN

4.1. Introduction	81
4.1.1. General features	81
4.1.2. Electrode potential	82
4.1.3. Sensitivity of the potentiometric sensors	85
4.1.4. Selectivity of the potentiometric sensors	87
4.2. Membranes	88
4.2.1. General features	88
4.2.2. Glass membranes	89
4.2.3. Crystallized inorganic membranes	95
4.2.4. Polymeric membranes	97
4.3. Current developments in potentiometric sensors	99
4.3.1. All-solid-state sensors	99
4.3.2. All-solid-state microsensors	105
4.4. Bibliography	109

Chapter 5. Amperometric Sensors — 115
Alain WALCARIUS, Chantal GONDRAN and Serge COSNIER

5.1. Sensors based upon chemically modified electrodes	115
5.1.1. Introduction	115
5.1.2. Fabrication and characterization	119
5.1.3. Fundamental principles and examples of application	127
5.2. Amperometric biosensors	138
5.2.1. Introduction	138
5.2.2. Immobilization of biomolecules	140
5.2.3. Amperometric biosensors, principle and description	142
5.3. Bibliography	160

Chapter 6. ISFET, BioFET Sensors — 173
Nicole JAFFREZIC-RENAULT and Claude MARTELET, translated by
Claude MARTELET

6.1. Structure of ISFET sensors	173
6.1.1. Introduction	173
6.1.2. MOS (metal-oxide semiconductor) structure	174
6.1.3. EOS (electrolyte-oxide semiconductor) structure	177
6.1.4. MOSFET	178
6.1.5. ISFET	179
6.2. Techniques used for ISFET fabrication and operation	180

6.2.1. ISFET fabrication.................................. 180
6.2.2. ISFET Measurement set-up.......................... 182
6.3. ISFET membranes...................................... 183
  6.3.1. Detection of H^+ ions.......................... 183
  6.3.2. Detection of other ions.......................... 184
6.4. Detection of molecular species....................... 187
  6.4.1. Metabolic biosensors............................. 187
  6.4.2. The enzymatic field-effect transistor (ENFET) principle .. 188
  6.4.3. Some ENFET examples.............................. 189
  6.4.4. ENFET and inhibition mechanisms.................. 192
6.5. BioFETs.. 193
  6.5.1. Systems based on affinity mechanisms............. 193
  6.5.2. BioFET based on cells and living organisms....... 195
6.6. Commercial devices................................... 197
  6.6.1. pH ISFETs.. 197
  6.6.2. Multidetection systems........................... 200
6.7. Conclusion and perspectives.......................... 201
6.8. Bibliography... 202

Chapter 7. Biosensors and Chemical Sensors Based Upon Guided Optics. 209
Jean-Pierre GOURE and Loïc BLUM

7.1. Introduction... 209
7.2. Definitions.. 210
  7.2.1. Luminous wave.................................... 210
  7.2.2. Optical fibers................................... 211
  7.2.3. Planar guides.................................... 212
7.3. Principles of optical microsensors................... 213
  7.3.1. Definition....................................... 213
  7.3.2. Modulation of the optical signal................. 214
  7.3.3. Techniques....................................... 218
  7.3.4. Refractometry.................................... 219
7.4. Optical fiber biosensors............................. 220
  7.4.1. Configurations of optical fiber biosensors....... 221
  7.4.2. Chemical sensors integrated in optical fiber sensors 221
  7.4.3. Optical fiber enzymatic biosensors............... 223
  7.4.4. Biosensors with non-catalytic bioreceptors (affinity biosensors).. 225
  7.4.5. Chemiluminescence and bioluminescence detection sensors 227
7.5. Perspectives and conclusions......................... 229
7.6. Bibliography... 229

Chapter 8. Sensors and Voltammetric Probes for *In Situ* Monitoring of Trace Elements in Aquatic Media 233
Marie-Louise TERCIER-WAEBER and Jacques BUFFLE

   8.1. Introduction. 233
   8.2. Basic principles of the voltammetric techniques and of their applications to analysis of water. 235
      8.2.1. Components and principles 235
      8.2.2. Influence of the transport properties of the electroactive species on the voltammetric signal 239
      8.2.3. Influence of the speciation of the electroactive compounds on the voltammograms 241
   8.3. Voltammetric techniques used for the analysis of trace elements in waters 244
      8.3.1. Sensitivity limit of the voltammetric techniques. 244
      8.3.2. Various voltammetric techniques 245
      8.3.3. Voltammetric determinations of natural samples in the laboratory or on the field 245
   8.4. Development of reliable submersible voltammetric probes. 247
      8.4.1. Working electrodes. 247
      8.4.2. Reference electrodes 252
      8.4.3. Voltammetric cells 255
      8.4.4. Interference due to the dissolved oxygen 257
      8.4.5. Interferences due to the adsorption of organic and inorganic compounds 259
      8.4.6. Gel-integrated microsensors 260
   8.5. Submersible voltammetric probes reported in the literature ... 264
      8.5.1. Continuous-flow probe based on a microelectrode and a pre-treatment of the sample. 264
      8.5.2. Continuous-flow probe based on a macro- or a microelectrode with no sample treatment 264
      8.5.3. Probes based on direct immersion of the electrodes. ... 265
      8.5.4. Continuous-flow probe based on a gel-integrated microsensor with no sample pretreatment: the VIP system. 267
   8.6. Conclusion 273
      8.6.1. Calibration of voltammetric procedures. 274
      8.6.2. Development of robust and reliable sensors and probes . 275
   8.7. Bibliography 275

Chapter 9. Chemometrics 287
Philippe BREUIL

   9.1. Introduction. 287
      9.1.1. The Problem of multivariate analysis 288

    9.1.2. Example: Beer-Lambert law of light absorption. 289
    9.1.3. General method. 290
  9.2. A particular case: the linear case 290
    9.2.1. Notations and preliminary considerations 290
    9.2.2. Simple least square methods 291
    9.2.3. Factor analysis 296
  9.3. Least squares methods: non-linear case 302
    9.3.1. Case when transformations can reduce the problem to linear
    functions. 302
    9.3.2. PLS can model non-linear phenomena 303
  9.4. Neural networks 303
    9.4.1. General structure of the network 304
    9.4.2. Learning (i.e. calibration) 304
    9.4.3. Prediction 305
  9.5. Conclusion 305
  9.6. Bibliography 306

Chapter 10. Impedancemetric Sensors 307
Jacques FOULETIER and Pierre FABRY

  10.1. Introduction. 307
  10.2. Fields of application. 307
  10.3. Conductivity of liquid media. 310
    10.3.1. Theoretical basis. 310
    10.3.2. Effect of temperature 312
  10.4. Impedance of first kind cell (direct measurement) 313
  10.5. Cell configurations and sources of error 317
    10.5.1. Types of conductivity cells 317
    10.5.2. Characteristics – specifications 321
  10.6. Second kind cells. 326
  10.7. Summary of practical precautions. 328
  10.8. Bibliography 329

List of Authors 331

Index 335

Foreword

The present book, devoted to sensors for ions and gaseous species in solution, deals with the theoretical basic concepts which are necessary for a better understanding, and consequently, for a better use of such sensors, and also presents the state-of-art and current developments. In order to provide a comprehensive overview of the subject, the authors are expert researchers and industrialists in each field.

Chapter 1 is devoted to a classification of chemical sensors, with a particular emphasis on electrochemical sensors. It provides some definitions and basic concepts, focusing on metrological characteristics and on major problems, specific to chemical sensors, which must be taken into account by the users.

In Chapter 2, the essential aspects of the industrial development of new sensors are discussed. The designers are not usually familiar with their complexity, as, for example, in the case of the qualification procedure of a new product, with regard to the specifications required by the future users.

Chapter 3 deals, from a general point of view, with electrochemical sensors for which the electrochemical reaction ensures the selectivity and the sensitivity of the device, but also the transduction to an electrical signal. The fundamental aspects are discussed and the methods of determination of the detection limit and of the selectivity coefficients are developed.

The two following chapters are devoted to a description of potentiometric and amperometric sensors used for the analysis of ionic species and dissolved gases. In Chapter 4, the main characteristics of specific electrodes are presented. The emphasis is on ion-sensitive membranes, which are ionic conductors in crystallized phases, glasses, or polymers. The main current developments of such sensors are also discussed. Amperometric sensors, which are the subject of Chapter 5, are used for the detection of dissolved species (using chemically modified electrodes) or for the analysis of biomolecules by immobilization of a biologic species on a transducer.

Although ISFET, i.e. ion-sensitive field effect transistors, are advantageous in the case of the ionic detection in comparison with classical chemical sensors, only the pH-ISFET has had commercial success up to now. The current situation is described in Chapter 6, discussing the developments led by multi-detection requirements, for instance, in biomedical applications with a strong demand for disposable and miniaturized multisensors.

The wide range of possibilities offered by sensors using optical fibers are presented in Chapter 7. The needs in instrumentation are tremendous: robotics, automatic manufacturing, aeronautics, automotive industry, etc. They are essentially optical biosensors, which have been developed recently and will certainly be industrially developed in the near future.

Chapter 8 discusses the characteristics of sensors and voltamperometric gauges for continuous *in situ* measurements of concentrations and electrochemical properties of trace species, especially metallic species in natural media. It is a domain where electrochemistry yields very interesting performances.

Chapter 9 is devoted to chemometrics, which enables the extraction of relevant and useful information from several pieces of physicochemical raw data. It is a relatively recent tool, rarely used in applications where it is coupled with sensors, but it also promises to have real development potential.

Finally, impedancemetric sensor devices are described in Chapter 10, where the essential concepts are discussed to guide users in their own applications, and therefore, to avoid errors due to a bad choice of the measuring parameters. Such sensors are traditional; nevertheless, several examples will be given, illustrating recent research and developments.

Jean-Claude Poignet, past Professor in Electrochemistry (Grenoble Institute of Technology), has translated most of the chapters of the book. His contribution has been well beyond that of simple translation; above all, the book owes a great deal to stimulating discussions we have had with him over the years. Finally, Véronique Ghetta is warmly acknowledged for her assistance in the preparation of the manuscript.

Jacques FOULETIER and Pierre FABRY

Chapter 1

General Features

1.1. Definitions

Before focusing on chemical sensors, and particularly on chemical microsensors, it is worth trying to define, or at least clarify the notion of a sensor and the qualities expected from such a device (which are the same whichever the measured property may be, either physical or chemical).

1.1.1. *Sensors*

The most frequently encountered definitions of a sensor are based on that given by ISA (Instrument Society of America at the time, now Instruments, Systems & Automation Society) in its ANSI MC6.1 standard of 1975, "Electrical Transducer Nomenclature and Terminology" [ISA 75], that is to say that a transducer, or sensor, is "a device that provides a usable output in response to a specified measurand". This standard specifies that the property is physical and the output electrical. This definition is generalized by including chemical properties and outputs usable by modern measuring chains (which, at term, may mean an optical output). Other definitions are possible, such as that of the AFNOR NF X 07-001 standard (December 1984) [AFN 84], in which the sensor is "a part of a measurement apparatus or of a measuring chain to which the property to be measured is directly applied". The ISA definition, which is much less restrictive, will be used here, but it is important to keep in mind that the definition of the sensor is not really unambiguous.

Chapter written in French by Bernard MICHAUX.

If this definition is rather easy to use in the field of physical properties (pressure, temperature, forces, acceleration, etc.), where the sensing element measures a well-defined property, it is far less easy in the field of chemical properties (concentration or activity of a chemical species in solution, partial pressure of a gas in air) because the influence of parameters different from the property to be measured (*matrix effect*) is both more important and more difficult to appreciate (therefore, to correct for) than in the case of physical properties.

In addition, a second difficulty arises from the existence and use, for similar aims, of more complex devices, generally called analyzers (chromatographs, spectrometers, etc.), which are able to measure several parameters simultaneously. However, it is difficult to define a clear limit between sensors and analyzers: the reason for this is that, because of the importance of matrix effects (which must be corrected for), the sensor will often also comprise measuring devices for additional parameters, different from the principal parameter, which would provide it with the qualities of an analyzer, according to the definition given previously.

For the needs of this chapter, the term chemical property sensor (improperly called chemical sensor for the sake of simplicity) will mean a simple device intended to identify and measure a single chemical species in a more or less complex matrix. However, it is worth stressing that this definition is of variable geometry according to the measuring conditions and must always be used with caution.

1.1.2. *Qualities of measuring sensors*

In this section the characteristics that qualify the behavior of a sensor and determine whether it is a "good" or "bad" sensor will be discussed. Of course these judgments are not absolute and it is fundamental that the qualities of the sensor are valued considering the nature of the measured entity and the measurement conditions. It should be noted that some ambiguity does exist, caused by the frequent use of the same terms to qualify both the behavior of the sensor and the quality of the measurement delivered: here again, common sense helps in solving possible difficulties of terminology.

These qualities are defined in the same way for physical and chemical sensors. However, the latter present some specificities, which will be briefly addressed in this section and, for some of them, developed further in this chapter. The definitions given are adapted from the above-cited NF X 07-001 standard, which can be referred to for further details (the list given later does not claim to be exhaustive and concerns mainly the qualities to take into account chemical sensors in particular).

1.1.2.1. *Accuracy*

This is the narrowness of the difference between the result of the measurement and the true value (unknown, but taken by convention) of the measured property. It is sometimes wrongly called "precision". This parameter is interesting when a property is measured as opposed to simply monitoring the variation in a property.

The difference between the true value and the measured value is called "absolute error". The ratio of the absolute error to the true value is the relative error. It is obvious that the values of both parameters cannot be known as the true value is unknown. Therefore, we use their superior limit, called respectively "absolute uncertainty" and "relative uncertainty", denoted Δx and $\Delta x/x$.

Accuracy can vary during the lifetime of an instrument. According to the continuous or random nature of this variation, we use the terms "drift" or "instabilities" of the sensor (or of the measurement).

1.1.2.2. *Resolution*

Resolution is the smallest increment of the measured value that can be significantly measured by the sensor. This parameter (which is limited by the whole noise of the sensor and the measuring chain) is important mainly if the sensor is used for monitoring the variations of a property or as a "zero apparatus" (element of a regulation loop intended to maintain the measured property at a constant value).

1.1.2.3. *Fidelity (or repeatability)*

This parameter is probably the most important, for physical sensors, as well as for chemical sensors. It is defined in the standard as "the aptitude of an instrument to yield, in given conditions, very close responses to repeated applications of the same input signal". In other words, the fidelity determines the degree of confidence that in the sensor for a given use: a low or very low fidelity precludes the use of a device for measurement applications.

Another kind of repeatability, which is not taken into account by the standard (which rather concerns physical measurements), is significant in the case of chemical sensors: it is what could be called "repeatability in a batch", i.e. the repeatability between different sensors manufactured in the same way and presumed to have identical responses. This parameter is of high practical importance: it necessitates whether or not each sensor has to be calibrated separately, which obviously causes the purchase price or the cost-in-use to rise (depending on whether the calibration is performed by the manufacturer or by the user).

In the case of the chemical sensors, the importance of batch repeatability is linked to the increasing complexity (as compared with physical sensors) of the measurement principles implemented. The number of elements involved makes the achievement of the "all things being equal", (i.e. composition of the medium and all parameters) more difficult, which is an indispensable condition for good repeatability for each sensor, but also between sensors.

The problems linked to the importance and to the difficulties of calibration of the sensors will be addressed later in this chapter.

1.1.2.4. *Sensitivity*

The sensitivity is the ratio of the increase in the response of an instrument to a corresponding increase of the input signal.

1.1.2.5. *Hysteresis*

Hysteresis is the property of a measuring instrument whereby the instrument has a response to a given input signal, and this response depends on the sequence of the previous input signals. Such a property is, quite obviously, highly undesirable. In the field of sensors, the inverse property, called reversibility, is often used, whereby neither the sensor's zero nor its sensitivity are modified by the parameter to be measured (in a range defined by the manufacturer).

Hysteresis (or lack of reversibility) can be ascribed to the measured property, but also to an important modification of influence parameters.

1.1.2.6. *Selectivity*

Selectivity is the ability of the sensor to measure only one quantity, or, in other words, to be as immune as possible towards influence parameters, which are not to be measured, but act upon the output of the sensor.

In the case of physical sensors, where the effects of various parameters can be easily separated, this parameter is generally good. It is generally much worse for chemical sensors because of the complexity of the measurement principles and of the numerous approximations and assumptions underlying the models used. This is actually the reason why the term "specific", formerly used to designate the measuring electrodes (and which suggested the measure of a single quantity), has been abandoned, and replaced by "selective". Therefore, the degree of selectivity of a chemical sensor is a very important parameter as far as its application domain is concerned. It will be examined in detail when dealing with the Nikolskii-Eisenman equation.

1.1.3. *Chemical and biochemical sensors*

The purpose of this section is to clarify some specificities of chemical and biochemical sensors with respect to sensors in general.

An initial way to define these sensors is to refer to the measured parameter: a chemical quantity would be referred to if the sensor measures the concentration (or the partial pressure) of a definite chemical species. Similarly, a biochemical sensor can be defined as measuring the concentration of species of biological importance (antigens, antibodies, haptens, etc.).

The second way, which is not incompatible with the previous way, is to consider as chemical or biochemical sensors those with a measuring principle that is based on a chemical or electrochemical interaction (physisorption, chemisorption, or reaction) between the sensor and the measurand, in opposition to physical interactions involved in physical sensors. According to this definition, the classical scheme of a chemical sensor is given in Figure 1.1.

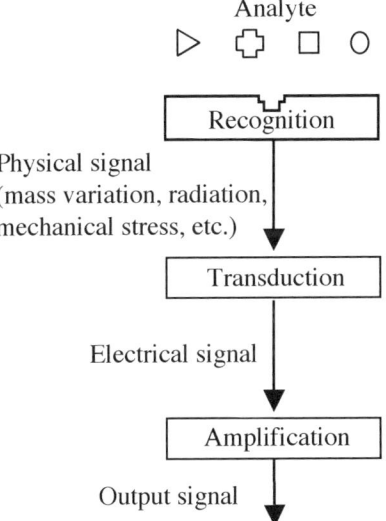

Figure 1.1. *Schematic drawing of the principle of a chemical sensor*

However, the difficulty (mentioned in section 1.1.1) in discriminating the fields of sensors and analyzers increased recently with the development of a new concept named µ-TAS (Micro Total Analysis System). In µ-TAS, pre-treatment and treatment of the samples is often necessary to ensure accurate measurement.

6   Chemical and Biological Microsensors

Therefore, the tendency now is to rely on microelectronics techniques (micropumps, channel machining, microdetectors, etc.) for elaborating very small devices ensuring the functions of an analyzer, with the goal generally assigned to sensors, which is the determination of a specific quantity. Another "definition jammer" is the artificial nose, which comprises a set of rather simple sensors, but which essentially makes use of a software treatment for a result similar to that expected from an analyzer.

Regarding the improvements of manufacturing technologies, it must be noted that the evolution of chemical sensors is clearly slower than that of the physical sensors, which has been considerable. At the present time, the achievements of the development are essentially limited to the glass electrode (pH) and its by-products (ionometry), as well as the Clark electrode (pO_2) and Severinghaus (pCO_2) electrodes. It is probably due to the fact that the models describing physical sensors (which generally involve simple principles: Wheatstone bridge, capacitance or inductance measurement, Hall effect, etc.) are much more robust than those used in chemical sensors (power laws for gases, Nernst law, or Nikolskii-Eisenman equation for liquids) which are merely approximate laws. It entails an intrinsic lack of reproducibility, both for a given sensor and for a series of sensors, and the expected advantages resulting from the evolution in technology have not been fruitful. The field of biosensors (enzyme electrodes and others) is the only field that appears to evolve (though rather slowly).

1.2. Classification

This classification, based on the principles of transduction[1], more specifically concerns chemical sensors. The specificities of biosensors are described in the relevant chapters.

1.2.1. *Mass variation*

The basic principle of these sensors is the variation of the frequency (*bulk acoustic wave*, BAW) or of the velocity (*surface acoustic wave*, SAW) of an acoustic wave generated in a piezoelectric medium (lithium niobate $LiNbO_{3-\delta}$, quartz, etc.) by a pair of electrodes, due to the adsorption of the target species on a convenient sensitive coating deposited on the piezoelectric substrate. It generally works on a differential mode, with a "sensitized" transducer and a second inert transducer. In the case of the SAW, the velocity difference is converted into frequency by means of an oscillator obtained by looping both transducers together.

---

1 For a more complete treatment of this topic, see [GÖP 91].

In the case of gases, this variation is clearly due to the mass increase (mass loading) following adsorption. The mechanism is much less clear in liquids and, as the possibility of using this method for species in liquid solution has been demonstrated [BAS 86], there has controversy as to whether it is a mass effect or a modification of the acoustic impedance via the adsorbed layer.

Apart from the question of their working principle, the main weakness of these sensors is their lack of selectivity: this property is supposed to rely on the chemical or biochemical selectivity of the adsorbing layer towards the target species. However, it is very difficult to prevent the competitive fixation (deposition of other species that are also able to react with the coating, this effect being used in some biochemical sensors) and the non-selective deposition: any species depositing on the substrate causes mass variation. It is practically impossible to prevent these unwanted deposits without pre-treating the solution. In addition, when the target species has a low mass (antigens, haptens), the interference (for example, dust) may have a much bigger effect than that of the target species.

For the time being, these sensors remain on the laboratory scale, without real development.

1.2.2. *Optics*

In the case of optics, the signal supplied by the transducer is an optical nature. The two main categories of principles used are deported spectrometry (usually by optical fibers) and the use of the modifications of the refraction index at the surface (use of the *evanescent wave*).

In the first category, we find either *spectrometry* in the visible range (color change of the solution), *luminescence* spectrometry (fluorescence and fluorescence extinction by oxygen [PET 84] or halide ions) or *Raman* spectrometry (in this case, because of the very weak intensity of the lines, laser excitation is generally used), in which the fiber only drives the excitating and re-emitted radiations, or fibers modified by the grafting of colored or photoluminescent indicators. This whole family of sensors is generally named *optodes* (from Greek meaning "vision path") or, more improperly and by assimilation with electrodes, *optrodes*, some consensus appearing on *opt(r)ode*. Visible and Raman spectrometry are generally used more often in chemical sensors, and luminescent markers in biochemistry (with no absolute character in this distinction). These methods remain at the laboratory scale and, in spite of their apparent interest (deported measurement, no sensitivity to electromagnetic interferences, large variety of possible indicators), they also have numerous drawbacks (a response curve that is often complex and, in the case of grafting, a loss of the indicator in solution or eventually photobleaching by the excitating radiation).

In the second category, we find interferometric methods (for instance, Mach-Zehnder integrated interferometer with one of its two branches passivated and the other covered by the adsorption layer), and the method of surface plasmon resonance (SPR). The latter deserves a more detailed description, as it is the only optical method that has given rise to an industrial development (BIAcore™, Pharmacia Biosensor, Sweden [BIA]): the surface plasmons are linked oscillating electromagnetic modes (polarons), which can be excited in a thin layer of a noble metal (silver or gold) deposited on a dielectric material (glass). The intensity of the field decreases exponentially in the direction normal to the layer. At resonance, which is obtained for a particular angle, the intensity of the reflected light decreases suddenly. The resonance conditions depend, via the evanescent field, on a possible adsorption on the metal layer. Therefore, any adsorption modifies the resonance angle. In the industrial equipment, the selectivity is obtained by a thin layer deposit of a porous polymer (dextran) on which the ligand is fixed (thus, non-selective fixation is avoided). However, this apparatus, by its size and its complexity as well as its price, falls more in the category of analyzers rather than sensors, and even less in the microsensors category.

1.2.3. *Electrical parameters*

Sensors using electrical transduction are very important in practice, as they are, with a few exceptions, the only chemical or biochemical sensors to have had real industrial development, and, therefore, to be used in industry. The methods used can be divided into three categories: conductimetry, potentiometry, and amperometry.

1.2.3.1. *Conductimetry*

This method consists of measuring the electrical conductivity (its unit being siemens per centimeter, S/cm) of the solution. It is performed by applying an alternating current (with an appropriate frequency) so that electrode polarization should be avoided. It is, of course, reserved to electrolyte solutions in which the species to be analyzed is the only species that varies (due to its principle, this method has no selectivity at all), the matrix remaining unchanged.

In spite, or because of its simplicity, this method is important from an industrial point of view: in fact, although it is not selective, it can be made as sensitive as necessary (size and nature of the electrodes, applied voltage, etc.) and gives reproducible results (as long as the conditions of use mentioned earlier are fulfilled). Therefore, conductimeters are particularly suitable for monitoring the concentration of a species (in a reasonable range) and for use as zero apparatus.

1.2.3.2. Potentiometry

This method has great practical importance, as a model for the working of most *selective electrodes*. This term stands for a whole set of membrane electrodes with a more or less selective response to a given ionic species in the presence of several other species.

The term membrane is taken here within its largest acceptation. It is a thin material, electricity conductive, separating two different solutions between which a voltage difference develops. Measuring this potential difference enables the activity to be determined (see section 1.2.3.2.1 for the definition of this parameter) of the ions present in one of the solutions, if this parameter is fixed in the other solution.

Potentiometry, as *amperometry*, belongs to a discipline called *electrochemistry*. Some fundamental elements of this discipline must be noted for a good understanding of the working and utility of these sensors.

1.2.3.2.1. Elements of thermodynamics and electrochemistry

The following thermodynamic quantities are defined for a given chemical system:

- *internal energy* $\quad U = W + Q$
- *entropy* $\quad S_{AB} = \int_{AB} \frac{dQ}{T}$
- *enthalpy* $\quad H = U + PV$
- *free enthalpy* $\quad G = H - TS$
- *chemical potential* $\quad \mu_i = \left(\frac{\partial G}{\partial n_i}\right)_{T,P,n_j}$

(where W stands for work, Q for heat quantity, T for temperature in Kelvin, P for pressure, and V for volume).

The chemical potential of a constituent i is given by:

$$\mu_i = \mu_i^o + RT \ln a_i \qquad [1.1]$$

where μ_i^o is the standard chemical potential, R is the ideal gas constant, and a_i the activity of species i. The activity is linked to the molar fraction x_i of i by:

$$a_i = \gamma_i \cdot x_i \qquad [1.2]$$

where γ_i is the activity *coefficient* of i.

An electrochemical system differs from a chemical system by the presence of active charges (free charges or dipoles). Electrical energy $\mathcal{E}$ is therefore present, and the corresponding electrochemical parameters (internal electrochemical energy noted $\tilde{U}$, etc.) can be derived by adding $\mathcal{E}$ to the thermodynamic parameters of the chemical system.

The electrochemical potential is then defined by:

$$\tilde{\mu}_{i\alpha} = \mu_{i\alpha} + n_i F \phi_\alpha \qquad [1.3]$$

where n_i is the charge number of ion i and F is the Faraday constant, i.e. 96,486 C/mol, and ϕ_α the *Galvani potential* of the α phase (the work necessary to bring a unit charge from the infinite to the bulk of the phase considered).

1.2.3.2.2. Electrode classification

Electrodes of the first kind

These are constituted by a metal dipped in a solution of one of its salts. At equilibrium we can write:

$$X^{n+} + n e^- \leftrightarrow X$$

At equilibrium, it can be easily demonstrated that the electrical voltage difference between the metal and the solution obeys the following equation[2]:

$$E_{abs} = E^o_{abs} + \frac{RT}{nF} \ln a_{X^{n+}} \qquad [1.4]$$

This equation is known as the *Nernst equation*. The quantity E^o_{abs} is called the *standard potential* of the electrode (it is the potential of this electrode in the presence of a solution of one of its ions with a unit activity). One important case is the hydrogen electrode (platinum foil – covered with platinum black – placed in a hydrogen gas stream and immersed in an acidic solution). The equilibrium is then written:

$$H^+ + e^- \leftrightarrow \frac{1}{2} H_2$$

---

[2] The demonstrations of the various equations are not given in detail in order to avoid making this chapter too cumbersome. The interested reader will refer to classical electrochemistry books (for example, [MIL 69]).

and the Nernst equation becomes:

$$E = E^o + \frac{RT}{F} \ln \frac{a_{H^+}}{(P_{H_2})^{1/2}} \qquad [1.5]$$

In the case of a normal solution of a strong acid ($a_{H^+} = 1$) and for hydrogen pressure of 1 bar, equation [1.5] gives:

$$E = E^o \qquad [1.6]$$

Such an electrode, the potential of which is the standard potential, is called the *standard hydrogen electrode*. Its potential is posed equal to zero at all temperatures, by convention.

Electrodes of the second kind

These are constituted of a metal covered by one of its weakly soluble salts, the whole being immersed in an electrolyte that contains the anion corresponding to the salt. Such an electrode is described by:

X (solid) / XA (solid) / A^{k-} (solution)

and the electrode reaction at the equilibrium is:

$$X + nA^{k-} \leftrightarrow XA_n + nke^-$$

Using the Nernst equation and the solubility product, we can write:

$$E_{abs} = E'^o_{abs} - \frac{RT}{kF} \ln a_{A^{k-}} \qquad [1.7]$$

The electrode–solution voltage difference depends only on the activity of the A^{k-} anions, which explains the use of these electrodes as reference electrodes, using a saturated solution of a salt containing the suitable anion as the electrolyte. The most widely used are the *calomel electrodes*:

Hg (liquid) / Hg_2Cl_2 (solid) / KCl (solution)

and the *silver–silver chloride electrodes*:

Ag (solid) / AgCl (solid) / KCl (solution)

Electrodes of the third kind

These are redox electrodes, with a noble metal (platinum) immersed in a solution containing salts of an element with two different oxidation numbers. The reaction equation is:

$$rOx + ne^- \leftrightarrow sRed$$

which yields the electrode potential:

$$E_{abs} = E^o_{abs} + \frac{RT}{nF} \ln \frac{a^r_{Ox}}{a^s_{Red}} \quad [1.8]$$

An important example is the *quinhydrone* (equimolar mixture of benzoquinone and hydroquinone) electrode which enables the measurement of the activity of hydrogen (H^+) ions. The reaction is:

$$benzoquinone + 2H^+ + 2e^- \leftrightarrow hydroquinone$$

which gives an electrode potential of the form:

$$E_{abs} = E^o_{abs} + \frac{RT}{2F} \ln \frac{a_{benzoquinone} \cdot a^2_{H^+}}{a_{hydroquinone}} \quad [1.9]$$

The standard potential (with respect to the standard hydrogen electrode) of this electrode is equal to 0.69976 V at 25°C.

1.2.3.2.3. ISFET (ion-sensitive field-effect transistor)

The ISFET is a microsensor (active surface of a few µm²), invented by P. Bergveld in 1970 [BER 70], which can be assimilated to potentiometric devices, but based on a principle different from that of the electrodes described above. It is a field-effect transistor (MOSFET, metal oxide semiconductor field-effect transistor) where the metallic gate is "replaced with" the whole set of reference electrode-solution-insulating surfaces (SiO_2 or, preferably, Si_3N_4). The electric field measured through this oxide depends on the charged species fixed on the exchange sites of the surface of the insulating material (the surface of which can be modified by grafting). It must be noted that this interface is blocking, which contradicts the case of potentiometric electrodes. Sensors of other ions, of gases (GASFET), or enzymes (ENFET), grouped under the generic name of CHEMFETS, have been developed, starting from the original device (sensitive to pH). These devices will be dealt with in Chapter 6, and will not be discussed in more detail here.

1.2.3.2.4. Scale of potentials

From all the previous information, it appears that absolute potentials cannot be measured: only potential differences are measurable. Therefore, it is essential to fix a reference (the value of which is not necessarily known but must be reproducible). By convention, *the standard hydrogen electrode* is generally selected, and its potential is taken as zero.

This electrode, which is not easy to use in practice, is often replaced by the *calomel electrode* (E^o = 0.2444 V at 25°C for a saturated KCl solution) or by the *silver–silver chloride electrode* (E^o = 0.22234 V at 25°C for a molar HCl solution; 0.1989 V for a KCl saturated solution).

Ionicity is then measured from the potential differences between the elements of a battery constituted by an indicator (or measuring) electrode and a reference electrode (whose voltage difference with the solution must be kept as constant as possible).

1.2.3.3. *Amperometry*

This method consists of measuring the current flowing between two electrodes (measuring and counter-electrode) dipped in a solution, as a function of the voltage difference applied to these electrodes. In the presence of electroactive species (liable to get oxidized or reduced) *1*, *2*, etc., the curve $i = f(E)$ presents the plateaux for voltage values E_1, E_2, etc., which are characteristic of each species and which depend on its oxidation or reduction potential. Moreover, the value i of each plateau-current depends on the concentration of the electroactive species. Figure 1.2 presents the corresponding *voltammogram*: it shows that, if the half wave potential corresponding to a given species is known, this species can be dosed, in principle, by polarizing the indicator electrode at this potential and measuring the current.

Amperometry is discussed in Chapter 5 of this book, and will not be treated in more detail here. However, it is worth making a few remarks at this stage. The most important one is that amperometry is a dosage method, and not an analysis method (it is necessary to know the target species and the matrix prior to dosing it). The reason is that any species present in the solution, with a potential smaller than that of the target species, will contribute to the observed current. Therefore, this method is selective only to the species responsible for the first plateau of the voltammogram. Besides, the dosed species is consumed by the reaction at the indicator electrode, and must, therefore, be renewed in the vicinity of the electrode by stirring. Also, the reaction must not cause a modification of the indicator electrode surface state (modification of the potential "seen" by the solution), therefore, noble metals (gold or platinum), which are stable in principle, are used. Lastly, as it is necessary to know or control the potential of the indicator electrode, either the potential of the counter-electrode is kept constant when the current varies (use of non-polarizable electrodes) or a three-electrode system is used (a reference electrode, with a fixed voltage versus the solution, is inserted). It is then possible to negate the problem caused by any variations of the potential of the counter-electrode.

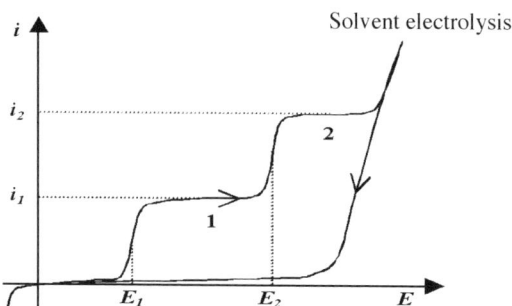

Figure 1.2. *Current-potential diagram. The arrows indicate the direction of the potential scan increasing, and then decreasing potential*

The most widely known and most ancient sensor, based on amperometry, is the oxygen (gas or dissolved) electrode, known as the *Clark electrode* [CLA 56]. It is worth noting that amperometry is used as a detection method in several types of biochemical sensors: the first enzymatic glucose sensor to be produced on an industrial scale [YSI] used amperometric detection of hydrogen peroxide (H_2O_2). To conclude, a three-electrode hypochloride ("dissolved chlorine") microsensor, developed by Microsens (Switzerland) and Cylergie (Lyonnaise des Eaux, France) was recently commercialized successfully [GRI 94].

1.3. Specific problems of chemical sensors

These will be studied by describing two examples: measurement of pH (defined as the decimal logarithm of the inverse of H^+ ion activity) using the glass electrode and *ionometry*[3].

1.3.1. *pH measurement: Nernst equation for the glass electrode*

This example was selected because of the importance of pH measurement by chemical sensors.

---

3 Regarding the use of ion-selective electrodes and the associated problems, it is possible to refer to [COV 79] and Chapters 3 and 4.

A *glass electrode* is made of a thin, special glass wall (silica and alkaline or alkaline earth metal oxides) separating an internal reference solution (*buffer*) from the external solution to be measured (see Chapter 4). During the first glass hydration (electrode conditioning), gelatinous layers (*hydrogels*) are formed on both sides of the glass wall, due to the exchange of sodium (Na^+) ions (glass) and H^+ ions (solution): these two hydrated layers are responsible for the sensitivity of the membrane to pH (exchange of H^+ with the reference and measured solutions). The bulk of the glass, which is not altered, only plays the role of an ionic conductor (by sodium ions). This electrode is associated with a reference electrode to constitute the measuring electrochemical chain. The electrode model is given in Figure 1.3.

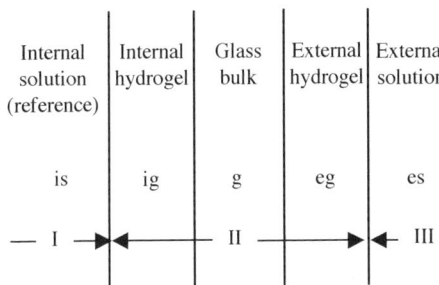

Figure 1.3. *Model for the glass electrode*

Let us calculate the potential difference existing, through the membrane, between the external solution and the internal reference solution. This quantity is given by:

$$E_{I/III} = E_{I/II} + E_{II/III} \quad [1.10]$$

The $E_{I/II}$ voltage is equal to the Galvani potential difference between the internal solution and the internal hydrogel. Writing the equality of the electrochemical potentials at equilibrium, using the corresponding expressions of these potentials, leads to:

$$E_{I/II} = \frac{\mu_{ig}^o - \mu_{is}^o + RT \ln \frac{a_{ig}}{a_{is}}}{F} \quad [1.11]$$

$E_{II/III}$ is calculated in the same way, using the activities in the external solution and in the external hydrogel:

$$E_{II/III} = \frac{\mu^o_{es} - \mu^o_{eg} + RT \ln \dfrac{a_{es}}{a_{eg}}}{F} \qquad [1.12]$$

The reference state is the same for both solutions and both gels (hence, the equality of the corresponding standard potentials μ^0) and the H^+ ions present in the hydrogels are supposed to be firmly fixed, which enables us to assume that their activity in these layers does not depend on the neighboring solution. It is also assumed that the activities in the external and internal hydrogels are equal ($a_{eg} = a_{ig}$) and we can write $a_{es} = a_e$ and $a_{is} = a_i$. With these assumptions, the potential difference between the two solutions is now:

$$E_{I/III} = \frac{RT}{F} \ln \frac{a_e}{a_i}. \qquad [1.13]$$

The activity of the H^+ ions in the reference solution is constant, which enables us to note $E^o = -(RT/F) \ln a_i$, and, by renaming a_{H+} the activity a_e of the H^+ ions in the external solution (which is to be measured), we finally obtain the potential difference through the membrane:

$$E = E^o + \frac{RT}{F} \ln a_{H^+}. \qquad [1.14]$$

This expression is known as the Nernst equation of the glass electrode. The value of the coefficient of the decimal logarithm is 0.05916 V at 25°C.

Two assumptions have been made for applying the Nernst equation to the glass electrode: the quality of the pH measurement made using a glass electrode is conditioned by the Nernst equation, that is to say by the validity of these two assumptions.

The first assumption is the *equality of the activity of the H^+ ions in the internal and external hydrogels*. In fact, due to the fabrication processes, these two activities are not equal (though having fixed values in the normal conditions of use). This results in a so-called *asymmetry potential*, which varies from one electrode to the other:

$$E_a = \frac{RT}{F} \ln \frac{a_{ig}}{a_{eg}} \qquad [1.15]$$

Therefore glass electrodes cannot be used for absolute measurements, and a preliminary calibration of the whole measurement chain using standard solutions is necessary.

The second assumption is that the *ionic activities in the hydrogels do not depend on the solution*. It can be considered as verified only if the solutions are sufficiently diluted, which is the case for the internal solution, which is freely adaptable, but not necessarily for the external solution.

Two situations can then be met: if the pH *value is high* (very basic solution), and, in particular, in the presence of sodium ions, some of the H^+ ions in the hydrogel may be replaced with alkaline ions of the solution. This entails a decrease of a_{eg} ($a_{eg} < a_{ig}$), which creates a positive ΔE shift with respect to the Nernst law. *The measured pH is then lower than its real value* (*alkaline error*). Conversely, if the pH of the solution is very acidic (concentrations over 1 mol/l), water loss in the hydrogel and replacement of alkaline ions in the glass with H^+ ions from the solution may both occur. Both phenomena contribute to increasing a_{eg} ($a_{eg} > a_{ig}$) and creating a negative ΔE shift with respect to Nernst law. This results in a too high pH reading (*acid error*). These effects are important (about one or two pH units), which, consequently, limits the use of the glass electrode to pH values between 1 and 12.

The influence of the external solution on the ion activities in the hydrogel is also the cause of memory phenomena (*hysteresis*): an electrode that has been dipped into a too concentrated acidic or basic solution has its external gelatinous layer modified, but will slowly return to its previous condition when it is immersed again in a diluted solution.

Lastly, some products may have a chemical action upon the hydrogel (e.g. surfactants, greases, proteins, etc.) and the glass itself (e.g. hydrofluoric acid, sodium hydroxide, etc.), which may also modify the electrode response.

The response of the sensor provides a value of the H^+ ions activity, but the experimenter is generally interested in knowing the concentration of the ions. The *activity coefficient*, γ, linking both quantities is equal to 1 only for infinitely diluted solutions. In fact, it was introduced as a link between the theory of real solutions and that of ideal solutions – no interaction between the species, i.e. an infinitely diluted solution, which is itself derived from the theory of ideal gases. The activity of the ions is based on the same principle as fugacity in gases, and *depends, in real solutions, on the ionic strength, which is related to the interactions between all the charged species present*. Consequently, it is not calculated, but determined experimentally.

The Nernst law can, therefore, only be considered as an approximation, giving the response of the glass electrode in diluted solutions. In addition, the existence of the asymmetry potential clearly indicates the necessity of individual calibration. The nature of standard solutions and the quality of the calibration procedure have important consequences on the quality of the measurement.

1.3.2. *Ionometry*

Starting with the glass electrode, the development of electrodes sensitive to other chemical species, in particular ions, has been attempted more or less successfully. Taking into account the number and the variety of the solutions investigated, the International Union of Pure and Applied Chemistry (IUPAC) recommends the following classification as far as selective electrodes are concerned:

1.3.2.1. *Primary electrodes*

– Crystalline membrane (two types):

- homogenous membrane (e.g. lanthanum fluoride or a mixture of silver iodide and silver sulfide);

- heterogenous membrane, where the active substance is mixed with an inert matrix (e.g. silver sulfide in a PVC membrane).

– Non-crystalline membrane (two types), supported or not (porous glass, Millipore filter, polyvinyl chloride (PVC), etc.):

- rigid matrix (for instance, glass electrodes sensitive to sodium);

- mobile carriers, either neutral or charged positively or negatively (e.g. quaternary ammonium salts, tetraphenylborate ion, valynomycin).

1.3.2.2. *Sensitized electrodes*

– Gas electrodes, using a gas-permeable membrane, or an air bubble separating the solution to be measured from a thin film of an intermediate solution.

– Enzyme electrodes, where an enzyme layer reacts with the solution, producing a species to which the electrode is sensitive.

Only primary electrodes will be considered here.

The lack of selectivity of the membranes used has been mentioned previously. It is, therefore, important to study their cross-sensitivity to various ions. By generalizing the reasoning made for the glass electrode, we can calculate the potential difference through the membrane, due to the response to a main ion M^{n+} by using the Nernst equation, which gives in this case:

$$E = E^o + \frac{RT}{nF} \ln a_{M^{n+}} \qquad [1.16]$$

where n is the charge of the ion and $a_{M^{n+}}$ the activity of the M^{n+} ion in the measured solution, the other face of the membrane being wetted by a solution in which the activity of this ion is constant.

However, this formulation does not take into account the possible response to interferent ions. However, except for the pH (glass) electrode and for the fluoride electrode (monocrystalline lanthanum fluoride LaF_3), these interferences are important. In such a case, the *Nikolskii-Eisenman equation* is used. For an electrode that is sensitive to a monovalent ion, i, and to a monovalent interferent, j, this equation is:

$$E = E^o \pm \frac{RT}{F} \ln (a_i + k_{ij} a_j) \qquad [1.17]$$

where a_i and a_j are the activities of both ions and k_{ij} the coefficient of selectivity of i with respect to j. The sign used is plus if i and j are cations, minus if they are anions. The most general form of the Nikolskii-Eisenman equation is:

$$E = E^o + \frac{RT}{n_i F} \ln (a_i + \sum_j k_{ij} a_j^{n_i/n_j}) \qquad [1.18]$$

where n_i and n_j are the valencies of the main and the interfering ions respectively.

It is very important to note that the Nikolskii-Eisenman equation, in its general form, merely corresponds to an empirical law that keeps the formalism of the Nernst equation. In particular, we must be careful not to assign an absolute value to the selectivity coefficients k_{ij}. *These coefficients are not constants* (in opposition to what is suggested by the term, *selectivity constants*, used previously), but they vary with temperature, with the ionic strength, the mobility and the diffusion coefficients of the various ions. Therefore, each measurement condition is unique, and the coefficients k_{ij} will have to be determined experimentally, and not taken from a table of data. Even so, *it is wise to consider these coefficients merely as indicating the probable performances of the electrode, and the highest quantity of interfering ions compatible with its satisfactory operation.*

The conclusion of the whole development given above is that ionometric measurements, in spite of their importance, are difficult to perform and require, in order to be performed satisfactorily, a large number of precautions.

It is essential to note that the measurements of the activity of an ion in solution are *relative in nature* and are determined by comparison with the electrode response to the same ion in a standard solution. Therefore, all the parameters of the measuring device, physical (temperature, stirring, flow, pressure, etc.) as well as chemical (matrix of the solution), must be kept as constant as possible. It is necessary to control the operating conditions rather than to try to correct the results. In practice, this means a pre-treatment of the solution (for instance, the use of an ionic strength buffer such as TISAB or other). If this is not possible, it is necessary to attempt to evaluate the order of magnitude of the error introduced by any discrepancy between the calibration and measurement conditions.

1.3.3. Conclusion concerning chemical sensors

The two examples treated above illustrate the major problems posed by chemical sensors to the user. These are of two kinds:

– *Lack of robustness of the models*: in most cases, the models used for solutions are idealistic rather than realistic, and are linked to the theory of ideal gases via a certain amount of mathematical and physical acrobatics, (see section 1.3.1). Their strict application needs the knowledge of numerous parameters that are ill-defined and difficult to measure (e.g. ionic strength, activity coefficient, etc.). For these reasons, theory will yield, in the best cases, a mathematical model of the response (e.g. logarithmic response for the Nernst equation). But, in order to determine the numerical values of the parameters to be used, the whole electrochemical measurement chain (selective and reference electrodes) will, in practice, have to be calibrated in conditions (concentrations, matrix, etc.) as close as possible to those of the considered utilization. Therefore, the experimental results are dependant on the quality of the calibration standards. These standards are, in fact, buffer solutions and, as far as pH is concerned, the best standards (commercialized by the National Institute of Science and Technology – NIST, USA) are given with two significant decimals. Consequently, whatever the indicator electrode is (glass, quinhydrone, IFSET), the measurement of pH (a parameter whose monitoring may be performed with a much better resolution) will be done, at best, with one significant decimal (it is generally admitted that the quality of a measurement is inferior to that of calibration by a factor of 10). The repeatability of these sensors is not very good (*hysteresis* due to measuring conditions). For all these reasons, as far as the actual measurement is concerned, chemical sensors are rather inferior to physical sensors, for fundamental reasons. However, they are well-adapted (because of their sensitivity) to monitoring the evolution of a parameter or to being used as a zero apparatus.

– *Lack of selectivity*: the example of *ionometry* (equation of Nikolskii-Eisenman) shows clearly that the difficulties *grow very quickly with the number of interfering ions*. This lack of selectivity, which is also inherent for fundamental reasons, has severely penalized the use of chemical sensors outside the laboratory (and in particular in industrial applications), except for a few successes (pH, pO_2, pCO_2). Regarding the application of chemical sensors in the medical field (and, *a fortiori*, biochemical), it is the monitoring rather than measuring quantities that are utilized, and the drawbacks of these sensors are, therefore, less penalizing.

After these somewhat negative considerations, it remains that these sensors have a *fundamental advantage*: they do *exist* and, in numerous instances, they are the *only* way to obtain the required information, despite the incomplete or approximate nature of the information received. Therefore, the above remarks must be considered as a "*caveat*" for the attention of the potential user when using these devices, as far as precautions are concerned.

1.4. Advantages and drawbacks of chemical microsensors

In the preceding text, the aspect of microsensors has not been fully elaborated, because a microsensor is generally a small-scale reproduction of a classical sensor, taking advantage of the use of technology offered by microelectronics (in particular, the micromachining of silicon and other materials). (The ISFET is a noticeable exception to this assertion.) Therefore, it generally has the same advantages and drawbacks. However, the small or very small size of the sensitive zone of microsensors provides them with some specific assets and/or drawbacks in the domain of chemistry:

– *Advantages*: they are, of course, capable of being inserted in areas of reduced accessibility (which is important for medical applications). In addition, in cases where reagents are consumed (by the sensor or for pre-treating the measured solution), the volumes are also reduced, which may be important for costly or dangerous reagents. However, the advantage offered by the small size is limited, in numerous instances, by the necessity of using the sensor together with a reference electrode (which must generally be placed close to the sensor). Also, theoretical considerations (blocking interface) preclude the use of an "*all solid state*" reference, and the existing miniaturized solutions are only applicable to a limited number of cases. The question of miniaturizing the reference remains one of the most important in the field of chemical microsensors. The expectations of low-cost, mass production of devices, such as ISFET, have, so far, been dashed (because the market is too small). It is worth noting that, beside microelectronic devices, miniaturized classical electrodes have existed for a long time (in particular for pH).

– *Drawbacks*: the small size of the measuring area of these sensors makes them very sensitive to accidental pollution (e.g. dust), leading possibly to *signal vanishing* and/or *sensor destruction*. Therefore, it is advisable to use them in very clean media. They are also sensitive to local variations in composition (unavailability of the "averaging effect" existing over the sensitive surface of a macroscopic sensor).

1.5. Perspectives

Although they do not fundamentally modify the problems linked to the nature of chemical measurement, miniaturization techniques have enabled improvements in two fields:

– *integrated optics*: in comparison with the micromaching of silicon or other materials for making chemical sensors, the techniques of integrated optics (developed mainly for telecommunications – couplers, multiplexers, etc.) have been less developed. These sensors have *very high sensitivity*, in particular, in sensors using the interaction with the evanescent wave. There are techniques that are already used for biosensors (for instance competitive fixation), which have partially overcome the problems caused by non-selective deposits. A good example of such techniques is the immuno-sensor called FCFI (Fluorescence Capillary Fill Immunosensor) developed, but not commercialized, by the Serono Diagnostics Society in the nineties [ROB 90];

– *μ-TAS*: these systems, which have already been mentioned, are miniaturized analyzers in that they achieve the necessary pre-treatment and conditioning of the solution to be measured, thus creating conditions suitable for a satisfactory operation of the chemical sensor. Here, the advantage offered by miniaturization is, apart from maintaining a reasonable size for the device, *to considerably decrease the quantity of reagents* used for each measurement, compared with an equivalent macroscopic device. It is then possible to consider an autonomous operation (without replacement of the reagents) which is compatible with industrial requirements. These devices, which are just beginning to appear, are mainly supported by the pharmaceutical industries (in pharmacy, the advantage is the use of aqueous solutions and, therefore, the possibility of using electrophoresis as the driving force of the charged species in the system). The problem is much more complicated in the field of chemistry, where movement of fluid is achieved using micropumps, which are difficult to fabricate. Nevertheless, this domain is certainly under development, with interest from the manufacturers and several government programs. Interested readers can refer to the historical description of this domain written by its creator [WID 96] and, more generally, to the proceedings of conferences dealing with this subject.

1.6. Bibliography

[AFN 84] ASSOCIATION FRANÇAISE DE NORMALISATION, *Vocabulaire international des termes fondamentaux et généraux de métrologie*, NF X 07-001, 1984.

[BAS 86] BASTIAANS G.J., GOOD C.M., "Immunoassay utilizing a piezoelectric surface acoustic wave mass sensor", *Proc. 2nd International Meeting on Chemical Sensors*, Bordeaux, p. 618-621, 1986.

[BER 70] BERGVELD P., "Development of an ion-sensitive solid-state device for neurophysiological measurements", *IEEE Trans. on BME*, BME-17, vol. 1, p. 70-71, 1970.

[BIA] BIAcore™, *Pharmacia Biosensor AB*, Uppsala, S-75182, Sweden.

[CLA 56] CLARK L.C., "Monitor and control of blood and tissue O_2 tensions", *Trans. Am. Soc. Artif. Int. Organs*, vol. 2, p. 41-48, 1956.

[COV 79] COVINGTON A.K., *Ion-selective Electrode Methodology*, vol. 1 & 2, CRC Press Inc., 1979.

[GÖP 91] GÖPEL W., HESSE J., ZEMEL J.N. (eds), *Sensors, A Comprehensive Survey*, vol. 2 & 3, *Chemical and Biochemical Sensors, I & II*, VCH, 1991-1992.

[GRI 94] GRISEL A., ARCHENAULT M., LACOMBE P., MONDIN G., "Integrated chlorine sensor for drinking water distribution security systems: micro total analysis systems (µTAS)", *Analusis*, vol. 22, p. M13-M15, 1994.

[ISA 75] INSTRUMENT SOCIETY OF AMERICA, *Electrical Transducer Nomenclature and Terminology*, ANSI Standard MC6.1-1975 (ISA S37.1) Research Triangle Park, 1975.

[MIL 69] MILAZZO G., "Bases théoriques, applications analytiques, électrochimie des colloïdes", *Electrochimie*, vol. 1, Dunod, 1969.

[PET 84] PETERSON J.I., VUREK G.G., "Fiber-optic sensors for biomedical application" *Science*, vol. 224, p. 123, 1984.

[ROB 90] ROBINSON G.A., "Optical immunosensing systems — meeting the market needs", *International Symposium on Miniaturized Total Analysis Systems, µTAS96, Basel, 19–22 November 1996, Proceedings of the 1st International Conference on Biosensors*, Singapore, Elsevier, 2 – 4 May 1990.

[WID 96] WIDMER H.M., "A survey of the trends in analytical chemistry over the last twenty years, emphasizing the development of TAS and µTAS", *Proceedings of the 2nd International Symposium on Miniaturized Total Analysis Systems*, µTAS96, Basel, 19–22 November 1996.

[YSI] Model 23 Glucose Analyzer, Yellow Springs Instruments, USA.

Chapter 2

Chemical Sensors: Development and Industrial Requirements

2.1. Introduction

2.1.1. *Overview of the various types of chemical sensors*

Numerous classifications for chemical sensors are found in the literature:

– according to the transduction mode: for example, thermal, mass (or piezoelectric), optical, or electrochemical sensors;

– according to the interaction mode between the medium and the sensor (surface or volume interaction);

– according to the type of use, i.e. measurement and control of pollution (air and water), health and security at work, detection of danger (explosive or flammable gases), process control, combustion control, biomedical field, etc.

This chapter is devoted to the industrial development of electrochemical sensors of the "potentiometric", "conductimetric" or "amperometric" type, used for chemical analysis in aqueous media.

Generally, it must be stressed that whichever the measuring device, the sensor remains always the core of the analysis device. The signal processing, however sophisticated it may be, will thus never cancel the drawbacks of a poor-quality sensor.

Chapter written in French by Jacques FOULETIER and Pierre FABRY, based on discussions with Jacques FOMBON, *Radiometer Analytical SAS.*

2.1.2. *General trends*

In the present situation, the potential users expect applications rather than sensors by themselves. This implies the design of sophisticated tools with the sensor environed by much automation, so that the quality of the results should not depend on the user, who may not be a specialized chemist.

For example, for environmental applications, which are undergoing extensive development, low-cost systems should be supplied for analyzing a target species or a group of species in a given medium and for a given concentration range. Moreover, the current demand is for accurate rather than sensitive sensors, which are easy to use and, if possible, have a simple calibration procedure.

Thus, the path is long from the research laboratory, where the idea was conceived and the first feasibility tests completed, to the commercial product providing full customer satisfaction.

The conscientious industrialist will firstly have to evaluate the suitability of the laboratory-developed sensor to the industrial needs that he is trying to inventory. He will attempt to adapt the products coming from research. In a second step, he will develop a new form of the sensor and of its instrumental environment, aiming at a given application field. The development will often be directed towards a "ready for use" procedure.

The aim of this chapter is to recall the essential features of the industrial development of a new sensor, as well as the importance of its qualification, with respect to the specifications expected by the future users.

2.2. Modern research and development (R&D) management methods applied to sensors

2.2.1. *Preliminary phases in the industrialization of a new sensor* [FOM]

2.2.1.1. *From the original idea to industrial design: research and technological development monitoring*

The modern tools for scientific and technological survey

Scientific survey allows the industrialist to detect new concepts developed in the field of sensors by research laboratories, particularly in public laboratories, and to monitor the progress of the work done at the international scale. It also helps to avoid pursuing unproductive pathways, such as re-inventing what has already been invented. Scientific survey necessitates monitoring scientific publications, participation in scientific conferences, or at least skimming through the congress proceedings. It must not only consider the field of sensors, but also the areas concerning the product before or after its development, more specifically the field of sensitive materials and their potential applications. At this level, modern tools, in particular, automated bibliography via the web, are widely used.

Technological survey is more precise, with the aim of keeping abreast of innovations by monitoring the technological improvements made by specialists in the field and rival companies. This survey involves monitoring publications, patents and their maintenance and their extensions. Moreover, monitoring the reaction of users offers marketing opportunities to provide new products that will respond better to user requirements.

The situation of the laboratories in fundamental and applied research

The industrialist himself is not always able to study the fundamental concepts of new sensors and will often have contacts in public laboratories. Usually, the laboratories contact the manufacturers for the development of their ideas and for support. When the idea is considered interesting, a partnership is established on a contractual basis, with or without the participation of a regional or state organization. The industrial contract must specify from the start of the collaboration a certain number of clauses concerning, for instance, industrial property, authorization to publish the results, confidentiality of thesis, partitioning the work between the partners, etc.

Instrumentation firms are at the interface between the researchers and the final user. They have the initial inciting role. They carefully follow the development of the research and take part in the feasibility tests of models and pre-prototypes in close collaboration with the research laboratories. If the results of the tests are positive, and if all the techno-economical elements necessary for industrialization are gathered, they take on the development of the sensor, then its production process and, finally, its commercial dissemination. The collaboration ceases, most of the time, as soon as the product is commercialized. However, the laboratory may also take part during the commercialization phase, especially in the training of the clients when the sensors have very specific applications (for instance, sensors for "*in vivo*" measurements).

Running feasibility tests

The feasibility tests are run on models fabricated by the laboratory or the industrialist. They are not yet prototypes, but only non-optimized devices as far as fabrication is concerned. In certain cases, the feasibility study can be performed in the frame of a thesis because it enlightens aspects that deserve detailed study.

28   Chemical and Biological Microsensors

Through the tests performed, the industrialist must acquire a full knowledge of the characteristics of the sensor:

– *Response law*: a function of the concentration of the species. This characteristic will bear consequences upon the optimal choice of the associated instrumentation.

– *Response range and practical detection limit*: the most important characteristics of a sensor, because they will define its suitability to potential applications. These characteristics are determined from the calibration curves, but with extremely clear definitions, for instance, from the recommendations of the International Union of Pure and Applied Chemistry (IUPAC) [IRV 77].

– *Selectivity*: species that may interfere with the response of the sensor, particularly in the frame of a planned application, must be systematically studied, and their relative importance must be estimated by determining the interference coefficients K_{ij} with a clearly specified method (see Chapter 3).

– *Repeatability, stability, reliability, and lifetime*: as sensors are often exposed to severe environmental conditions, it is wise to determine the variability of their response and of their selectivity, by a series of measurements repeated many times, over several months. Several identical sensors will be fabricated and be submitted to the same tests with the same protocol.

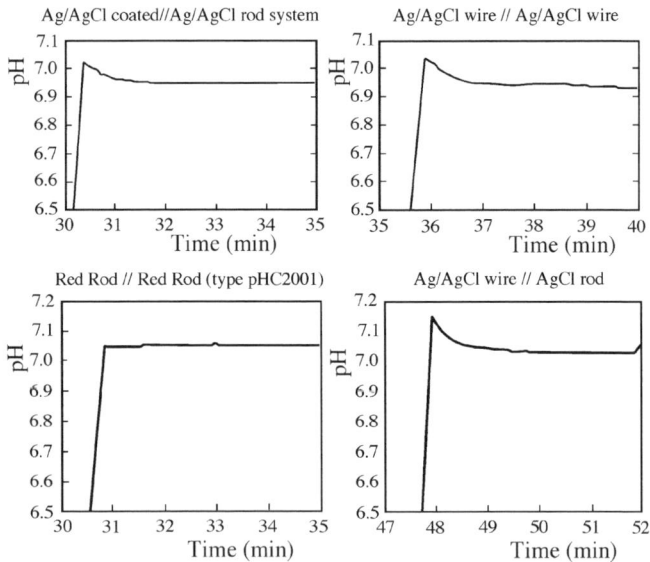

Figure 2.1. *Examples of response curves of various combined electrodes to simultaneous temperature (20 to 50°C) and pH steps*

– *Uncertainty*: this is a metrological quantity that allows the quantification of a margin of results (for instance, 95%), during measurements performed on standards or on *certified* reference samples. The traceability of the results with respect to the corresponding primary standards can thus be demonstrated unambiguously, and comply with the current quality standards.

– *Influence parameters*: are the remaining parameters that may manipulate the response, such as temperature, hydrometric degree, etc. The preliminary quantification of these potential influences will avoid unpleasant surprises during the commercial distribution of the sensor. Note that the working conditions in industrial workshops will have few similarities with those existing in the initial research laboratory.

– *Maintenance and conservation constraints*: these constraints concern, for instance, the influence of light, the storage in dry or conversely humid atmospheres, the advised positioning of the sensor, the solvents that may be harmful towards the life of the components of the sensor, etc. The industrialist will determine and describe them carefully, because they will contribute to the optimal use and to the lifetime of the sensor.

2.2.1.2. *Role of the Marketing Department: tuning to the customer and market survey*

"The success of a product does not arise from its own characteristics only, but also from the characteristics of the market ..."

Once the model or pre-prototype has been tested, it is necessary to make sure that the need is sufficiently important to ensure a potential market that justifies the investment for industrial development. From the knowledge of the operating characteristics of the model, a subtle equation taking into account the desirable performances, the feasible performances, and the corresponding market is established by the marketing specialist. In achieving this task, he must bear in mind the fact that the performances–market curve may present sudden breaks in certain "sensitive" zones. Let us consider, for example, the maximum level of nitrate in tap water, currently fixed at 50 mg/l; the potential market for a new nitrate sensor will obviously depend strongly on the fact that this concentration will be included or not in its response range.

The estimation of the potential market will also be influenced by the projected price, which is itself conditioned by the estimated production cost, which depends on the size of the production series, the size of which is conditioned by the flow of orders, therefore, by the market, etc. It is the well-known looped system, which must be properly mastered before marketing a new sensor!

The department of marketing is devoted to survey tasks, essentially by keeping in touch with clients, which enables the creation of an inventory of the requirements of the client. It is here that the commercial network in contact with the clients plays an important role. The "voice of the client (VOC)" thus helps to outline a potential market. The marketing survey also concerns the following up of products proposed by competitors by identifying their characteristics (drawbacks and qualities), in order to identify their potential position in the market. The collaboration between the marketing and R&D departments is thus permanent, with the common objective of evaluating the level of suitability between "desirable" and "feasible".

The marketing department also determines a projected selling price considering the situation of rival products and the VOC. It can then estimate a projected turnover for the coming years. A cash-flow simulation is then done in view of deciding whether the project should continue or not. This simulation takes into account the production costs, the costs for introducing the product to the market, for training the clients, etc.

If adequacy between the expenses and the receipts cannot be established on the basis of the projected techno-economic study, the project will be abandoned. This point is not always well-understood by the laboratory researchers who were at the origin of the sensor and consider that the innovation of their "baby" is indisputable.

If the operation is continued, the industrial development is engaged on the basis of the pre-project.

2.2.1.3. *The pre-project*

The pre-project constitutes a synthesis of the whole set of documents described above. Furthermore, it tackles an evaluation of the technological or human risks, such as a failing of the specialized human staff, of a supplier or a subcontractor (is it necessary to stock so-called strategic components, what will be the cost of this stock, etc.).

The development project is divided into additional steps each of which undergoes an evaluation of the risk. This evaluation phase is difficult, and needs experienced staff.

The specifications describe all the technical aspects of the development of the sensor: detailed drawings of the various constitutive elements, associated calculations, design aspects on its final form, and installations necessary for its fabrication.

2.2.2. Development and industrialization

During this stage, the product is conceived in its final form, with the definition of the machines necessary for its fabrication, the required competences, the suppliers and subcontractors.

Prototypes (sensors in their final form) are prepared and submitted to laboratory tests of conformity with respect to the specifications, but also to tests in actual in-use conditions, possibly at a standard user's place ("β-tests") whose opinion is often very useful. Once the validation is done, the production files are established.

Project management software helps planning and running the successive tasks, as well as the commercial launch date, the anticipated knowledge of which makes it possible to plan the other tasks linked to introducing the sensor on the market.

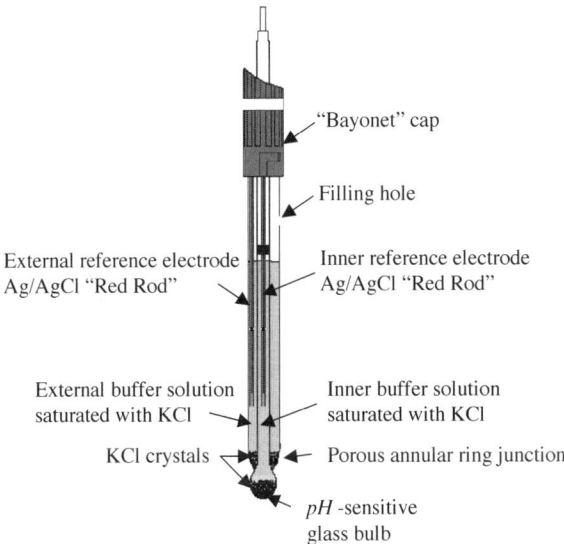

Figure 2.2. *Combined pH electrode, known as "red rod" (Radiometer Analytical)*

At the same time, quality-control tools are installed. A sequence of tests certifying the quality of the product is finalized. The objectives of the quality department are the following:

– quantifying the satisfaction of the clients;

– performing a statistical treatment of the claims and returns;

– establishing an assessment of the internal and external audits;

– developing an assistance service to the clients;

– controlling the conformity of the procedures and instructions taking place in production and control.

For a certified company, the ISO 9001 standard (in its recent version ISO 9001/2000) is used in this respect. The "quality handbook" gathers all the procedures to be respected by the staff, the forms for recording the progress of fabrication, the controls, the client requests, the complaints, etc.

The numerous development stages, which are rather long and costly, are justified by the fact that a new product must be perfectly known and present a guaranteed reliability when it is put on the market.

2.2.3. *Information and establishment of the commercial network*

The commercialization of a new sensor necessitates a certain number of actions, internal or/and external, for better information of the client on this new product, among with:

– training of commercial agents and distributors, when the sensor involves new concepts;

– writing the directions for use, the commercial brochure, and the specifications sheet;

– launching a promotion campaign prepared by the marketing department, including participations in national and international exhibitions.

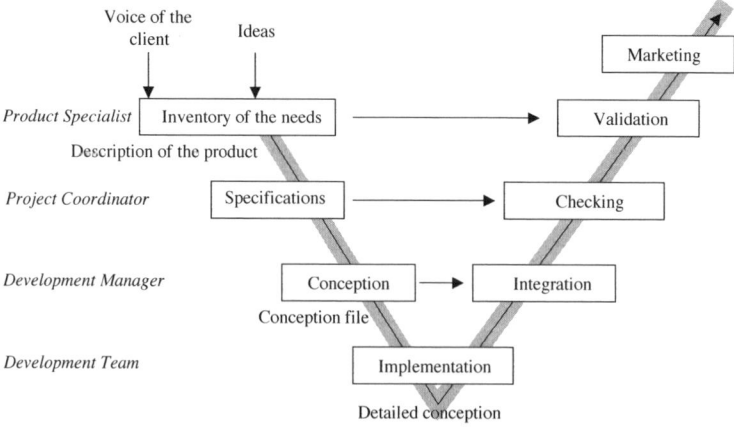

Figure 2.3. *"V" cycle*

To sum up the various stages during the development of a sensor, from the discovery of the concept to the commercialization, we can refer to the "V cycle" schematized in Figure 2.3.

2.3. Applications and inventory of the needs

The needs for chemical sensors are quite diverse. They can be expressed according to two main types of uses:

– discontinuous use in the laboratory;

– industrial use on line.

The biomedical applications, which are markedly different from the others, will be discussed separately.

2.3.1. *pH sensors*

pH sensors are, by far, the most widely used chemical sensors in liquid medium, with over 2 million per year sold in the world. These sales are divided in approximately equal parts between the two types of use indicated above.

Table 2.2, presented below, clearly shows the diversity of the domains in which the need for pH measurements exists. As in the case of the other sensors, the current demand focuses on reliable sensors, requiring as little maintenance as possible. Currently, these needs are not fully satisfied, because of the constraints of calibration and of maintenance of the liquid junctions of the reference electrodes, which are essential for good-quality measurement. Of course, additional needs, which entail additional technological constraints, appear according to the projected use of the sensors; symmetrical measurement chains are thus necessary when the temperature of the samples may vary (for example, combined electrodes "with red tube").

2.3.2. *Potentiometric selective electrodes*

The needs for selective electrodes are also quite diverse; their concept, as in the case of pH sensors, permits their adaptation to laboratory use or to industrial monitoring. However, most have to be calibrated frequently to enable total confidence in the quality of the results; their implementation in analysis systems on line is consequently more complicated and limited. The world sales of selective electrodes of all types, may be evaluated as a few thousand per year, which is much less than those of pH sensors, but constitutes a significant turnover, because their production and control costs make them 5 to 10 times more expensive than the pH sensors. Table 2.1 gives an approximate description, in decreasing order, of the relative importance of the most frequently proposed requirements and application domains for selective electrodes. Some examples are shown in Figure 2.4.

Automated titrations are a good example of the use of chemical sensors (see Figure 2.5).

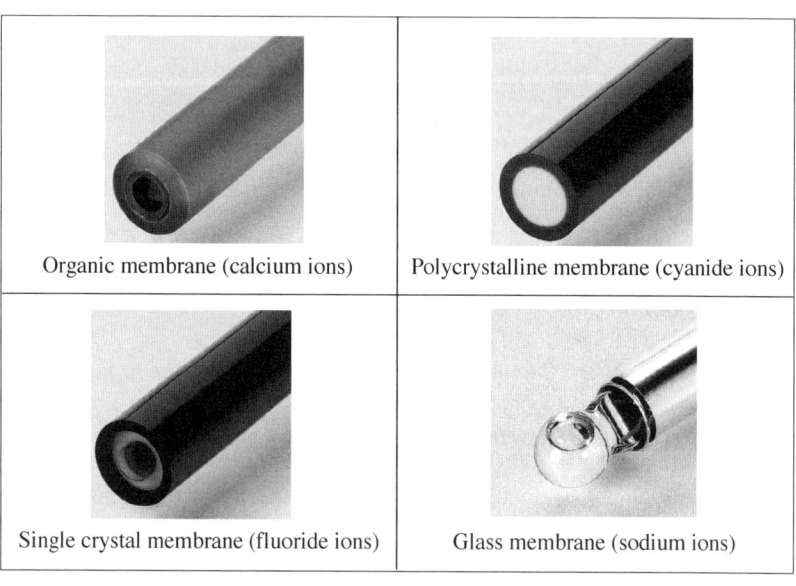

Figure 2.4. *The various types of sensitive elements of selective electrodes*

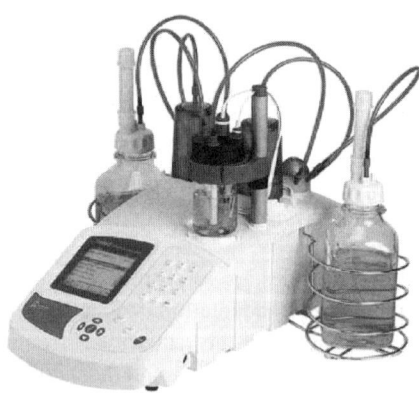

Figure 2.5. *Automated type RIM 865 titrator (Radiometer Analytical)*

Type of selective electrode	Relative demand	Application domains
Fluoride ions	30%	Environment, water industry, chemical and pharmaceutical analysis, surface treatments
Chloride ions	10%	Environment, water industry, food processing analysis, power stations, protection against corrosion
Calcium ions*	10%	Environment, water industry (water hardness measurement)
Sodium ions	8%	Water industry, power plants, food processing industry
Nitrate ions	8%	Environment, water industry, food processing industry
Ammonia/ ammonium ions[†]	8%	Environment, water industry, food processing industry
Cyanide ions	5%	Surface treatments, environment
Lead ions[‡]	5%	Water industry, environment
Potassium ions	5%	Food processing industry
Sulfide/silver ions[§]	5%	Industry and photographic laboratories, paper-making industry, environment
Iodide ions	3%	Food processing industry, environment
Bromide ions	2%	Industry and photographic laboratories
Cupric ions[‖]	1%	Surface treatments, environment
Others	1%	

*Using this selective electrode, the selective titration of calcium and magnesium ions by ethylenediaminetetraacetic acid (EDTA) is also possible.

[†]A gas diffusion electrode sensitive for ammonia (dissolved gas) exists, as well as an organic membrane electrode sensitive for ammonium ions.

[‡]The selective electrode for Pb^{++} ions, with a crystalline membrane, is, in fact, mainly used as an indicator electrode during the titration of sulfate ions. The reason is that its sensitivity is too low for the direct determination of lead at the legal toxicological level (50 µg/l); moreover, in its most frequent crystalline form, it suffers from strong interference from copper ions.

[§]The same type of electrode can be also used for determining sulfide or silver ions.

[‖]This electrode is also used as indicator electrode in the complexometric titration of other metallic ions in solution (cadmium, zinc, nickel, etc.).

Table 2.1. *Main application domains of the various types of selective electrodes*

2.3.3. Amperometric specific electrodes

These sensors differ from the previous ones by their working principle, which is presented elsewhere in this volume. Over 90% of the needs concentrate on dissolved oxygen sensors, or the Clark electrode, widely used in domains where dissolved oxygen consumption, or its residual concentration, yield essential static or dynamic information on the equilibria (chemical or biochemical) in the media examined. Several thousands of oxygen sensors are sold each year for laboratory or *in situ* industrial measurements.

Amperometric sensors have also been used for a few years for the determination of chlorine in its various oxidized forms in solution (ClO^-, $HClO$, ClO_3^-, ClO_2^-, ClO_2). Some are used for direct measurement; however, the precise values of the concentration of the various oxidized species (or of total chlorine) are preferably obtained by titrations involving amperometric sensors with two platinum electrodes.

2.3.4. Conductimetric sensors

These sensors are commonly classified among chemical sensors, although they yield the measurement of a physical parameter: the electrical conductivity of solutions (see Chapter 10). This parameter is directly linked to the presence of mobile electric charges constituted by all the ions present in these solutions.

The need for conductimetric sensors appears every time the salinity constitutes an important parameter, i.e. in oceanography, in the food processing industry, and in all situations involving the use of salt-containing solutions.

Conductimetric sensors are also used for testing the purity of water in the pharmaceutical field (injectable solutions) and the vapor exchangers of thermal and nuclear power plants.

2.3.5. Biosensors

The glucose sensors, lactate sensors and some others are different from previous ones in that they involve biochemical transductions between the measured species and the species to which the chemical sensor itself responds, which is generally amperometric or potentiometric. They are thoroughly described in this volume.

These sensors, well-studied in the laboratories for over 20 years, have not, as yet, produced an industrial application to meet industry requirements, because of their relative fragility, and of the necessary calibration and maintenance constraints.

However, the recent dramatic rise of biochips, examples of which are DNA probes, and the considerable investment in applied research for their optimization, lead us to expect considerable development in biosensors in the pharmaceutical, biomedical, and environmental domains.

2.3.6. *Biomedical applications*

Biomedical applications constitute a particular field in which the needs expressed are uniquely angled the "analysis of living," with all the implications related to the nature of the samples. The concepts used in theses sensors are, of course, the same as those for the other applications. However, the profile of the users and the needs expressed are so particular that the reasoning of the biomedical instrumentation suppliers constitutes an independent job, with its own specificity.

In the hospital environment, chemical sensors are key for certain essential diagnosis (such as analysis of gases in blood, oxygen (O_2) and carbon dioxide (CO_2), with the pH analysis). The sensors themselves are concealed in a very sophisticated hydraulic and data-processing environment, which is essential for the final information given to the hospital doctor to be 100% reliable and usable in real-time. However, it must be noted that the sensors always remain, as in any other application, the main component controlling the quality of the result.

Biomedical analysis laboratories use multiparameter automatons which make relatively little use of chemical sensors, and are more delicate than optical sensors. As indicated elsewhere, this tendency may be reversed in the next decade because of the development of biochips.

The chemical sensors market in the biomedical field is thus relatively small (a few thousands per year), but amounts to several hundred million Euros per year if the whole instrumental systems are included.

2.4. New needs and industrial applications

As mentioned several times in this chapter, there are several areas in which end-user requirements have not been met. These can be summed up in four essential aspects.

38   Chemical and Biological Microsensors

2.4.1. *Creation of new sensors*

In some cases, the lack of a suitable chemical sensor impedes the measurement of a substance, which is required by an end-user. For instance, the field of environment science expresses specific needs that remain unsatisfied. For example, there is a requirement for simple sensors of sulfate, phosphate, various metal ions as traces, and, generally speaking, sensors capable of performing in the field or in an automatic measurement station.

These determinations currently necessitate the transfer of samples to the laboratory, and the mobilization of heavy and costly instrumentation (ionic chromatography, inductively coupled plasma mass spectrometry (ICP-MS), high-performance liquid chromatography (HPLC)).

The biomedical field also demands selective sensors suitable for measurement of for example, lithium ions, nitrogen monoxide, neurotransmitters, etc, in biologic fluids (blood, plasma, etc.).

2.4.2. *Reliability and absence of maintenance*

The applications concerning continuous monitoring would probably develop considerably if the chemical sensors were as simple to use as physical sensors (pressure, temperature). The efforts of the laboratories must, therefore, concentrate on the development of robust and stable sensors and improve performance.

2.4.3. *Miniaturization and robustness*

These two characteristics, antagonists as they might be, are gathered in the latest generation of semiconductor sensors, the ion-sensitive field-effect transistor (ISFETs), the working principles and recent developments of which are described in Chapter 6.

Considering the viewpoint of the industrial suppliers, the ISFETS, with the same production means as the electronic components, necessitate heavy equipment investments that will be justified only by the importance of their commercial diffusion.

The improvements of the reliability of these sensors are important, but their usage and calibration constraints remain too heavy for a consumer application to be envisaged.

2.4.4. *Lowering the production and use costs*

Chemical sensors have generally quite reasonable costs as compared with heavy analysis means, such as spectrometric or chromatographic techniques. However, their price is still too high for consumer applications in the electrical appliance field or in the field of processed foodstuffs or biomedical controls, in which the sensors would be used only once.

When the production costs and the maintenance-less reliability reach the desired level, the market scale will change and the field of chemical sensors for solutions will probably undergo a revolution with an amplitude similar to that which occurred for chemical gas sensors, applied to the car industry for optimizing the combustion process by continuous control of exhaust gases.

2.5. The sensor in the measuring chain

2.5.1. *General features*

The knowledge of the future environment of the sensors in their application conditions is very useful for making sure that the sensor will operate correctly and will have optimal use. In view of assistance to future users, the supplier must specify the compatibility of the products of his catalogue with the various envisaged using conditions. As an example, Table 2.2 gives recommendations about the choice of the most appropriate pH electrode type for the aimed application.

Of course, the user will not be exempt from carrying out correct maintenance of his sensors, performing at regular intervals the necessary calibrations, and undertaking quality checks by using reference standard samples delivered by specialized organizations.

For example, in the case of pH measurements in milk, which can be considered a colloidal suspension, the supplier will have to propose sensors capable of avoiding the blocking of liquid junctions and thus ensuring optimal reliability of the measurement, in order to fulfill the requirements of the end-user.

Applications	Available combined electrodes	Available glass electrodes	Available reference electrodes
Hydrofluoric acid			
Photographic baths	For each electrode in the catalog and according to the application to which it is aimed, the following indication is mentioned:		
Sea water, swimming pools			
Emulsions			
Teaching	• *recommended*		
Grease, cosmetics	• *possible*		
Hops, beer	• *not recommended*		
Oil	• *deterioration risk*		
Milk			
Lacquers			
Paint			
Penetration (cheese)			
Blood			
Alkaline solutions			
Suspensions			
Soil			
Yoghurt, curdled milk			

Table 2.2. *Selection mode of the appropriate pH electrode for a given application*

2.5.2. *Quality and metrology of the standards and of the instruments*

2.5.2.1. *Traceability chains and calibration procedures*

In principle, a primary standard should be universal and stored in drastic conditions. As far as standard solutions are concerned, the problem is more delicate as, to be used as reference, these solutions must be manipulated, which entails risks of modification of their properties. The "primary" standards are developed in only a few laboratories in the world. Their specific properties (conductivity, pH) are determined with high precision and serve as references. So-called "secondary" standards are then prepared in similar conditions and their properties are compared on the ground with those of the primary standards, which entails a markedly wider uncertainty range on their value.

The secondary standards are then dispatched, in various forms (phials, powders) to laboratories or suppliers who will be able to adjust their own production by comparison, before proposing them to their clients, with an uncertainty level that is, in turn, significantly higher than those of the secondary standards. The whole set of stages stretching along this transfer of reference measurements constitutes the traceability chain.

Figure 2.6 presents the two possible types of traceability chains applied to the reference materials constituted by pH buffer solutions elaborated, for instance, by Radiometer Analytical SAS.

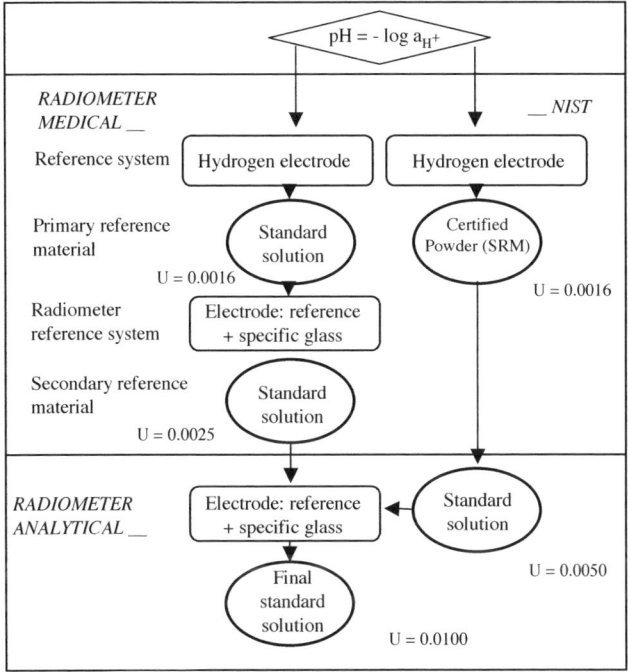

Figure 2.6. *Traceability chain of the pH buffer solutions certified by Radiometer Analytical*

The reference laboratory of *Radiometer Analytical* has been accredited by the COFRAC[1], which guarantees the traceability of the products prepared and standardized in this laboratory.

---

1 Comité Français d'Accréditation (French Committee of Accreditation).

As far as the pH electrodes are concerned, the quality guarantee is based upon a strict control of every electrode produced. The following characteristics are specifically examined using a control bench involving widely the signal processing of the results:

– slope of the response as a function of pH,

– response time,

– influence of stirring: this criterion is essential for testing the combined electrodes because it allows checking the absence of liquid junction potential[2].

Conductivity measurements

The primary standards are elaborated in a few certified laboratories. The calibration is operated in thermostatic cells ($\pm 10^{-3}$°C), by impedance measurements of the standard solution contained in calibrated tubes with various lengths. The electrical uncertainty of the measurements is less than 0.005%. In the same way, starting from the primary standards, secondary standards are prepared and dispatched to the suppliers who wish to link their standards to an indisputable traceability chain.

The certified standards can also be used for determining cell constants, the results of which will be traceable to the primary standards. *Radiometer Analytical SAS* proposes to certify certain types of conductivity cells used for testing the chemical purity of the water used in pharmaceutical industry (USP25/645 standard).

Selective electrodes

In principle, the samplings are not performed using traceable solutions because such solutions (particularly the most diluted ones) would not have enough stability. According to the domain of use of the electrodes, three to five measurement points are done with solutions having various concentrations (use of an automatic bench with signal processing) [FOM 91]. The overall response curve is calculated by adjusting the experimental points to the equation of Nikolskii. The control sheet delivered together with each electrode includes the calibration curve, the value of the response slope, the value of the potential for unit activity, and the experimental detection limit, in conformity with the recommendations from IUPAC (International Union of Pure and Applied Chemistry).

---

[2] This test consists of measuring the pH with and without stirring. The difference must be less than 0.5 mV at pH = 7.

2.5.2.2. Calibration of the measuring instruments

Numerous users also demand a normalized certification of the measuring instruments. For that, electric standards are used, the traceability of which is also defined metrologically. The calibration certificates are delivered upon request with the delivery of the instrument. They contain various pieces of information among which, for example, is the linearity level in a given range of measurements.

2.6. Conclusions and prospects

The field of chemical sensors is doubtlessly multidisciplinary and in full development. Improvements in the field of materials elaboration, of the control of their bulk or surface properties, and the support of the latest technologies have widely contributed either to the optimization of sensors based on new concepts, or to their potential applications in new fields. However, the development of such sensors on an industrial scale necessitates close collaboration between laboratories and specialized instrumentation companies.

2.7. Bibliography

[FOM] FOMBON J.J., PAUZON J.J., "Capteurs ioniques et moléculaires. Tests de validation et développement industriels", in JAFFREZIC N. *et al.* (eds), *Les Capteurs Chimiques*, CMC2, p. 123-130, 1991.

[FOM 91] FOMBON J.J., Procédé et appareil pour la mesure potentiométrique de la concentration d'une espèce chimique chargée, ou d'un gaz dissous, French patent no. 9107276.

[IRV 77] IRVING H.M.N.H., FREISER H., WEST T.S., *Compendium of Analytical Nomenclature*, IUPAC, Pergamon Press, p. 168, 1977.

Books dealing with chemical sensors

[KOR 91] KORYTA J. (ed), *Ions, Electrodes and Membranes*, John Wiley & Sons, 1991.

[MOR 81] MORF W.E., *Studies in Analytical Chemistry, vol. 2, The Principle of Ion-Selective Electrodes and of Membrane Transport*, Elsevier, 1981.

[GOP 89] GÖPEL W., HESSE J., ZEMEL J.N. (eds), *Sensors, A Comprehensive Survey*, vol. 1, *Fundamental and General Aspects*, VCH, Winheim, 1989.

[GOP 91] GÖPEL W., HESSE J., ZEMEL J.N. (eds), *Sensors, A Comprehensive Survey*, vol. 2, *Chemical and Biochemical Sensors*, VCH, Winheim, 1991.

[SEI 88] SEIYAMA T. (ed), *Chemical Sensor Technology,* vol. 1 & 2, Kodansha Ltd./Elsevier, Amsterdam, 1988 & 1989.

[YAM 91] YAMAZOE N. (ed), *Chemical Sensor Technology,* vol. 3, Kodansha Ltd./Elsevier, Amsterdam, 1991.

[YAM 92] YAMAUCHI S. (ed), *Chemical Sensor Technology,* vol. 4, Kodansha Ltd./Elsevier, Amsterdam, 1992.

Chapter 3

Sensitivity and Selectivity of Electrochemical Sensors

3.1. General concepts

As mentioned in Chapter 1 of this volume, the function of a sensor is to capture a physical characteristic, such as activity, concentration, or a partial pressure in the case of chemical sensors, then to transform it into an electric signal via a function generally called a transducing function. The main advantage of electrochemical sensors is that they ensure both the sensitivity towards the measured quantity, and the transduction into an electric signal, which is easily and directly adaptable, in most cases, to a measuring instrument.

These sensors make use of the ability of materials to exchange matter with the analyzed medium, by an electrochemical process, either involving ionic species or molecular entities to be necessarily transformed into ionic species at the interface. For that, sensitive materials must be ionic conductors (in the case of membranes) or electronic conductors (in the case of electrodes).

When the device exhibits higher sensitivity towards a given species (or towards a given class of species) it is characterized by the additional dimension of selectivity. Selectivity is an extremely important property in the case of chemical analyses, especially in aqueous solutions where a high variety of species are simultaneously encountered. Such specificity can be obtained, in particular, with materials possessing the properties of molecular recognition, which develop, on the molecular

---

Chapter written in French by Pierre FABRY and Jean-Claude MOUTET, and translated by J.C. POIGNET and P. FABRY.

scale, one or several types of specific interactions with the target species: coordination phenomena, electrostatic interactions, hydrophobic interactions, hydrogen bonds. These recognition properties can be utilized when a molecular receptor is integrated in an electrochemical device, for instance, by fixation on an electrode, or by incorporation into a material deposited on the electrode. The interactions with the target species of the analyzed medium and this host species take place at the interface with the solution, or in the bulk of the material if it allows penetration of the target species.

The term membrane is often reserved for materials having ionic conduction, where the mobile ions are able to exchange with the analyzed medium, i.e. ionic or ion-sensitive membranes. This term is also used in the case of porous or microporous membranes that are permeable to neutral species in a more or less selective way (perm-selective membranes).

3.1.1. *Various kinds of electrochemical sensors*

As discussed in the literature [OEH 91] and in Chapter 1, electrochemical sensors can be classified according to their transduction modes.

3.1.1.1. *Potentiometric transduction*

This term concerns devices that deliver a potential signal at zero current. The sensing material (electrode or membrane) reaches an equilibrium state with the environment, via exchange of matter, and, therefore, undergoes fluctuations of its internal electric potential. These potential variations are transmitted to the measuring device via an electrochemical chain. Ion-selective electrodes (ISE) belong to this category [COV 79, MOR 81, KOR 83]. Of course, a reference point with a constant voltage (reference electrode) must be available. In fact, a potentiometric sensor is equivalent to an electrochemical half-cell, the remaining half-cell being the external reference electrode [IVE 61]. Figure 3.1 shows an example of a complete electrochemical chain.

Me	Ag	AgCl	Internal solution Cl^-, Na^+	Na^+ glass	Na^+ solution to be analysed	External reference	Me

Figure 3.1. *Electrochemical chain corresponding to a Na^+-sensitive ISE (glass membrane) dipped into an aqueous solution, with an external reference electrode*

It is important to consider here that the quantity of matter exchanged between the membrane and the measured medium during the establishment of equilibrium is extremely minute, because very few charges (ions) are sufficient to modify the potential significantly. According to the interface exchange properties, the membrane will be sensitive towards a particular species or not.

3.1.1.2. *Amperometric transduction*

In this case, the membrane exchanges matter with the medium under the action of an electric current (matter consumption generally occurs in the measured medium). The experimental conditions are selected in such a way that the activity (or concentration, or pressure) is zero at the interface. In this limiting condition, the current reaches a limit fixed by the facility of matter to move and feed the interface. The Clark electrode for measuring dissolved oxygen [CLA 59, OEH 91] is the most famous example.

Some gas sensors are based on this principle but, because of a dissymmetric conception, they themselves generate an electric potential, and the battery thus constituted, once short-circuited, provides the necessary current. These sensors are of the "fuel cell" type [MOD 91, GOU 08]. They are essentially used for the analyses of gaseous species, but a device working on this principle could be designed for analysis in liquid solutions.

At high temperatures, the so-called "coulometric sensors" work by integration of the quantity of current necessary to empty or fill a closed vessel, by the effect of electrochemical pumping [HAA 77, KLE 91, FAB 97]. There is no equivalent in the case of ionic or gaseous species in solution, although nothing fundamentally precludes such a conception.

3.1.1.3. *Conductimetric transduction*

In this case, an electric signal, generally alternating with a frequency selected in order to minimize the electrode polarization effects, is applied to the sensor. The resistance, which is nearly proportional to the overall ionic concentration in solution, is then measured. In the case of ionic sensors, the resistance involves all the ions present in the solution and, as it also depends on the values of the ionic mobilities. There is no selectivity, unless we can separate the species, as a function of time for instance (chromatography), by a construction artifact (see Chapter 10).

Solid materials can be used if their conductivity is modified by the surrounding medium. In this case, the sensitive effect is indirect (second kind). Some gas sensors are based on this principle, at high temperature, using oxides with a mixed conductivity [FAB 97]. The latter devices are by no means suitable for measurements in solutions. However, organic conductors offer a similar approach for which a conductimetric detection is viable. This is particularly the case for electronic conducting polymers, the electrochemical properties of which are extremely sensitive to the environment [SKO 98]. Their interaction with chemical species, either in gaseous phase or in solution, can induce a very large variation of their conductivity. A large increase in conductivity can be observed in the presence of oxidizing or reducing species, which, in this case, play the role of n or p dopants [HEI 90].

The conductivity of such polymers is also extremely sensitive to changes in morphology [WAR 89] and to the creation of defects in the polymer [BER 87] induced by interaction with chemical species, which cause a decrease of the bond lengths and/or of the electronic communication between the polymer chains. Electrostatic [AND 83] and solvation effects [JAV 88, MCD 94] can also cause significant modifications of conductivity of these polymers, because of their influence on the mobility of the charge carriers in the polymeric matrix. It is important to note that the charge transport properties in a conducting polymer, which reflect the overall characteristics of the system, can be markedly altered by a small number of localized events. In other words, the transport of a large number of charges in a conducting polymer, i.e. its conductivity, can be markedly influenced by merely one or a few interactions with an analyte, which results in amplification of the detection [SWA 98]. Moreover, grafting or encapsulating functional molecules in conducting polymers allows elaboration of electro-active materials presenting selective interactions and responses towards target species [GAR 89].

Lastly, recent results reported in the literature have demonstrated the possibility of detecting electro-inactive metallic cations, using impedance spectroscopy, at gold electrodes modified by a self-assembled monolayer of selective ligands, such as crown-ethers [FLI 98, FLI 99, BAN 00] or nitrolotriacetic acid [STO 97], via the variation of the capacitance or of the charge transfer resistance of the ligand monolayer.

3.1.1.4. *Voltammetric transduction*

In this case, an electronic conducting electrode is used that exchanges electrons with the species in solution and modifies their oxidation numbers. This operation is made by applying a potential sweep, and measuring the resulting current waves or peaks that enable the identification of the species (potential values of the peaks or the waves) and its concentration (height of the peak or of the wave). The voltammetric methods developed, such as polarography, rotating disc voltammetry, cyclic voltammetry or differential pulse voltammetry are efficient and polyvalent analysis techniques (see Chapter 8). So far, they have hardly been used in the field of sensors because they necessitate more complex signal control and data processing devices than in the case of usual sensors. Such devices rather fall within the field of electrochemical analysis instrumentation. Nevertheless, along with the improvements of microcomputing science, they might undergo interesting developments in the future.

It must be noted that voltammetry with anodic stripping at a mercury electrode is certainly one of the most sensitive techniques for trace analysis of metals in the environment [WAN 94a] (see Chapter 8).

Because of toxicity problems and some limitations linked to the use of mercury, recent research in this field has shifted towards the development of non-toxic materials, such as, for example, electrodes made from doped diamond layers [WU 96, FER 02] or molecular materials allowing pre-concentration of metals by complexation [BEE 98]. These novel interfaces can also increase the sensitivity and selectivity, with the use of ligands that complex specifically targeted metallic cations [BAK 02]. Electro-active materials, such as, conducting polymers functionalized by complexing agents, can ensure transduction of the electrochemical response of the sensor to an analyte [BAN 00].

A very recent approach in voltammetric analysis concerns redox-active receptors, created by covalent assembling of a selective molecular receptor with a stable and reversible redox group, such as a metallocene or an organic redox group [BEE 98]. These systems are meant to recognize charged or neutral species, or even non-electro-active species, provided that the interaction of the target species with the receptor perturbs its intrinsic electrochemical activity. However, studies of these systems remain at the fundamental level, and currently, no sensor based on this principle has been developed [BAK 02].

3.1.2. *Interference and selectivity*

The problem of interference is inherent to chemical sensors. In the case of electrochemical sensors, it is caused by non-selective exchange phenomena at the interface between the membrane and the analyzed medium which, in the presence of interfering species (noted i), leads to an electrical response that overlaps that generated by the target species, which will be called the "primary species" (denoted p). The sensor responds then to the concentration, C_p, of the primary species, but, at the exploitation level, one obtains an approximating value given by:

$$C_{app} = C_p + \Delta C \qquad [3.1]$$

where ΔC stands for a systematic error, generally in excess. In fact, this expression arises from several phenomena and is expressed in the following form:

$$C_{app} = C_p + \sum_i K_{p,i} C_i \qquad [3.2]$$

where $K_{p,i}$ are the selectivity coefficients of the primary species in the presence of the interfering species i. The lower these coefficients are, the more selective the system is. Notice here that the terminology adopted is not very judicious from a logical point of view. It would be wiser to use interference coefficients of ions, i, on the primary species, p. The value of the coefficient would thus be the smaller as interference would be weaker, the logic would then be direct instead of reversed. Nevertheless, to concur with a standard frame, it is better to surrender and accept the terminology adopted by the International Union of Pure and Applied Chemistry (IUPAC).

Chemometrics rests on a whole set of measurements using several sensors and a mathematical processing of the results which allows acceptance of the major part of this difficulty (see Chapter 9). It is mostly used for the gas sensors (electronic nose), and rather scarcely developed, until now, for species in solution using electrochemical devices (electronic tongue) [OTT 85, VLA 97, BAR 00].

Among the other methods able to cope with the interference problems are those consisting of devices which have good selectivity properties, i.e. with a very low $K_{p,i}$ coefficient. Several methods enable this:

– use of selective membranes (inorganic membranes with selective conduction);

– introduction of highly complexing molecules into the sensitive material;

– use of catalysts to accelerate reactions involving the primary species to the detriment of the interfering species;

– introduction of perm-selective membranes to slow down, or even block, the flux of interfering species at the interface.

Obviously, the interfering species could be eliminated using chemical reactions, for instance, by a reactant causing precipitation. However, this rather efficient method can only be applied to very specific cases, for instance, when adding a reactant that does not disturb the analyzed medium (analysis of samples in the laboratory).

3.1.3. *Nature and shape of materials*

The materials used in the fabrication of the ISE (pH, pNa) were initially bulb-shaped glass membranes. Subsequently, numerous ISE were built from disc-shaped membrane pellets, for example, monocrystalline (LaF_3 for pF measurement) or polycrystalline for the insoluble salts-based ISE. They are fixed to an electrode body made of a material that is inert towards the analyzed solutions. ISE based on liquid or polymeric membranes have similar configurations, but the nature of the membrane is specific (hydrophobic solvent containing complexing agents, with enrichment in ionic carriers for lowering the electric impedance, see Chapter 4).

Over the last 20 years, copious research has been carried out in the field of membrane manufacture using microelectronic (or micro-ionic) processes, to obtain miniaturized sensors with thin or thick-layer configurations (thermal evaporation, cathodic sputtering, screen-printing, etc, see relevant chapters).

Grafting techniques of membranes or of complexing molecules, often based on organic compounds, are also under development for making ISFET (see Chapter 6) or modified electrodes.

3.2. Models for the sensitivity and selectivity of potentiometric sensors

3.2.1. Basic concepts

Prior to describing the phenomena governing selectivity, it is worth noting the principles underlying the sensitivity towards an ionic species, because it plays a major role in interference phenomenon.

As already mentioned, the fundamental principle of potentiometric sensors relies on the establishment of an equilibrium between the conducting material (called the membrane) and the external medium, via the electrochemical exchange of species. The description of this equilibrium can be made using classical electrochemical thermodynamic concepts (see Chapter 1). From the equilibrium reaction

$$\sum_{reactants} v_r Reactants \Leftrightarrow \sum_{products} v_p Products \qquad [3.3]$$

we obtain the following relation between the electrochemical potentials (or chemical potentials for neutral species):

$$\sum_{reactants} v_r \tilde{\mu}_r = \sum_{products} v_p \tilde{\mu}_p \,. \qquad [3.4]$$

Let us note that $\tilde{\mu}_A$ is the electrochemical potential of species A (see Chapter 1), which can be written:

$$\tilde{\mu}_A = \mu^\circ{}_A + RT \ln a_A + z_A F \varphi_\alpha \qquad [3.5]$$

where φ_α is the electrical potential of phase α containing A (with charge z_A), $\mu°_A$ is the chemical potential of A taken in its standard state.

Two cases will be distinguished, whether the ionic exchange is direct or involves a change of oxidation state (electrode reaction).

3.2.1.1. Single ionic exchange

This case corresponds to an interface between a membrane (denoted phase α) and the analyzed solution (phase β). The exchanged species is supposed to be an ion, A^{z+}, with the algebraic charge, z. The equilibrium can then be simply written:

$$A^{z+}{}_\alpha \Leftrightarrow A^{z+}{}_\beta \qquad [3.6]$$

A basic calculation that makes use of the equality of the electrochemical potentials, yields the relation:

$$\varphi_\alpha - \varphi_\beta = \frac{\mu°_\beta - \mu°_\alpha}{zF} + \frac{RT}{zF} \ln \frac{a_{A,\beta}}{a_{A,\alpha}}. \qquad [3.7]$$

This potential difference, called Galvani voltage, plays a fundamental role in the function of a potentiometric sensor, because it varies with the activity of the A^{z+} ion in the solution, provided that the activity of A^{z+} in the membrane remains constant. For instance, it is the case of an ionic conducting membrane in which A^{z+} is highly concentrated. This condition is sufficient, but not necessary in the case of an electrochemical chain with several interfaces. The fundamental assumption is always that the membrane by itself is at equilibrium, i.e. that the electrochemical potential of A^{z+} does not depend on its position in the membrane. Only the difference between both ends of the chain is taken into account in the calculation. This will be illustrated later on in the case of a sensor of the first kind.

More generally, when several exchanges can occur at an interface, an interference phenomenon intervenes. The measured voltage is then marred by mistakes, as previously mentioned. In the case of an ionic interference, Nikolskii proposed an empirical law [NIK 37] quantifying these phenomena:

$$E = E° + \frac{RT}{z_p F} \ln(a_p + \sum_i K^{pot}_{p,i} a^{z_p/z_i}) \qquad [3.8]$$

where a_p and a_i are respectively the activities of the primary and interfering ions, z_p and z_i their respective charges and $K^{pot}_{p,i}$ the potentiometric interference coefficients of species i, in the presence of the primary ion p.

It may be noticed that, in the absence of any primary species p, the law appears *a priori* complex, however, if the membrane is sensitive to a major interfering species (indexed j), a similar relation is obtained, but with a shift in the value of $E°$ due to the coefficient $K_{p,j}^{pot}$.

The pH electrode is based on such a property. The ionic membrane is a glass doped by alkaline ions (A^+) that have extremely strong interference with the protons ($K_{A,H}^{pot}$ of the order of 10^{13}). The Nikolskii law, applied to an alkaline sensitivity, would give the expression:

$$E = E° + \frac{RT}{F} \ln(a_A + 10^{13} a_H) \qquad [3.9]$$

which can also be written:

$$E = E'° + \frac{RT}{F} \ln(a_H + 10^{-13} a_A). \qquad [3.10]$$

In this use as an inverse function, i.e. for sensitivity to pH, the interference towards the alkaline ions gets extremely weak. However, it causes a limitation of the use of such a membrane in the case of very basic media, were the alkaline ions are markedly predominant with respect to the protons.

3.2.1.2. *Exchange at an electrode*

This case corresponds either to a metallic electrode in contact with a solution (dissolved redox couple), or at the head of a chain, between the metallic connector M and an ionic phase (membrane or intermediate phase in the internal reference system, phase γ). One can for instance write the following reaction:

$$M_{metal} \Leftrightarrow M^+{}_\gamma + z\,e^-{}_{metal} \qquad [3.11]$$

which leads to the relation

$$\varphi_{metal} - \varphi_\gamma = \frac{\Delta\mu°}{zF} + \frac{RT}{zF} \ln \frac{a_{M^{z+},\gamma}}{a_{M,metal}} + \frac{RT}{F} \ln a_{e,metal}. \qquad [3.12]$$

This expression contains the activity of electrons in the metal (electrons are then considered as an electrochemical species), but, in fact, as the electron concentration in a metal is very high, the corresponding term is constant. Nevertheless, as there are always two electric connectors in a measurement chain (a reference electrode is necessary), this term vanishes in the overall value of the measured potential difference. It is to be noted that the activity of M is equal to 1 if M is a pure metal.

In the case of equilibriums at metallic electrodes or, more generally speaking, at materials partly conducting electrons (n or p type), if several redox couples intervene, each couple takes part in the definition of the potential. The resulting value is intermediate between the thermodynamic potentials of the two couples (in the simplest case), and is called "mixed potential", which is well known in electrochemistry, particularly in the field of corrosion [SAR 91]. One of the couples plays an oxidizing role, while the other plays a reducing role. An example is schematically given in Figure 3.2. Therefore, the total current ($I = I_{ox} + I_{red}$), resulting from a positive oxidation current (I_{ox}) and a negative reduction current (I_{red}), both having the same amplitude, is zero. The value of E_m depends on the equilibrium potentials (E_1 and E_2) of the couples, but also on the kinetics of the electrode reactions. The fastest kinetics will be that with the most flattened $E(I)$ curve in the selected system of coordinates, i.e. with the smallest variation of the potential as a function of the current. It tends to impose its potential (as in the case of reaction 2 in Figure 3.2). The problem is generally complex and cannot be easily modeled.

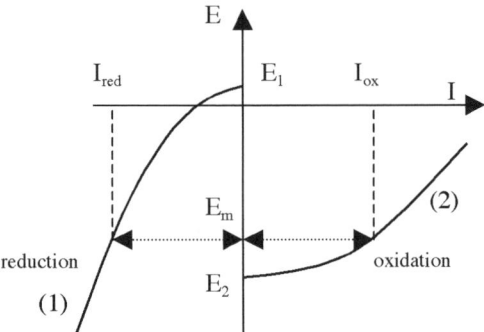

Figure 3.2. *Typical behavior of an electrode with a mixed potential*

The redox electrode used for monitoring the oxidizing (or reducing) property of water works on this principle. This kind of electrode is not a selective sensor, it only indicates a global state; in practice, this is interesting for determining, for instance, the oxidizing power of water, or of an effluent, or for doing laboratory redox titrations.

The phenomenon of mixed potential plays a harmful role in the functioning of an ionic conducting membrane because it too has electronic conduction, as the interface can be the support for several exchanges. This situation corresponds to an interference phenomenon, particularly if the analyzed solution contains a dissolved redox couple, for example, dissolved oxygen, which fixes a rather high electrode potential (O_2/H_2O couple) in water. The Nikolskii law no longer applied in this case, modeling is complex because the value of the mixed potential, situated between E_1 and E_2, depends on the equations ruling the kinetics of both reactions (Butler Volmer type regime or limiting diffusion type, etc.) [SAR 91]. However, if the exchange kinetics of the primary ion with the membrane are very fast (quasi-constant potential whatever the current), it will tend to fix the potential, and the deviation due to this type of interference will possibly be low. This is the reason why membranes of silver sulfides or selenides, which possess some electronic conduction, are weakly sensitive to the presence of oxygen in the analyzed solution.

3.2.2. *Ionic conducting membranes of the first kind*

3.2.2.1. *Sensitivity*

In the case of ISE of the first kind, the ionic conduction of the membrane is insured by the same ionic species as that which is to be measured, for example, the glass membrane, which conducts using Na^+ ions, is used to measure Na^+ concentration. The equilibrium at the interface between the membrane and the solution fixes a value of the electrical potential of the membrane, which must be picked by an electric connexion (metal). It is necessary to set up an electrochemical chain between these two phases which is as reversible as possible (very fast exchanges at the interfaces) to obtain stable and quick responses. The chain can be symbolized by the diagram given in Figure 3.3.

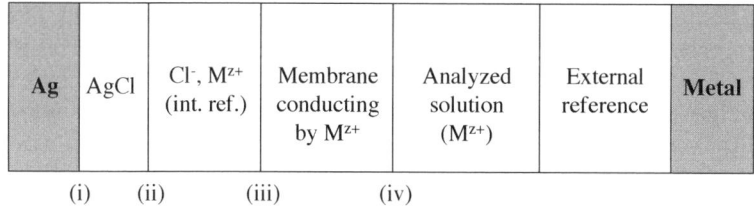

Figure 3.3. *Electrochemical chain corresponding to an ISE of the first kind dipped in a solution*

The equilibriums at the interfaces (i), (ii), (iii) and (iv) are respectively:

$$Ag \Leftrightarrow Ag^+_{(AgCl)} + e^-_{(Ag)} \qquad [3.13]$$

$$Cl^-_{(AgCl)} \Leftrightarrow Cl^-_{(int)} \qquad [3.14]$$

$M^+_{(int)} \Leftrightarrow M^+_{(memb)}$ [3.15]

$M^+_{(memb)} \Leftrightarrow M^+_{(sol)}$. [3.16]

For each equilibrium, the equality of the electrochemical potentials of the species intervening at the interface is written. At equilibrium, the electrochemical potential of the mobile species is constant in all points of the material, and the difference of the electric potentials of silver (Ag) and the analyzed solution can be calculated. It will be noted that the potential of the solution is also related to that of the metal used at the reference electrode. The following relation is then obtained:

$$\varphi_{Ag} - \varphi_{ref} = E° - \frac{RT}{F} \ln[a_{Cl^-,int}(a_{M^{z+},int})^{1/z}] + \frac{RT}{zF} \ln a_{M^{z+},sol} \quad . \quad [3.17]$$

The logarithmic variation of the delivered voltage is of the Nernstian type. It is noticeable that the second term of this expression contains the activities of the ions in the internal system (also called the internal reference or internal ionic bridge, or internal contact, depending on the authors). This expression readily shows that the voltage depends on the salt concentration in this internal system. This concentration is not, in principle, modified in the course of the measurements, provided that no electric accident, such as a short circuit, occurs and causes enrichment or depletion, depending on the direction of the current flowing through the system.

3.2.2.2. Eisenman's model for describing interferences

Eisenman's model [EIS 68] was established to account for the interference mechanism of the glass membrane. It is based on the diffusion of the interfering ion across the membrane. The interface with the solution then becomes sensitive towards the primary ion and to the interfering ion via a double exchange. Eisenman had conceived his model for a Na^+ ion-sensitive membrane, where H^+ was considered as interfering. But we know, at the present time, that H^+ ions do not penetrate the glass and that the interference is a surface reaction phenomenon (due to silanol groups, see Chapter 1). It turns out that the example selected was an ill-founded choice; however, the model holds its validity for the study of the interference of other alkaline ions.

Internal solution Na^+	m Na^+	m' Na^+, Li^+	Analyzed solution Na^+, Li^+

Figure 3.4. *Electrochemical chain for a Na^+-sensitive ISE dipped in a solution containing interfering Li^+ ions*

To illustrate this model, we will take the example of a membrane sensitive to Na^+, with Li^+ as interfering ion. The electrochemical chain can be represented by the diagram of Figure 3.4. A junction potential occurs at the interface between the polluted zone (m') and the non-polluted membrane (m), due to the fact that Li^+ and Na^+ ions, submitted to concentration gradients, tend to diffuse towards the bulk of the membrane or inversely. Concentration profiles are qualitatively shown in Figure 3.5.

The potential difference appearing across this non-homogenous zone (which can be very thin and even limited to the interface) can be evaluated using the Henderson relation when the initial activity a_{Li} in the membrane is zero [HEN 07]:

$$E_{Henderson} = \varphi_{m'} - \varphi_m = \frac{RT}{F} \ln \frac{D_{Na} \times a_{Na}^m}{D_{Na} \times a_{Na}^{m'} + D_{Li} \times a_{Li}^{m'}} \qquad [3.18]$$

where D_{Na} and D_{Li} are the respective diffusion coefficients (which could be replaced by the electrochemical mobilities) of the Na^+ and Li^+ ions in the membrane (both are assumed to keep the same value in m and m'), and a_{Na} and a_{Li} their activities in the phases indicated by superscripts m or m'.

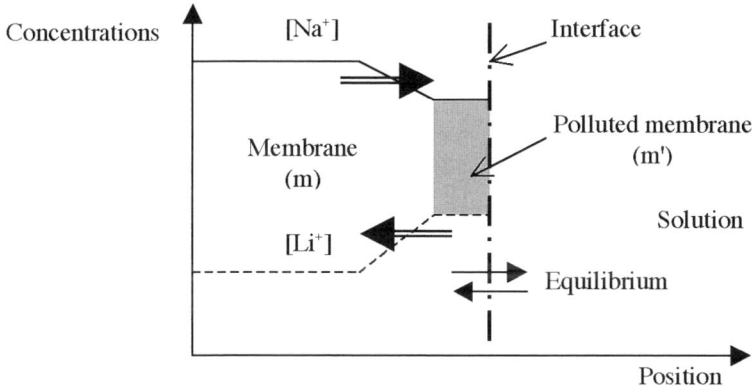

Figure 3.5. *Concentration profiles of (primary) Na^+ ions and (interfering) Li^+ ions in an ionic conducting membrane, caused by diffusion of the interfering ion from the solution*

The equilibrium established at the interfaces between the solutions (analyzed solution and internal reference) is:

$$Na^+_{sol} \Leftrightarrow Na^+_{memb} \qquad [3.19]$$

A similar equilibrium can also be written for Li^+ ions at the interface between the solution and the m' part of the membrane. Moreover, taking into account the exchange equilibrium at the interface where interference occurs

$$Li^+_{sol} + Na^+_{m'} \Leftrightarrow Na^+_{sol} + Li^+_{m'} \quad [3.20]$$

with the equilibrium constant K_{exch}

$$K_{exch} = \frac{a^{sol}_{Na} a^{m'}_{Li}}{a^{sol}_{Li} a^{m'}_{Na}} \quad [3.21]$$

and by combining the various contributions of the potential difference between the metal of the internal connection and the solution (and, therefore, that of the external reference electrode), we finally obtain

$$E = \frac{RT}{F} \ln \left(\frac{a^m_{Na} a^{sol}_{Na}}{a^{int}_{Na} a^{m'}_{Na}} \frac{D_{Na} a^{m'}_{Na} + D_{Li} K_{exch} a^{sol}_{Li} a^{m'}_{Na} / a^{sol}_{Na}}{D_{Na} a^m_{Na}} \right) \quad [3.22]$$

or, put in another way,

$$E = \frac{RT}{F} \ln \left(\frac{1}{a^{int}_{Na}} \right) + \frac{RT}{F} \ln \left(a^{sol}_{Na} + \frac{D_{Li}}{D_{Na}} K_{exch} a^{sol}_{Li} \right) \quad [3.23]$$

and, setting

$$K^{pot}_{Na,Li} = \frac{D_{Li}}{D_{Na}} K_{exch} \quad [3.24]$$

we obtain an expression of the form:

$$E = E° + \frac{RT}{F} \ln \left(a^{sol}_{Na} + K^{pot}_{Na,Li} a^{sol}_{Li} \right). \quad [3.25]$$

This expression is equivalent to the Nikolskii relation (also called the generalized Nernst law). According to this model, the selectivity coefficient is a function of the ratio of the mobilities of the interfering and primary ions. Moreover, it depends on the equilibrium constant of the exchange between the primary and the interfering ions between the solution and the membrane. This coefficient will be weaker as the membrane hardly accepts interfering ions in its bulk and as these ions are less mobile in the membrane. Therefore, steric effects acting on conduction properties can play a major role in the selectivity of this type of membrane. In this respect, experimental results were obtained in agreement with this point in the case of a crystalline membrane such as NASICON [MAU 98].

This demonstration is obviously based on a simple case, where both ions have a single charge. In the case of ions with different charges, this equation is no longer valid, but the physical basis of the model can be retained.

This kind of model seems to indicate that the selectivity coefficient is a constant, the value of which depends on the membrane properties. In fact, it has been observed that this value depends on the concentration of the primary ion and on that of the interfering ion [CRE 97a]. This is the reason why the experimental conditions in which the values of $K_{p,i}^{pot}$ have been determined must always be specified [REN 99].

Moreover, it has been observed, for instance, using ISE in flow injection analysis (FIA cells: analysis under flux in dynamic regime), that the values of the selectivity coefficients in transient mode vary with time. Thus, if the interfering ion is slower to reach equilibrium than the primary ion, the selectivity coefficient can appear improved [CRE 97b], but the inverse situation can occur [DOR 02]. This time-dependence of the selectivity coefficients could become very interesting for the development of chemometrical methods.

3.2.3. *Ionic conducting membranes of the second kind*

In the case of sensors of the second kind, the membrane is conducts using an ion that reacts with that of the solution to be analyzed and forms an insoluble phase. A typical example is the AgX membrane used for measuring the concentration of a halide ion (X^-).

3.2.3.1. *Sensitivity*

In the abovementioned case, the potentiometric sensor can be schematized as in Figure 3.6.

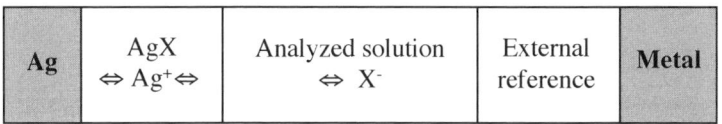

Figure 3.6. *Electrochemical chain of an ISE with an Ag^+ ions-conducting membrane sensitive to X^- ions (second kind)*

The global equilibrium between silver and the analyzed solution is

$$Ag + X^-_{solution} \Leftrightarrow AgX + e^-_{metal} \qquad [3.26]$$

which can be split into :

$$Ag \Leftrightarrow Ag^+_{internal\ interface} + e^-_{metal} \qquad [3.27]$$

$$Ag^+_{internal\ interface} \Leftrightarrow Ag^+_{external\ interface} \qquad [3.28]$$

$$Ag^+_{external\ interface} + X^-_{solution} \Leftrightarrow AgX \qquad [3.29]$$

Calculation of the voltage yields the same result whatever the reaction path used. The difference in writing the reactions merely allows better understanding of the transduction mechanism through the electrochemical chain. We finally get:

$$\varphi_{Ag} - \varphi_{sol} = \frac{1}{F}(\mu^\circ_{AgX} + \mu_e - \mu^\circ_{Ag} - \mu^\circ_{X^-,sol}) - \frac{RT}{F}\ln a_{X^-,sol} \qquad [3.30]$$

and, with respect to an external reference electrode:

$$E = E^\circ - \frac{RT}{F}\ln a_{X^-,sol}. \qquad [3.31]$$

3.2.3.2. Example of an electrode sensitive to heavy metals

The usual ISE sensitive towards transition and heavy metal ions are generally based on silver sulfides or selenides doped with sulfides or selenides of these metals, either as polycrystals, or as glasses (see Chapter 4). In this case, sensitivity is based on a similar principle. The free M^{2+} ions in the analyzed solution reach equilibrium with the S^{2-} ions of the membrane, which conducts using Ag^+ ions. The interface can be schematized as shown in Figure 3.7.

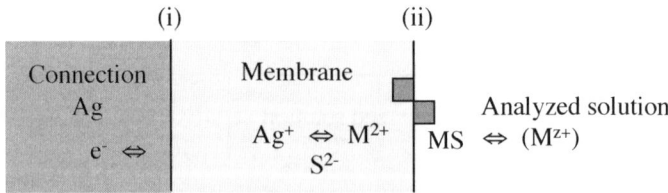

Figure 3.7. *Schematic description of an ISE made sensitive to M^{2+} ions by doping an Ag_2S membrane*

In this type of membrane, the Ag^+ ions are mobile, whereas M^{2+} and S^{2-} are not. The Ag^+ ions can partially be replaced by M^{2+} ions in the immediate vicinity of the interface (shaded squares in the figure) and form some MS groups causing the MS grains to grow or decrease slightly (if MS is in a separate phase) or causing the MS activity to increase if this species is dissolved in Ag_2S. The following equilibrium relations can then be written:

– at the interface (ii) $\quad M^{2+}_{sol} + S^{2-}_m \Leftrightarrow MS$ [3.32]

– within the membrane $\quad Ag_2S \Leftrightarrow 2\, Ag^+_m + S^{2-}$ [3.33]

– at the interface (i) $\quad 2\, Ag^+_m + 2\, e^-_{Ag} \Leftrightarrow 2\, Ag$ [3.34]

and the overall equilibrium is:

$$M^{2+}_{sol} + Ag_2S + 2\, e^-_{Ag} \Leftrightarrow 2\, Ag + MS.$$ [3.35]

Using the electrochemical potentials of the species involved in the equilibrium [3.35], we obtain the relation:

$$\varphi_{Ag} - \varphi_{sol} = \frac{1}{2F}(\mu°_{Ag_2S} + 2\mu_e + \mu°_{M^{2+}_{sol}} - 2\mu°_{Ag} - \mu_{MS}) + \frac{RT}{2F}\ln(a_{M^{2+}_{sol}}).$$ [3.36]

A Nernstian response to M^{2+} ions in the solution is again seen. In this expression, all the terms included in the first brackets are constants at a given temperature, apart from the μ_{MS} term, which is constant only if MS forms a distinct phase (i.e. $a_{MS} = 1$), or if MS forms a solid solution with Ag_2S that is sufficiently buffered to avoid significant fluctuations in composition. If this condition is not fulfilled, the response is not Nernstian.

3.2.3.3. Simplified interference model

The simplest model accounting for the response of this kind of sensor towards X^- ions in the presence of an interfering ion, Y^-, is based on the solubility equilibriums of AgX and AgY. In this way, if the sensor responds selectively to X^-, its voltage can be written:

$$E_1 = E°_X - \frac{RT}{F}\ln a_{X^-,sol} = E°_X - \frac{RT}{F}\ln K^s_{AgX} + \frac{RT}{F}\ln a_{Ag^+,sol}$$ [3.37]

and, if it responds preferentially to Y^-

$$E_2 = E°_X - \frac{RT}{F} \ln K^{pot}_{X,Y} a_{Y^-,sol} \qquad [3.38]$$

or, put in another way:

$$E_2 = E°_X - \frac{RT}{F} \ln K^{pot}_{AgX} - \frac{RT}{F} \ln K^s_{AgY} + \frac{RT}{F} \ln a_{Ag^+,sol}. \qquad [3.39]$$

When $E_1 = E_2$, the value of $a_{Ag,sol}$ is the same (considering that this membrane is of the first kind towards Ag^+) and, by identification, we can write the simple relation:

$$K^{pot}_{X,Y} = \frac{K^s_{AgX}}{K^s_{AgY}}. \qquad [3.40]$$

This relation can be expressed using the precipitation constants of the salts (with the relation $K^{preci} = 1/K^s$).

Equation [3.40] shows that the selectivity is determined by the ratio between the solubility products of the silver salts considered. The more soluble the interfering ion is in the analyzed solution (high value of K^s or low value of K^{preci}), the weaker the selectivity coefficient is. A few values of pKs for silver salts are given in Table 3.1. This table shows that a silver iodide membrane will show low sensitivity to Cl^- in the presence of I^-, but conversely, a silver chloride membrane will be very strongly influenced by the presence of I^- in the solution. There is good agreement between the experimental values and the theoretical ones deduced from the solubility products. Slightly more complex models are described in the literature [MOR 81].

AgX	AgSCN	Ag_2CO_3	Ag_2S	AgCl	AgBr	AgI
pKs	12	11	49	9.8	12.3	16

Table 3.1. *Values of pKs for silver salts in water (weak ionic strength)*

A similar model could be derived for the interference of N^{2+} with a membrane sensitive to M^{2+} ions, even though the latter is based on silver sulfide. From a practical point of view, the agreement between this simple theoretical model and the true values is only semi-quantitative.

3.2.4. *Liquid and organic membranes*

The use of organic molecules possessing a recognition center has been widely developed in the field of potentiometric sensors (see Chapter 4). In such cases, the molecule is included in a liquid medium, or in a polymer having hydrophobic properties. The exchange by complexation equilibrium takes place at the interface between this medium and the aqueous solution analyzed, which contains the target element (ion A^+). If, for instance, the molecule complexing the A^+ cation is L_m, we can write the reaction:

$$A^+{}_{sol} + L_m \Leftrightarrow LA^+{}_m \qquad [3.41]$$

and the calculation of the sensitivity, carried out with the same method as above, shows that L_m and $LA^+{}_m$ must have constant activities for the response to be Nernstian.

The most famous example is that of valinomycin, which strongly complexes potassium, the size of the complexation site being well fitted to that of K^+. Li^+ and Na^+ ions are too small, and the Cs^+ and Rb^+ ions too big, for a stable complex to be formed. The crown-ethers are additional examples of this type: the complexation constant for a given alkaline ion depends strongly on the size of this ion and on the number of oxygen atoms of the crown. In this respect, the steric effect is evident. The literature, which is very rich in the domain of ionophores for selective electrode membranes, has been extensively reviewed [BUH 98].

Fundamentally, the modeling of interference phenomenon is not similar to that of inorganic membranes with insoluble salts, because the activity of complexes cannot be considered as constant. Using Eisenman's model, it has been shown elsewhere that the selectivity coefficient can be written [FAB 08]:

$$K^{pot}_{A,B} = \frac{\tilde{u}_{LA}}{\tilde{u}_{LB}} \frac{K^{cpl}_{LB}}{K^{cpl}_{LA}} \qquad [3.42]$$

where $\tilde{u}_{LA}$ and $\tilde{u}_{LB}$ are the electrochemical mobilities, K^{cpl}_{LA} and K^{cpl}_{LB} the complexation constants of the complexes LA and LB, respectively. Nevertheless, because the complexation constants are generally very different whereas the mobilities are not, the selectivity coefficient can be assimilated, in a first approach, to the ratio of the complexation constants. In this way, it is possible to obtain highly selective membranes, for instance, with selectivity coefficients of about 10^{-4} or even less.

3.3. Case of amperometric sensors

3.3.1. *Principle of sensitivity*

If a species A is consumed at an interface, e.g. an electrode, a feeding flux establishes and tends to homogenize the solution with respect to the depleted zone. This is the diffusion phenomenon, as shown in Figure 3.8 (one-dimensional example). The flux in the zone that is not homogenous in concentration is governed by the general relation:

$$J_A = -C_A \tilde{u}_A \frac{\partial \tilde{\mu}_A}{\partial x} \qquad [3.43]$$

where $\tilde{u}_A$ is the electrochemical mobility of A, which is linked to the diffusion coefficient by:

$$\tilde{u}_A = \frac{D_A}{RT}. \qquad [3.44]$$

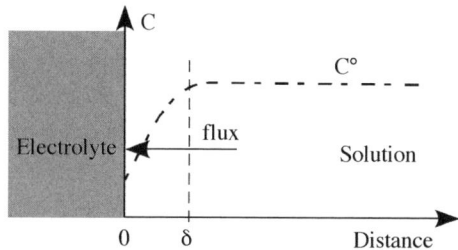

Figure 3.8. *Schematic diagram of the concentration profile of a species that is consumed at an electrochemically reactive interface*

If the solution is highly conducting, the electrical potential is constant and the flux relation is simplified:

$$J_A = -D_A \frac{\partial C_A}{\partial x} \qquad [3.45]$$

This relation is Fick's first law. The resulting current is proportional to the interfacial flux of the A species, the charge of which changes by n (algebraic) in the electron exchange reaction:

$$I = nFS\, J_A \qquad [3.46]$$

where S is the section area. If the bulk concentration of A is $C°$ at a distance δ from the interface, beyond the diffusion layer, and gets lowered to zero at the interface due to the reaction, the current reaches a limiting value:

$$I_{lim} = nFSD_A \frac{C°}{\delta}. \qquad [3.47]$$

This relation shows that the limit current is proportional to the concentration (which is considered as the activity) of the target species A. The proportionality coefficient is a function of the limiting mass transport parameters: diffusion or porosity in the inserted material, orifice in an insulating wall, diffusion-convection layer imposed by a particular hydrodynamic regime (rotating disc electrode, thin layer, or wall jet cells etc.). The Clark electrodes are based on this principle (see Chapter 5).

If the device is capable of filtering the species, either at the interface level (selective interface) or by a specific additional material, a selectivity effect can be obtained. The use of ionic membranes could help conceiving a rather large variety of new sensor devices.

3.3.2. Selectivity model

3.3.2.1. General expression

When several species (A and B for instance) can be simultaneously consumed at an interface at a given value of the potential, the overall current is the sum of the corresponding partial currents, provided that the reactions are independent:

$$I = I_A + I_B = FS\left(\frac{n_A D_A C_A}{\delta_A} + \frac{n_B D_B C_B}{\delta_B}\right) \qquad [3.48]$$

or

$$I = n_A FS \frac{D_A}{\delta_A}\left(C_A + \frac{n_B}{n_A}\frac{D_B}{D_A}\frac{\delta_A}{\delta_B} C_B\right) \qquad [3.49]$$

setting

$$\alpha_A = n_A FS \frac{D_A}{\delta_A} \quad [3.50]$$

and

$$K_{A,B}^{amp} = \frac{n_B D_B \delta_A}{n_A D_A \delta_B} \quad [3.51]$$

the sensitivity response is:

$$I = \alpha_A (C_A + K_{A,B}^{amp} C_B) \quad [3.52]$$

where α_A is the sensitivity towards the primary ion A and $K_{A,B}^{amp}$ is the selectivity coefficient of the sensor (or the interference coefficient of B on the measurement of the concentration of A).

3.3.2.2. *Insertion of a selective membrane*

Improving the selectivity is made possible by depositing a membrane material possessing properties of selectivity on the electrode surface (see Chapter 5). In this case, two diffusion phenomena occur: one in the solution, the other one through the membrane, as shown in Figure 3.9. The concentration at the electrode is zero when the current reaches the limiting value. Let C_{int} be the concentration value at the membrane–solution interface. The conservative flux remains constant whatever the position, leading to the equation:

$$nFS \frac{D}{\delta}(C_{sol} - C_{int}) = nFS \frac{D_m}{d} C_{int} \quad [3.53]$$

where D and D_m are the diffusion coefficients of the electro-active species, respectively, in the solution (with a thickness δ), and in the membrane (with a thickness d).

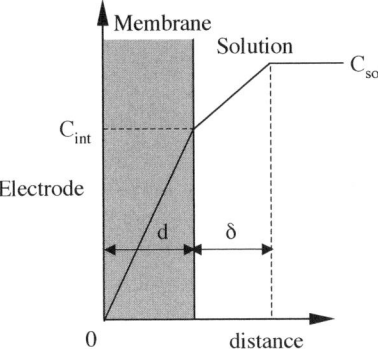

Figure 3.9. *Concentration profiles at an electrode covered by a membrane*

By setting

$$P_{sol} = D/\delta \quad \text{and} \quad P_m = D_m/d \qquad [3.54]$$

for the respective permeabilities of the solution and of the membrane, we obtain

$$C_{int} = C_{sol} \frac{P_{sol}}{P_{sol} + P_m} \qquad [3.55]$$

and, as

$$I_{lim} = nFS \frac{D_m}{d} C_{int} \qquad [3.56]$$

it generates:

$$I_{lim} = nFS \frac{D}{\delta} C_{sol} \frac{1}{1 + P_{sol}/P_m}. \qquad [3.57]$$

If the electrode is sensitive towards two species and if the reactions are independent (the limiting currents are additive), the following relation is obtained:

$$I_{lim} = n_1 FS \frac{D_1}{\delta_1} \frac{1}{1 + P_{sol,1}/P_{m,1}} (C_{sol,1} + \frac{n_2 D_2 \delta_1 (1 + P_{sol,1}/P_{m,1})}{n_1 D_1 \delta_2 (1 + P_{sol,2}/P_{m,2})} C_{sol,2}) \qquad [3.58]$$

and a selectivity coefficient can be defined as follows:

$$K_{1,2}^{amp} = \frac{n_2 P_{sol,2}(1 + P_{sol,1}/P_{m,1})}{n_1 P_{sol,1}(1 + P_{sol,2}/P_{m,2})} \qquad [3.59]$$

Therefore, the choice of the membrane controls selectivity improvement. Wang has derived and applied this model to amperometric sensors [WAN 94b]. He showed that the interference of ascorbic acid and uric acid towards glucose was very different when a PPM (phenylenediamine) or a NAFION membrane was used in an amperometric biosensor using glucose oxydase.

3.4. Molecular recognition and sensors

Molecular recognition is based on the ability of a host molecule (receptor) to complex a target species selectively, thanks to molecular interactions, especially: coordination, hydrogen bonds, electrostatic, and hydrophobic interactions, etc. This property can be exploited for the detection and the electrochemical dosage of chemical species, provided that the recognition event can yield an electrical signal (variation of potential, current, resistance, or capacitance). Currently, the main approaches for electrochemical recognition have remained in the field of prospecting. They are based upon surface modifications of electrodes using molecular materials (adsorbed polymer films, grafted or self-assembled monolayers) possessing recognition properties (specific interactions with the target) and a well-defined redox activity.

A widely investigated approach makes use of electron conductor polymers functionalized by receptor sites. These conducting polymers offer many functionalization possibilities; they are stable and adhere well to various conductors or semiconductors. Moreover, their redox properties are very sensitive to even minute perturbations (electrostatic or hydrophobic interactions, steric stresses, etc.) arising from contact interactions with chemical species. The main examples listed in recent reviews are related to conductor polymers modified by ionophores, such as polyalkylether chains, crown-ethers, or polypyridinic ligands, for the detection of metallic cations [MCQ 00, FAB 98, RON 99]. The modification of conductor polymers by active optical functions is promising for an enantioselective electrochemical detection [MCQ 00]. Their functionalization by biological molecules, such as proteins, oligonuclides, enzymes, or antibodies, has already largely proved to be promising in the field of manufacturing bioelectrochemical sensors [MCQ 00].

Transition metal complexes, which are able to complex and/or selectively catalyze oxidation or reduction of various substances, such as anions (halides, cyanides, organic anions, nitrogen oxides, etc.) and gases (CO, CO_2, NO, etc.), are very promising for elaborating electrochemical interfaces in view of electroanalysis in aqueous solutions [EFI 90]. Detecting such species by voltamperometry is then possible, due to the setting of a redox couple that is specific to the complex associated with the analyte by a catalytic current corresponding to the selective oxidation or reduction of the target species.

For example, the first case can be illustrated by the detection of organo-halide polluting agents on carbon electrodes, modified by polymer films of cobalt porphyrines [RUH 87, ORD 00]. In such an example, the species {R-Co(III)} resulting from an oxidizing addition of RX on the cobalt(I) complexes are characterized by a reduction wave with a redox potential typical of the alkyl-halide target.

Numerous studies have been devoted to the development of electrochemical microsensors for the analysis of nitrogen monoxide (NO) in biological medium, based on its selective oxidation by metallic complexes [PON 00]. The prime electrode materials for these sensors are nickel porphyrine films, with a Nafion® membrane covering that limits the interference due to anions often concurrently present with NO in biological media [BED 96]. The use of ultramicroelectrodes (carbon fibers) with analytical techniques, such as differential pulsed voltammetry, allows great sensitivity and NO detection at very low concentrations (10^{-9} mol/l).

The voltamperometric detection of electro-inactive species or of species that cannot form electro-active complexes should benefit from the recent research in the field of active redox receptors [BAK 02]. Numerous prototype redox receptors have been studied in homogenous phases, especially macrocycle receptors, such as crown-ethers and cryptants, grafted on organometallic or coordination complexes for the detection of alkaline and alkaline earth cations [BEE 92]. Complexation of anions followed by electrochemical detection is also conceivable [BEE 00], with functionalization by redox groups of anion receptors recognizing these species via interactions (electrostatic, hydrophobic, hydrogen bound, or Lewis acid–base reactions).

The voltamperometric detection of target species by these receptors is made possible because the electro-activity of the redox center is modified by the complexation reaction. This process can be schematized by a square diagram, as shown in Figure 3.10, where E^0_1, E^0_2, K_0 and K_+, respectively, stand for the redox potentials of free and complexed systems, and the association constants between the receptor, L, and the substrate, S, relative to the oxidized and reduced forms [MED 92].

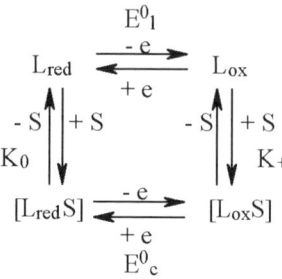

Figure 3.10. *Complexation equilibria and redox reactions in a redox-active receptor*

In other words, the electrochemical detection will be possible if the redox event causes significant stabilization or destabilization of the host–guest association, due to electrostatic interaction or repulsion forces between the redox center and the complexed substrate. The potential difference between the free and complexed forms (E_c^o) of the redox receptor will, therefore, depend on the respective stabilities of the complex in its oxidized and reduced states:

$$E_1^o - E_c^o = \frac{RT}{nF} \ln\left(\frac{K_o}{K_+}\right) \qquad [3.60]$$

Depending on the electrochemical perturbation obtained, theses receptors will be applicable to potentiometric detection (in the case of a mere progressive shift of the voltamperometric wave characteristic of the redox ligand) or to amperometric detection (if a redox system specific to the complexed receptor is formed).

The use of such redox receptors in electrochemical sensors will need improvements beyond the numerous current limitations. Particularly, most of the current studies have been performed in organic solutions and the development of sensors will obviously necessitate an immobilization of redox receptors without any denaturation and with a good stability in aqueous medium.

3.5. Characterization methods

3.5.1. *Definition and determination of the detection limit*

The definition of the detection limit is essential for a user because it fixes the sensitivity range of the sensor. Generally, the detection limit could be defined as the smallest value of the detectable measurand activity (i.e. the minimal concentration). It is considered as an intrinsic parameter of the sensor and is independent of the measurement instrumentation (resolution).

In the case of amperometric sensors, the detection limit is expected to be fixed by the resolution (i.e. the smallest value of the signal measurable by the device). In fact, at low values, the signal is blurred by the electric noise arising from several origins (among which is the electrochemical noise). So, the detection limit is generally defined as the smallest value of concentration giving an electrical signal (i.e. current intensity) at least equal to three times the noise amplitude.

The principle of potentiometric sensors is based on equilibrium between species present both in the membrane and in the analyzed solution. It is easily conceivable that, if there is no trace of the target species in the solution, the membrane will tend to relieve these species (the rich medium tends to give to the poor medium). For example, the final value will be imposed by the solubility of the membrane (or of any other part of the device) or by an interference phenomenon with the species present in the analyzed medium (e.g. H^+ or OH^- ions). Using simple models, in a first approximation, it can be shown that the response obeys the general law:

$$E_p = E°_p + \frac{RT}{z_p F} \ln(a_p + DL) \qquad [3.61]$$

where DL is the detection limit value of the sensor giving a potentiometric response E_p to the activity a_p of the species p.

For instance, the value of DL is equal to $\sqrt{K^s}$, where K^s is the solubility constant of the membrane chemical compound (for a 1-1 salt, e.g. AgX), or equal to $K^{pot}_{p,i} a^{1/z_i}$ if the detection limit of p is fixed by the interfering ion, i.

The shape of the corresponding curve is given in Figure 3.11. The detection limit, according to IUPAC recommendations, is given by the intersection of both asymptotes (Nernstian straight line response and non-response horizontal asymptote). A simple calculation shows that the difference between the response and the horizontal asymptote is equal to

$$\Delta E = \frac{RT}{z_p F} \ln 2 \qquad [3.62]$$

and, as this horizontal asymptote is fixed at

$$E_\infty = E° + \frac{RT}{z_p F} \ln(DL)$$ [3.63]

the value of *DL* could be determined from the value E_∞ for infinite dilution. But as the experimental points defining this asymptote are often rather scattered, (fluctuations, for very low activities, of the phenomena fixing the response), this method will not be accurate.

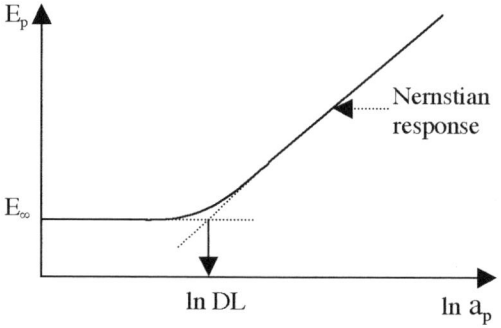

Figure 3.11. *Typical shape of a potentiometric response close to the detection limit, DL*

The method that consists of undertaking additions of a reference solution (with a known concentration), starting from infinite dilution up to concentrated solutions, is generally preferred. The use of buffered solutions (pH or ionic strength fixed using a non-interfering salt) is recommended. If the ionic strength is not fixed, the calibration curve must be established as a function of the activity, by the determination of the activity coefficient of the *p* species, using the Debye Hückel relation (which is always an approximation) [SAR 91].

The *DL* value is a very useful parameter, but the medium in which it was determined must be clearly specified. The simplest illustration is given by an ISE based on a glass membrane and sensitive to Na^+, because the interference to the proton is strong (K^{pot} can be about 100). With such an ISE, for pH = 8, the detection limit is about 10^{-5} or 10^{-6} mol/l and rises to 10^{-3} for pH = 3. A similar demonstration could be made with an ISE sensitive to F^- with a strong interference to OH^-.

3.5.2. Determination methods of selectivity coefficients

Although some suppliers give a few indications concerning the interferences of their sensors, in particular, in the case of ISE, these indications are often merely qualitative, or given without any clear indications of the medium or of the determination conditions. This is why the user must determine the selectivity coefficients in conditions close to the experimental situation under study. The most commonly used determination methods of these parameters will be given in the following section [MAC 96, REN 99, BAK 00].

3.5.2.1. Separate solutions method (SSM)

The SSM determination of selectivity coefficients is doubtlessly the fastest method, but it is lacks accuracy and only gives the order of magnitude. This method consists in dipping the sensor successively into two solutions. The first solution,, which contains only the primary species (target) with a well-known concentration (or activity), gives the response X_p. The other contains only the interfering species, also at a known concentration, and the response becomes X_i.

In the case of sensors with a linear response versus the concentration (e.g. amperometric sensors), such as

$$X = \alpha \times (C_p + K_{p,i} C_i) \qquad [3.64]$$

the selectivity coefficient emerges as:

$$K_{p,i} = \frac{X_i}{X_p} \times \frac{C_p}{C_i} \qquad [3.65]$$

This expression can be simplified if both concentrations are equal, or if these concentrations are selected so that the X_i and X_p responses are equal.

In the case of sensors with a logarithmic response, (as is the case in potentiometric sensors), the response can be written

$$E = E° + \alpha \ln(a_p + K_{p,i}(a_i)^\beta) \qquad [3.66]$$

and the coefficient $K_{p,i}$ can be shown to be:

$$K_{p,i} = \frac{a_p}{a_i^\beta} \exp\frac{E_i - E_p}{\alpha} \qquad [3.67]$$

Here again, the expression of the selectivity coefficient can be simplified if the a_i and a_p activities are equal or if these activities are selected so that both responses are equal. In the case of Nernstian responses, note that the equality between activity and concentration is only approximate and increases in accuracy as the concentration decreases.

The drawback of the SSM is that the sensor is not used in a real situation, i.e. there is no competition between the primary species, p, and the interfering ones, i, in the construction of the response.

3.5.2.2. Mixed solution methods (FPM and FIM)

Such methods are recommended by IUPAC. They are closer to real situations, i.e. when the sensor is in contact with the primary species in the presence of the interfering species, so competition between both species occurs and the kinetic phenomena play their role fully.

Two variants of this method exist. In the first, the activity of the primary species is kept constant (FPM stands for fixed primary method) during the addition of the interfering species. It means that the solution added has a high concentration of the interfering species and, as far as the interfering species is concerned, has the same concentration as in the initial solution. Of course if activities and concentrations are assumed equal, this method is valid only at low concentrations, more so if the ions are multicharged.

In the case of the linear response mentioned previously, the sensor response can be expressed as

$$X_i = \alpha \times C_p^\circ + \alpha \times K_{p,i} C_i \qquad [3.68]$$

which can also be written

$$X_i = X_p^\circ + \frac{X_p^\circ}{C_p^\circ} K_{p,i} C_i \qquad [3.69]$$

where X_p° is the response value before any addition of the interfering species (concentration C_p°) and X_i is the response when the interfering species is present with a variable concentration C_i. This expression can be used in different ways for the determination of $K_{p,i}$ (by linearization, etc).

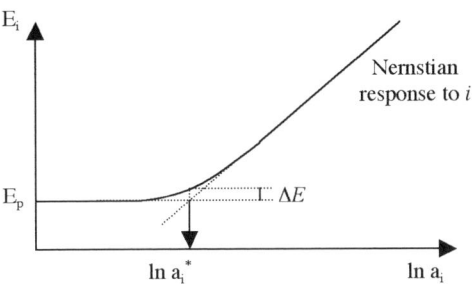

Figure 3.12. *Determination of $K_{p,i}$ by a mixed solution method (FPM)*

In the case of potentiometric sensors, the experimental method is identical and the exploitation of the response versus C_i (or a_i) can be made similarly to the determination of the detection limit *DL* (intersection of the asymptotes, see Figure 3.12).

The equations of the asymptotes are given by the following relations, at low activity of the interfering species, *i*, compared with that of the primary species, *p*, (relation [3.70]) and at low activity of *p*, compared with that of *i* (relation [3.71]) respectively:

$$E_p = E° + \frac{RT}{z_p F} \ln a_p \qquad [3.70]$$

$$E_{(i)} = E° + \frac{RT}{z_p F} \ln(K_{p,i}(a_i)^{z_i/z_p}) \qquad [3.71]$$

and the intersection for $a_i = a_i^*$ (Figure 3.12) yields the relation:

$$K_{p,i} = \frac{a_p}{(a_i^*)^{z_p/z_i}} \qquad [3.72]$$

In fact, this method cannot be applied for too low values of $K_{p,i}$, because it then necessitates too high concentrations of the interfering species, *i*, for the "Nernstian" asymptote related to this species to be obtained. Moreover activities and concentrations can no longer be considered as equal. Sometimes only the left part of the curve is used. The value a_i^* is then determined by the gap relative to E_p which is equal to:

$$\Delta E = \frac{RT}{z_p F} \ln 2 \qquad [3.73]$$

It is about 18 mV, at room temperature, if z_p is equal to 1. This is a second method for the exploitation of such a curve. Another method consists in using software tools for the determination of $K_{p,i}$ (by linearization of the Nikolskii equation). This method is more accurate.

The other variant of the mixed solution method consists of keeping the concentration (or activity) of the interfering species, i, constant during the addition of the primary species p (FIM: fixed interference method). The curves obtained are relatively similar to the previous ones. The apparent detection limit is then correlated with the interference phenomenon and the value of $K_{p,i}$ is determined by a method which is similar to that of the preceding case.

As a general remark and as mentioned previously, the values of the selectivity coefficients are unfortunately not constant, because they often are functions of the concentrations of the primary and interfering species. Generally, the values depend on the experimental conditions of the determination. For this reason, the experimental conditions corresponding to each determination should be specified systematically.

3.5.2.3. Matched potential method (MPM)

This method only deals with the experimental observations. In this frame, a concentration (or activity) of the primary species, p, is first chosen as a nominal value, and the method then consists of investigating which concentration increment of this species would produce the same signal as that obtained when the interfering species is introduced into the solution at the nominal concentration. In the case of a potentiometric sensor, the selectivity coefficient is then calculated by the ratio:

$$K_{p,i}^{MPM} = \left(\frac{\Delta[p]}{[i]} \right)_{\Delta E = cte} \qquad [3.74]$$

This definition, which is not often used, is clearly independent of the Nikolskii relation and can be generalized to any situation (for example, amperometric sensors). It offers the advantage of taking clearly into account the fact that the value determined depends on the experimental parameters, which must be specified in the definition of the selectivity coefficient value.

3.6. Bibliography

[AND 83] ANDRÉ J.J., BERNARD M., FRANÇOIS B., MATHIS C., "Depinning of charge carriers in n-type semiconducting polyacetylene", *J. Phys.*, vol. 44-C3, p. 199-202, 1983.

[BAK 00] BAKKER E., PRETSH E., BUHLMANN P. "Selectivity of potentiometric ion sensors", *Anal. Chem.*, vol. 72, p. 1127-1133, 2000.

[BAK 02] BAKKER E., TELTING-DIAZ M., "Electrochemical sensors", *Anal. Chem.*, vol. 74, p. 2780-2800, 2002.

[BAN 00] BANDYOPADHYAY K., SHU L., LIU H., ECHEGOJEN L., "Selective K^+ recognition at the interface during self-assembly of a bis-podand thiol on a gold surface", *Langmuir*, vol. 16, p. 2706-2714, 2000.

[BAR 00] BARET M., MASSART D.L., FABRY P., CONESA F., EICHNER C., MENARDO C., "Application of neural network calibrations to an halide ISE array", *Talanta*, vol. 51, p. 863-877, 2000.

[BED 96] BEDIOUI F., TREVIN S., DEVYNCK J., "Chemically modified microelectrodes designed for the electrochemical determination of nitric oxide in biological systems", *Electroanalysis*, vol. 8, p. 1085-1091, 1996.

[BEE 92] BEER P.D., "Transition metal and organic redox-active macrocycles designed to electrochemically recognize charged and neutral guest species", *Adv. Inorg. Chem.*, vol. 39, p. 79-157, 1992.

[BEE 98] BEER P.D., GALE P.A., CHEN Z., "Electrochemical recognition of charged and neutral guest species by redox-active receptor molecules", *Adv. Phys. Org. Chem.*, vol. 31, p. 1-90, 1998.

[BEE 00] BEER P.D., CADMAN J., "Electrochemical and optical sensing of anions by transition metal based receptors ", *Coord. Chem. Rev.*, vol. 205, p. 131-155, 2000.

[BER 87] BERCK F., BRAUN P., OBERST M., "Organic electrochemistry in the solid state overoxidation of polypyrrole", *Ber. Bunsenges. Phys. Chem.*, vol. 91, p. 967-974, 1987.

[BUH 98] BUHLMAN P., PRETSCH E., BAKKER E., "Carrier-based ion-selective electrodes and bulk optodes. 2. Ionophores for potentiometric and optical sensors", *Chem. Rev.*, vol. 98, p. 1593-1688, 1998.

[CLA 59] CLARK L.C., Electrochemical device for chemical analysis, U.S. Patent, 2 913 386, 17 November 1959.

[COV 79] COVINGTON A.K., *Ion Selective Electrode Methodology*, vol. 1, CRC Press, London, 1979.

[CRE 97a] CRETIN M., FABRY P., "Detection and selectivity properties of Li^+-ion-selective electrodes based on NASICON-type ceramics", *Anal. Chim. Acta*, vol. 354, p. 291-299, 1997.

[CRE 97b] CRETIN M., ALERM L., BARTROLI J., FABRY P., "Lithium determination in artificial serum using flow injection systems with a selective solid-state tubular electrode based on NASICON membranes", *Anal. Chim. Acta*, vol. 350, p. 7-14, 1997.

[DOR 02] DORNEANU S.A., POPESCU I.C., FABRY P., "NASICON membrane used as Na^+-selective potentiometric sensor in steady state and transient hydrodynamic conditions", *Sens. and Actuators B*, vol. 91, p. 67-75, 2002.

[EFI 90] EFIMOV O.N., STRELETS V.V., "Metal-complex catalysis of electrochemical reactions", *Coord. Chem. Rev.*, vol. 99, p. 15-53, 1990.

[EIS 68] EISENMAN G., "Similarities and differences between liquid and solid ion exchangers and their usefulness as ion specific electrodes", *Anal. Chem.*, vol. 40, p. 310-320, 1968.

[FAB 97] FABRY P., SIEBERT E., "Electrochemical sensors", Chapter 10 in GELLINGS P.G., BOUWMEESTER H.J.M. (eds), *The CRC Handbook of Solid State Electrochemistry*, CRC Press, p. 329-369, London, 1997.

[FAB 98] FABRE B., SIMONET J., "Electroactive polymers containing crown ether or polyether ligands as cation-responsive materials", *Coord. Chem. Rev.*, vol. 178–180, p. 1211-1250, 1998.

[FAB 08] FABRY P., GONDRAN C., *Capteurs Electrochimiques*, Ellipses, 2008.

[FER 02] FERREIRA N.G., SILVA L.L.G., CORAT E.J., TRAVA-AVIOLDI V.J., "Kinetics study of diamond electrodes at different levels of boron doping as quasi-reversible systems", *Diamond Relat. Mater.*, vol. 11, p. 1523-1531, 2002.

[FLI 98] FLINK S., BOUKAMP B.A., VAN DEN BERG A., VAN VEGEL F.C.J.M., REINHOUDT D.N., "Electrochemical detection of electrochemically inactive cations by self-assembled monolayers, of crown ethers", *J. Am. Chem. Soc.*, vol. 120, p. 4652-4657, 1998.

[FLI 99] FLINK S., VAN VEGEL F.C.J.M., REINHOUDT D.N., "Recognition of cations by self-assembled monolayers of crown ethers", *J. Phys. Chem. B*, vol. 103, p. 6515-6520, 1999.

[GAR 89] GARNIER F., "Functionalized conducting polymers – Towards intelligent materials", *Angew. Chem. Int. Ed. Engl.*, vol. 28, p. 513-517, 1989.

[GOU 08] GOURE J.P., FABRY P., TARDY P., "Capteurs de gaz basés sur d'autres principes", Chapter 11 in MENIL F. (ed), *Microcapteurs de Gaz.*, Hermes-Lavoisier, p. 337-377, 2008.

[HAA 77] HAALAND D. M., "Internal-reference solid-electrolyte oxygen sensor", *Anal. Chem.*, vol. 49, p. 1813-1817, 1977.

[HEI 90] HEINZE E., "Electronically conducting polymers", in E. STECKAN (ed), *Topics in Current Chemistry*, vol. 152, p. 1-47, Springer-Verlag, 1990.

[HEN 07] HENDERSON P., "Zur Thermodynamik der Flüssig Keitsketten", *Z.. Phys. Chem.*, vol. 59, p. 118-127, 1907.

[IVE 61] IVES D.J.G., JANZ G.J., *Reference Electrodes – Theory and Devices*, Academic Press, 1961.

[JAV 88] JAVADI H.H.S., ANGELOPOULOS M., MC DIARMID A.G., EPSTEIN A.J., "Conduction mechanism of polyaniline: Effect of moisture", *Synth. Met.*, vol. 26, p. 1-8, 1988.

[KLE 91] KLEITZ M., SIEBERT E., FABRY P., FOULETIER J., "Solid state electrochemical sensors", Chapter 8 in GÖPEL W., HESSE J., ZEMEL J.N. (eds)., *Sensors, a Comprehensive Survey*, vol. 2, Part I, VCH, p. 341-428, 1991.

[KOR 83] KORYTA J., STULIK K., *Ion Selective Electrodes*, Cambridge University Press, 1983.

[MAC 96] MACCA C., "Determination of potentiometric selectivity", *Anal. Chim. Acta*, vol. 321, p. 1-10, 1996.

[MAU 98] MAUVY F., CRETIN M., SIEBERT E., FABRY P., "Mechanism of selectivity in alkali sensitive membrane based on NASICON", *J. New Mat. For Electro. Syst.*, vol. 1, p. 71-76, 1998.

[MCD 94] MC DIARMID A.G., EPSTEIN E., "The concept of secondary doping as applied to polyaniline", *Synth. Met.*, vol. 65, p. 103-116, 1994.

[MCQ 00] MC QUADE D.T., PULLEN A.E., SWAGER T.M., "Conjugated polymer-based chemical sensors", *Chem. Rev.*, vol. 100, p. 2537-2574, 2000.

[MED 92] MEDINA J.C., GOODNOW T.T., ROJAS M.T., ATWOOD J.L., LYNN B.C., KAIFER A.E., GOKEL G.W., "Ferrocenyl iron as a donor group for complexed silver in ferrocenyldimethyl[2,2]cryptand – A redox-switched receptor effective in water", *J. Am. Chem. Soc.*, vol. 114, p. 10583-10595, 1992.

[MOD 91] MODELEY P.T., NORRIS J.O.W., WILLIAMS D.E., *Techniques and Mechanisms in Gas Sensing*, Adam Higler, 1991.

[MOR 81] MORF W.E., *The Principle of Ion Selective Electrode and of Membrane Transport*, Elsevier, 1981.

[NIK 37] NIKOLSKII B.P., "Theory of the glass electrode", *Zh. Fiz. Khim.*, vol. 10, p. 495-503, 1937.

[OEH 91] OEHME F., "Liquid electrolyte sensors: potentiometry, amperometry, and conductometry", Chapter 7 in GÖPEL W., HESSE J., ZEMEL J.N. (eds), *Sensors, a Comprehensive Survey*, vol. 2, Part I, VCH, p. 239-339, 1991.

[ORD 00] ORDAZ A.A., ROCHA J.M., AGUILAR F.J.A., GRANADOS S.G., BEDIOUI F., "Electrocatalysis of the reduction of organic halide derivatives at modified electrodes coated by cobalt and iron macrocyclic complex-based films: application to the electrochemical determination of pollutants", *Analusis*, vol. 28, p. 238-244, 2000.

[OTT 85] OTTO M., THOMAS J.R.D., "Model studies on multiple channel analysis of free magnesium, calcium, sodium, and potassium at physiological concentration levels with ion-selective electrodes", *Anal. Chem.*, vol. 57, p. 2647-2651, 1985.

[PON 00] PONTIE M., BEDIOUI F., "Design of electrochemical microsensors to monitor nitric oxide production in biological systems: a global compilation", *Analusis*, vol. 28, p. 465-469, 2000.

[REN 99] REN K., "Selectivity problems of membrane ion-selective electrodes – a method alternative to the IUPAC recommendation and its application to the selectivity mechanism investigation", *Fres. J. Anal. Chem.*, vol. 365, p. 389-397, 1999.

[RON 99] RONCALI J., "Electrogenerated functional conjugated polymers as advanced electrode materials ", *J. Mater. Chem.*, vol. 9, p. 1875-1893, 1999.

[RUH 87] RUHE A., WALDER L., SHEFFOLD R., "Modification of carbon electrodes by vitamin B-12-polymers", *Makromol. Chem., Makromol. Symp.*, vol. 8, p. 225-233, 1987.

[SAR 91] SARRAZIN J., VERDAGUER M., *L'oxydoréduction, Concepts et Expériences*, Ellipses, 1991.

[SKO 98] SKOTHEIM T.A., REYNOLDS J.R., ELSENBAUM R.L., *Handbook of Conducting Polymers*, 2nd edn, Marcel Dekker, 1998.

[STO 97] STORA T., HOVIUS R., PACHOUD M., VOGEL H., "Metal ion trace detection by a chelator-modified gold electrode: A comparison of surface to bulk affinity", *Langmuir*, vol. 13, p. 5211-5214, 1997.

[SWA 98] SWAGER T.M., "The molecular wire approach to sensory signal amplification", *Acc. Chem. Res.*, vol. 31, p. 201-207, 1998.

[VLA 97] VLASOV Y., LEGIN A., RUDNITSKAYA A., "Cross-sensitivity evaluation of chemical sensors for electronic tongue: determination of heavy metal ions", *Sens. and Actuators B*, vol. 44, p. 532-537, 1997.

[WAN 94a] WANG J., "Decentralized electrochemical monitoring of trace metals: from disposable strips to remote electrodes", *Analyst*, vol. 119, p. 763-766, 1994.

[WAN 94b] WANG J., "Selectivity coefficients for amperometric sensors", *Talanta*, vol. 41, p. 857-863, 1994.

[WAR 89] WARREN L.F., WALKER J.A., ANDERSON D.P., RHODES C.G., "A study of conducting polymer morphology – The effect of dopant", *J. Electrochem. Soc.*, vol. 136, p. 2286-2295, 1989.

[WU 96] WU J., ZHU J., SHAN L., CHENG N., "Voltammetric and amperometric study of electrochemical activity of boron-doped polycrystalline diamond thin film electrodes", *Anal. Chim. Acta*, vol. 333, p. 125-130, 1996.

Chapter 4

Potentiometric Sensors (Ions and Dissolved Gases)

4.1. Introduction

4.1.1. *General features*

Potentiometric sensors, which constitute the main part of the selective electrodes commonly called ISEs (or "ion-selective electrodes"), occupy a particular position among the class of chemical sensors. This term designates electrodes, generally of the membrane type, which respond selectively to an ionic species in the presence of other species, which may or may not interfere; therefore, the ISE are not specific of a given species. The principle on which the use of these electrodes is based has long been exploited for pH measurement using the glass electrode, which is a class of ionic electrode having a H^+ ion-selective membrane. These electrodes make use of the abilities of the materials from which they are made to exchange ionic or molecular species with the analyzed medium. These materials must be electrical conductors. The term membranes is generally used if the conduction is mainly ionic (for example, alumino-silicate glasses conducting via Na^+ ions for dosing sodium in solution); if the conduction is electronic (for example, a metallic electrode in a redox solution: platinum (Pt) wire in a solution containing iron (Fe^{3+} and Fe^{2+}) ions), the term electrode is used. The transduction mode is potentiometric, i.e. the open-circuit potential difference is measured (which can be related to the analyte concentration), which exists between the terminals of the electrochemical chain constituted by the

---

Chapter written in French by Annie PRADEL and Eric SAINT-AMAN.

ISE and the analyzed solution one side, and a reference point having a constant potential on the other (Figure 4.1).

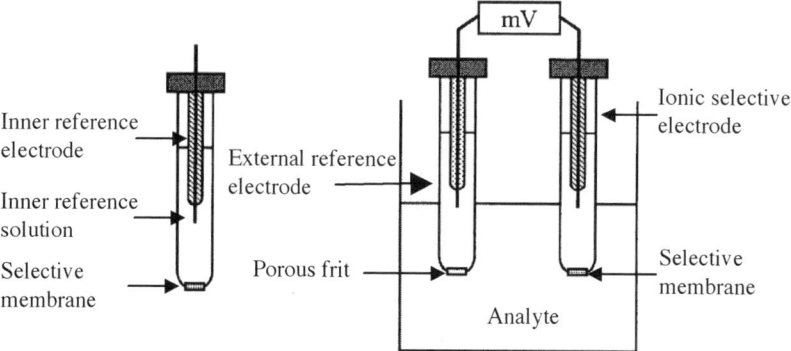

Figure 4.1. *Diagram of an ionic electrode and of the measuring chain*

The application field of the ISE is extremely wide, but is mainly restricted to aqueous media. Examples are used in the fields of environmental science (CN^-, F^-, S^{2-}, etc, in effluents and natural waters); agriculture (NO_3^-, NH_4^+, K^+, etc, in soils, fertilizers, plants, etc); food processing (salt in meat, fish, fruit juice); the detergent industry (Ca^{2+}, Ba^{2+}, F^- in residual waters); the paper-making industry; the explosive industry (F^-, Cl^-, NO_3^- in explosives and their combustion products); and biomedical science (Ca^{2+}, K^+, Cl^- in biological fluids, etc). Beyond the practical drawbacks of the analysis techniques (e.g. selectivity, reproducibility, fragility), theses applications are essentially motivated by the low cost of the ISE, their simple use, and their fast response.

Before describing the main characteristics of the materials constituting the potentiometric sensors, in particular, those of the ion-sensitive membranes, it is necessary to present more details of the working principle of an ISE and to define some of its characteristics.

4.1.2. *Electrode potential*

As mentioned in Chapter 3, the functioning of potentiometric sensors is based on creating equilibrium between the selective electrode and the *analyzed* medium by electrochemical exchange. This equilibrium is obtained when the values of the electrochemical potentials of the exchanged species (X^{z-} ion) in the *analyzed* solution (external phase, denoted "ext.") and in the membrane (phase denoted "mem.") are equal:

$$\tilde{\mu}_{X,mem.} = \tilde{\mu}_{X,ext.} \qquad [4.1]$$

with, in the j phase,

$\tilde{\mu}_{i,j} = \mu_{i,j} + Z_i F \varphi_j$: electrochemical potential of species i [4.2]

$\mu_{i,j} = \mu_{i,j}^o + RT \, Ln \, a_{i,j}$: chemical potential of species i [4.3]

$a_{i,j} = \gamma_{i,j} \, C_{i,j}$: activity of species i [4.4]

where $\mu_{i,j}^o$ is the standard chemical potential of species i, φ_j: is the electrical potential of phase j, γ_i is the activity coefficient of species i and $C_{i,j}$ is the concentration of species i.

A specific electrode can be represented by the electrochemical chain of Figure 4.2. For the sake of simplicity, the ionic species are supposed to be monovalent, and the simple case where the *analyzed* species, X⁻, is the only species that may exchange at the surface (specific membrane) is considered.

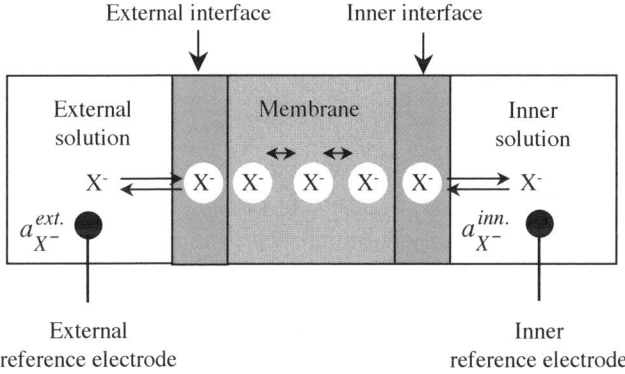

Figure 4.2. *Diagram of the working principle of an ISE*

The potential difference ΔE between the internal and external reference electrodes is:

$$\Delta E = E_{ref.,int.} + E_{int.} + E_{a.m.} - E_{ext.} - E_{ref.ext.} - E_j \qquad [4.5]$$

where $E_{ref.,int.}$ and $E_{ref.ext.}$ are potentials of the internal and external reference electrodes, $E_{int.}$ and $E_{ext.}$ are the potential differences between the internal or external electrolyte and the membrane, E_j is the junction potential of the external electrode,

$E_{a.m.}$ is the asymmetry potential (a few mV) due to the different physico-chemical behavior of the two interfaces. This potential, which depends on the electrode and on ageing, may be considered as constant at the timescale of the experiment.

In equation [4.5], $E_{ref.,int.}$, $E_{a.m.}$, $E_{ref.ext.}$ and E_j are nearly constant. Hence:

$$\Delta E = ct + E_{int.} - E_{ext.} \qquad [4.6]$$

$$\Delta E = ct + \Delta\phi_{mem.} \qquad [4.7]$$

where $\Delta\phi_{mem.}$ is the membrane potential.

Equilibrium at the interfaces and the homogenous state of the membrane imply that the electrochemical potentials of X^- in the membrane and in the internal and external solution have the same value:

$$\tilde{\mu}_{X,int.} = \tilde{\mu}_{X,ext.} \qquad [4.8]$$

Grouping expressions [4.2], [4.3], [4.7] and [4.8] leads to:

$$\Delta\phi_{mem.} = (\mu_{X,ext.} - \mu_{X,int.}) / Z_X F \qquad [4.9]$$

$$\Delta\phi_{mem.} = \frac{RT}{Z_X F} Ln \left(\frac{a_{X,ext.}}{a_{X,int.}} \right) \qquad [4.10]$$

The internal solution is permanent, which necessitates that $a_{X,int.}$ is constant. Therefore, relation [4.11] is established:

$$\Delta E = \Delta E° + \frac{RT}{Z_X F} Ln \left(a_{X,ext.} \right) \qquad [4.11]$$

This is the basic expression for the use of ionic electrodes, but it has been established for an ideal situation. In fact, specificity is never obtained and the electrode may react with one or more ions in the external solution. The effect of interference is to deliver a response, which will add to that of the dosed ion.

Equation [4.11] shows that the response of ionic electrodes is a function of the ionic activities and not of the concentrations. For pH measurements, this is precisely what is desired, but it is not a general case. It is then necessary to use standard solutions with a composition close to that of the *analyzed* solution, or buffers with sufficiently high ionic strength to make any variation of ionic strength during the titration negligible.

It must also be noted that the electrode response varies linearly as a function of the logarithm of the activity of the analyte. Consequently, the calculated concentration is highly dependent on the precision of the potential measurement. For example, an error of 1 mV entails an error of 4% in the determination of the concentration of a monovalent cation (8% for a divalent cation). In the case of the determination of $pH = -log(a_{H^+})$, an error of 5 mV corresponds to an error of 0.1 unit of pH.

The above calculation was carried out considering the ideal case of a first-order response. In such a case, the "mobile" ion is also the ion to be dosed, but ionic electrodes may also have a second-order (or even third-order) response if intermediate reactions intervene, when the mobile ion is not the ion to be dosed.

Let us consider, for example, an ionic electrode with a silver halide (AgX) membrane, which is conducting by Ag^+ ions. Its response is, therefore, of the first order with respect to Ag^+, but it is also sensitive to the halide X^-. This is because, due to some solubility of the membrane – characterized by the solubility product Ks_{AgX} – some activity of Ag^+ is set at the interface with the external solution:

$$Ks_{AgX} = a_{Ag,ext.} \cdot a_{X,ext.} \qquad [4.12]$$

the measured potential, ΔE, can be written:

$$\Delta E = \Delta E° + \frac{RT}{F} Ln\left(a_{X,ext.}\right) = \Delta E° + \frac{RT}{F} Ln\left(\frac{Ks_{AgK}}{a_{X,ext.}}\right) \qquad [4.13]$$

hence

$$\Delta E = \Delta E'° - \frac{RT}{F} Ln\left(a_{X,ext.}\right) \qquad [4.14]$$

Relation [4.14] expresses the link between the activity of X^- and the measured potential.

4.1.3. Sensitivity of the potentiometric sensors

The use of a specific electrode rests on the potential difference established between the presumably specific electrode and a reference electrode. This potential difference is ideally determined, as has just been mentioned, by a Nernstian law. Figure 4.3 gives the typical response of an electrode selective to the species i: $E = f(pX_i)$ with $pX_i = -log\ C_i$.

The curve has three distinct branches. The interesting zone is the intermediate zone, for $1 < pX_i < 5$ (values usually encountered), which corresponds to the linearity domain of the electrode response, which is a theoretical slope (called the "Nernst" slope) of 59.16 mV at 25°C for a monovalent cation ($Z_i = 1$). In the two zones corresponding to $pX_i > 6$ and $pX_i < 1$ the sensor has no sensitivity. These two zones are delimited by the high and low detection limits of the specific electrode; they correspond to the highest and the lowest detectable quantities of analyte. The International Union of Pure and Applied Chemistry (IUPAC) proposes two definitions for them. The first definition corresponds to the concentration defined by the point situated on the extrapolated Nernstian straight line at a distance (RT/F) log 2 (about 18 mV at 25°C) from the experimental curve. The second definition, most commonly accepted, corresponds to the intersection between the Nernstian response and the non-sensitivity asymptote (Figure 4.3).

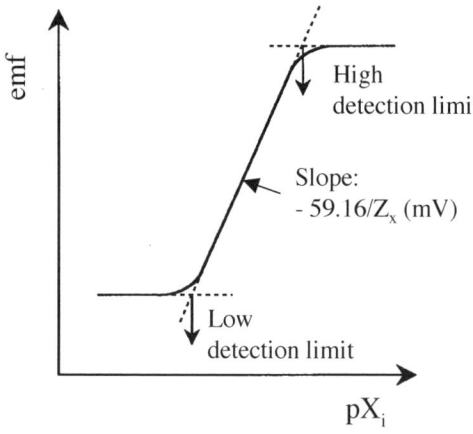

Figure 4.3. *Typical response of an ISE to the presence of a species, i*

Let us cite two phenomena, among those intervening in the definition of the detection limit: the presence of interfering ions and the membrane solubility. In the latter case, a simple example can be given. It concerns the use of a silver halide (AgX) membrane for dosing a solution of a MX salt (for example, NaX). The detection limit is directly linked to the solubility of AgX and equal to $(Ks_{AgX})^{1/2}$. It should be noted, however, that it is precisely this phenomenon that is used for dosing X^-, with a second kind response.

4.1.4. Selectivity of the potentiometric sensors

An ion-sensitive membrane is generally in contact with several ionic species, which may interfere with the response of the target ion. In this case, the generalized Nernst law, also called the Nikolskii-Eisenman relation, which takes into account the selectivity of the membrane to other species in solution, must be applied.

If n ions may interfere in the response of X^-, the following relation holds:

$$\Delta E = \Delta E° + \frac{RT}{Z_X F} \, Ln \left(a_X + \sum_1^n K_{X^-/Y_i} a_{Y_i}^{m_i} \right) \quad [4.15]$$

with a_X, a_{Y_i}: activities of the various species in the external solution, $m_i = Z_{Y_i}/Z_X$ and K_{X^-/Y_i}: coefficient of selectivity of X^- with respect to Y_i (also called potentiometric coefficient) which depends on the mobilities of X^- and Y_i in the membranes and on the exchange equilibrium constant:

$$m_i X^-_{mem.} + Y_{i,ext.} \Leftrightarrow m_i X^-_{ext.} + Y_{i,mem.}$$

K_{X^-/Y_i} also represents the ratio of the activities of X^- and Y_i that would cause the same electrode response. This coefficient permits quantifying the contribution of each ion taking part in establishing the electrode potential. Its determination is important because it yields, as mentioned earlier, the value of the detection limit of the sensor. For example, a Na^+-ISE with $K_{Na^+/H^+} = 100$ will have a detection limit of 10^{-5} mol/l if the pH of the solution is 7, and a value of 10^{-3} mol/l if the pH is equal to 5. This shows that the detection limit of a membrane is not an intrinsic parameter as it depends on the measuring conditions. The detection limit also takes into account the selectivity of the electrode. Thus, a AgBr-based membrane will be selective for chloride ions, characterized by: $K_{Br^-/Cl^-} = 10^{-2}$, but it will not be selective to iodide ions, which will interfere with a coefficient of the order of 10^4 (see Chapter 3, Table 3.1).

Several methods for determining the selectivity coefficients have been proposed and are described in Chapter 3. The mixed solutions method is the easiest to use and consists of fixing the concentration of the interfering ion in the solution, while the concentration of the primary ion is varied in a range identical to that used during the standardization. A typical curve is given in Figure 4.4.

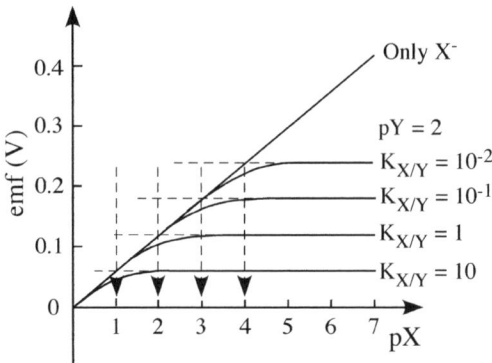

Figure 4.4. *Calibration curves, calculated using equation [4.15], illustrating the determination of the coefficient of selectivity by drawing the intersection point of the calibration curves obtained in the absence and in the presence of an interfering ion with a fixed concentration. The origin of the electromotive force values is arbitrary*

The potentiometric coefficients are determined from the experimental curves of the potential as a function of the primary ion concentration, for a fixed concentration of the interfering ion. The electromotive force measured in the presence of an interfering ion tends, at low X^- concentrations, towards a constant value given by $\Delta E = \Delta E° + \dfrac{RT}{Z_X F} Ln\left(K_{X^-/Y_i} a_{Y_i}^{m_i}\right)$. The intersection of the calibration curve drawn in the absence of interfering ion and the extrapolated plateau of the curve drawn in the presence of interfering ion yields the value of the halide ion activity:

$$a_{X^-} = K_{X^-/Y_i} a_{Y_i}^{m_i} \quad \text{or} \quad K_{X^-/Y_i} = \dfrac{a_{X^-}}{a_{Y_i}^{m_i}} \qquad [4.16]$$

In equation [4.16], a_{X^-} is the abscissa of the intersection point, and a_{Y_i} is fixed by the experimental conditions.

4.2. Membranes

4.2.1. *General features*

Let us focus now on the membranes. Two cases will be distinguished, whether the ionic exchange is direct or it involves a change of oxidation number. In the case of ionic exchange, two types of membranes will be differentiated:

– *ionic conducting membranes of the first kind*, in which ionic conduction is provided by the ion to be dosed in the solution (for instance Na^+/sodium aluminosilicate glass, or Na^+/NASICON (Na super-ionic conductor)),

– *ionic conducting membranes of the second kind*, in which the membrane conduction is provided by an ion that reacts with the ion to be dosed in the solution, forming an insoluble phase (for example, a AgX-type membrane, with conduction provided by Ag^+ ions to dose the halide ions, or membranes based on silver chalcogenides containing heavy metal ions for dosing the latter).

It is to be noted that the presence of some possible electronic conduction in the membrane material is not a major impediment if no electrode reaction occurs (in potentiometric sensors, the total current across the system is zero, in the absence of any electrode reaction i_e and i_{ion} are zero and the voltage delivered corresponds to the Galvani potential). Nevertheless, if the *analyzed* medium contains redox couples, the potential of the membrane will result from two exchanges: the ionic exchange and the oxido-reduction reaction due to electron exchange between the membrane and the redox couple of the *analyzed* solution, and the measurement will be spoilt by errors.

Let us now focus on the nature of the ion-selective membrane. These membranes can be classified in three large families: the glass membranes, the inorganic or crystallized inorganic (polycrystalline-ceramics or monocrystalline), membranes, and the polymer membranes.

4.2.2. *Glass membranes*

In addition to good chemical durability, glasses have the advantage of offering a wide choice of composition, which enables easy adaptation of their formulation. Two glass families are used: oxide glasses, which were among the first to be used for selective electrodes, and the chalcogenide glasses.

4.2.2.1. *Oxide glass-based membranes*

The electrode is constituted by a blown glass membrane (cationic conductor by Na^+, Li^+, etc) having a suitable composition. It usually has a spherical shape and is welded at the extremity to a robust glass tube or fixed to a plastic tube. This tube is filled with a reference solution in which an internal reference element, generally Ag/AgCl, is immersed. An electrochemical cell containing a glass electrode sensitive to M^+ ions is presented in Figure 4.5.

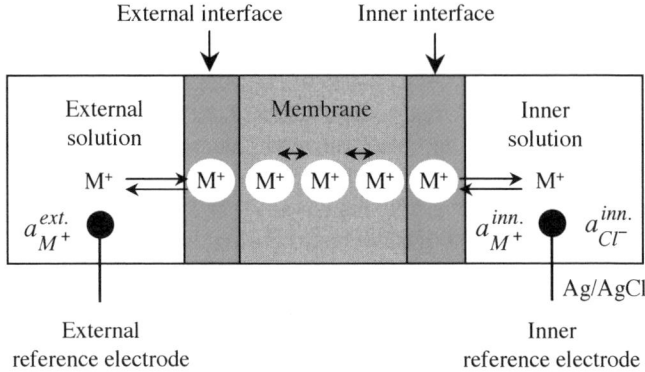

Figure 4.5. *Measuring chain of an ISE constituted of a glass-based membrane*

4.2.2.1.1. Example of the pH electrode

The ionic membrane is a glass with conduction provided by Na^+ ions, with an extremely high interference to the protons ($K_{H^+/Na^+} \sim 10^{13}$). Historically, the first glass used (Corning 015) had the molar composition 0.214 Na_2O-0.064 CaO-0.72 SiO_2. This glass membrane was not usable for pH >9. The current glass electrodes have far more complex compositions. They are Li^+ ion-conducting glasses, containing niobium and tantalum oxides and sometimes uranium or rare-earth oxides, which lower the electrical resistance while increasing the robustness of the electrodes. These electrodes and the corresponding measuring chain are schematized in Figure 4.6.

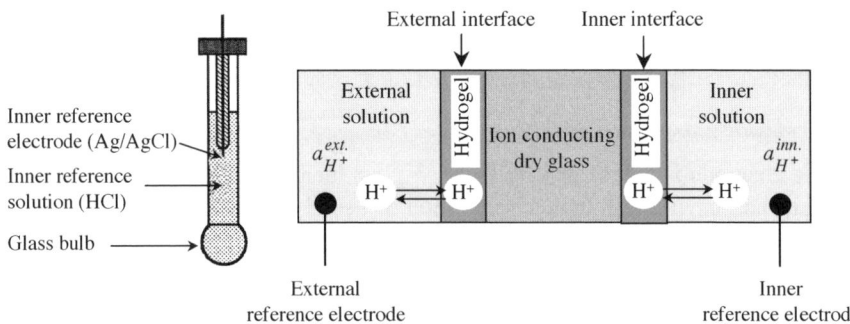

Figure 4.6. *Diagram of a pH glass electrode and corresponding measuring chain*

Before being used, the glass electrode is "hydrated" by immersion in an aqueous solution, which causes an ion exchange between certain components of the glass material (for instance, Na^+) and the solution (H^+). Thus, a hydrated layer (a hydrogel a few hundreds of nanometers thick) is formed on the surface, similar to grafted H^+ ions, which are available for future exchanges via the creation of an equilibrium between the silicate groups SiO^- of the glass (very basic, which explains the high selectivity of the electrode to H^+ in comparison with Na^+, for instance) and the -SiOH groups. These hydrogel layers are, therefore, responsible for the sensitivity to pH. However, H^+ is not the species ensuring the ionic conduction of the glass, as the bulk of this material remains dry. The ionic conduction is rather obtained by micromovements of highly mobile small ions (essentially Na^+) that migrate from site to site on the defects of the network.

At high pH values, which correspond to low H^+ activities in the measured solution and to high activities of alkaline cations, the exchange equilibrium alkaline cation/H^+ is reversed, i.e. H^+ ions from the hydrated layer are now replaced by cations from the alkaline solution: this phenomenon is known as alkaline error. These interferences can be decreased by an appropriate selection of glass composition, or conversely, they can be enhanced to cause the electrode to behave as an ionic electrode with a solid membrane sensitive to Na^+, K^+, or Li^+.

4.2.2.1.2. Example of the dissolved gas-electrode

The potentiometric sensors for dissolved gases are based upon a conventional selective electrode, but its response to its specific analyte gets modified by the interaction (which can be quantified) of a dissolved gas with that species. The most commonly *analyzed* gases, by this technique, are carbon dioxide (CO_2), ammonia (NH_3), and hydrogen sulfide (H_2S). The reason is that they have acido-basic properties, which will make their analysis possible via pH modification of the medium. The corresponding potentiometric sensor is then a mere pH electrode. An electrode selective to a dissolved gas is thus based upon a membrane that is permeable to the gas and separates the *analyzed* solution from an internal solution containing an appropriate electrolyte, and in which a pH electrode is dipped. In the case of NH_3 or CO_2 gases analysis, the electrolyte is a solution of NH_4Cl or of $NaHCO_3$, respectively (Figure 4.7).

In the case of the CO_2-selective electrode, the following acido-basic equilibrium prevails in the internal acidic solution:

$$CO_{2,aq} + H_2O \Leftrightarrow H^+ + HCO_3^-$$

It is characterized by its acidity constant:

$$Ka = \frac{a_{H^+} a_{HCO_3^-}}{a_{CO_2}} \qquad [4.17]$$

or by:

$$pCO_2 = pH + pHCO_3^- - pKa. \qquad [4.18]$$

In this medium, where the HCO_3^- activity is fixed, the response of the pH electrode as a function of pH is linear, and is, therefore, linear as a function of pCO_2 in the internal solution, which is proportional to the value of pCO_2 in the external solution.

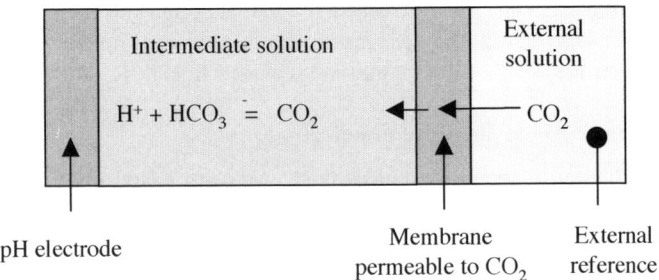

Figure 4.7. *Principle of the measuring chain of a CO_2-selective electrode*

The response of the ISE to the presence of CO_2 dissolved in the external solution is thus:

$$\Delta E = \Delta E° + \frac{RT}{F} Ln\, (a_{CO_2}^{ext.}). \qquad [4.19]$$

4.2.2.1.3. Example of electrodes for alkaline metals

As described in section 4.2.2.1.1, the membrane of the pH electrodes is sensitive to alkaline metal ions. Adjusting the glass composition, for instance, by incorporating alumina, causes further enhancement of these interferences to obtain an electrode that is sensitive to Na^+, K^+, or Li^+. Thus, a membrane having the composition 0.11 Na_2O-0.18 Al_2O_3-0.71 SiO_2 is suitable for preparing an Na^+-sensitive electrode, whereas glass with the composition 0.27 Na_2O-0.04 Al_2O_3-0.69 SiO_2 is used for measuring K+. Similar selective electrodes can be made by replacing sodium with lithium. It must be noted that these electrodes remain sensitive to pH, which must be fixed at a high value for a limitation of its influence, by the use of buffer solutions. Interferences with other monovalent cations must also be taken into account during the measurements [MID 91].

4.2.2.2. *Membranes based upon chalcogenide glasses*

The chalcogenide glasses were used for the first time as ion-sensitive membranes of selective electrodes in the early 1970s [BAK 71]: the chalcogenide glass with the formulation $Sb_{12}Ge_{28}Se_{60}$, doped by copper or iron, was capable of selectively detecting Cu^{2+} and Fe^{3+} ions, respectively, in aqueous solution. Following this discovery, many researches have been devoted to this class of materials, which, once doped with a metal element, can detect selectively a good number of ions of various natures. Table 4.1 lists chalcogenide glasses used as ion-sensitive membranes, specifying for each glass its composition and the main electrochemical characteristics of the detected ions.

A certain number of advantages have been put forward to justify the use of chalcogenide glasses as ion-sensitive membranes, particularly in comparison with the crystallized materials used for conventional electrodes. Among them, the following advantages can be cited:

– improved chemical stability in highly oxidizing [VLA 86] or acidic [VLA 92] aqueous solutions. These situations are encountered during analysis of media containing ionic species such as Fe^{3+}, Hg^{2+}, etc, which are stable only at very low pH values (pH <2);

– higher selectivity with respect to a number of interfering species, in particular heavy metal ions [VLA 94];

– as these materials generally have a large vitreous domain, the possibility exists of adjusting the chemical composition of the glass to optimize its performance and adapting it to meet different demands (specific media).

Z	Composition	Linearity (mol/l)	Slope (mV/pM)	Interferents $K^{pot}_{Z^{2+},j} > 10^{-1}$	Ref.
Cu^{2+}	Cu-As-S	$10^{-5} - 10^{-2}$	30		[JAS 74]
		$10^{-6} - 10^{-1}$	28		[OWE 80]
	Cu-As-Se	$10^{-6} - 10^{-2}$	30	Ag^+, Hg^{2+}, Fe^{3+}	[BOH 81]
	Cu-As-Se (thin film)	$10^{-5} - 10^{-1}$	27 – 30		[VLA 83]
	Cu-As-Se-Te	$10^{-6} - 10^{-2}$	29		[TOH 86]
	Cu-Ag-As-Se	$10^{-6} - 10^{-1}$	30		[BYC 95]
Pb^{2+}	Pb-Cu-As-Se	$10^{-5} - 10^{-1}$	30		[TOH 86]
	Pb-I-Ag-As-S	$10^{-6} - 10^{-2}$	28.9		[VLA 86]
	Pb-S-Ag-As	$10^{-6} - 10^{-1}$	20 – 25	$Cu^{2+}, Ag^+, Hg^{2+}, Fe^{3+}$	[VLA 88]
	Ag-S-As	$10^{-6} - 10^{-2}$	24		[VLA 92]
	Pb-S-Ag	$10^{-5} - 10^{-2}$	29.5		[TOH 83]
Fe^{3+}	Sb-Ge-Se-Fe	$10^{-5} - 10^{-1}$	57.6 ± 2.9		[BAK 71] [TOH 83]
Ag^+	Ag-As-S	$(1–5) \times 10^{-6} - 10^{-1}$	59		[VLA 88]
	Ag-As-Se	$(1–5) \times 10^{-6} - 10^{-1}$	59	none	[VLA 88]
	Ag-Br-S-As	$(2–5) \times 10^{-6} - 10^{-1}$	59.2–59.7		[VLA 88]
	Ag-As-Se-Te	$10^{-6} - 10^{-1}$	59		[VLA 88]
Cd^{2+}	Cd-S-Ag-As	$(1–5) \times 10^{-6} - 10^{-1}$	28		[VLA 87a, 88]
	Cd-S-Ag-I-Sb	$10^{-6} - 10^{-3}$	28		[GUE 98]
	Cd-I-Ag-As-S	$(1–5) \times 10^{-6} - 10^{-1}$	28	$Cu^{2+}, Pb^{2+}, Ag^+,$	[VLA 87a, 88]
	Cd-I-Ag-S	$10^{-7} - 10^{-1}$	25 – 28	Hg^{2+}, Fe^{3+}	[MIL 99]
	Cd-S-Ge-Se-Te	$10^{-5} - 10^{-1}$	30		[TOH 83]
Hg^{2+}	Hg-S-Ge-Te-Se	$10^{-6} - 10^{-2}$	30		[TOH 83]
		$10^{-6} - 10^{-2}$	24		[GUE 95]
	Hg-Te-Ge-Se	$10^{-8} - 10^{-6}$	30	Ag^+	[MIL 99]
	Ag-Br-As-S	$10^{-6} - 10^{-3}$	90 – 110		[MIL 99]
	Ag-Br-As-S	$10^{-3} - 10^{-1}$	60		[MIL 99]
	Ag-Br-As-S				
Cr^{VI}	?	$10^{-6} - 5 \times 10^{-4}$	90		[CHR 97]
Br^-	Ag-Br-S-As	$8 \times 10^{-7} - 10^{-1}$	59.8	I^-, CNS^-	[VLA 89]
	Ag-Br-S	$7 \times 10^{-6} - 10^{-1}$	57.5		[VLA 89]
S^{2-}	Cu-As-S	$10^{-10} - 10^{-5}$	40		[CAL 01]

Table 4.1. *Chalcogenide materials used for the detection of various ions and electrochemical characteristics*

4.2.3. *Crystallized inorganic membranes*

A small number of typical examples of selective electrodes are presented in this chapter to highlight the properties of the detecting membrane (low solubility, adapted structures, etc). The crystallized membrane materials are generally comprise insoluble inorganic salt crystals, and may be homogenous (one single crystal or a pressed polycrystalline powder, in most cases) or heterogeneous (with the crystals incorporated in a polymer matrix). Charge transport by diffusion and migration is ensured by defects in the network. A dopant is sometimes added to the inorganic salt constituting the membrane, to increase the defect concentration and, consequently, the conductivity of the membrane. For example, the ionic F^--sensitive electrode is constituted by a LaF_3 single crystal doped with EuF_2.

4.2.3.1. *Electrode based upon silver sulfides or selenides*

The properties of silver sulfide or selenide salts, including their low solubility, are suitable for preparing electrodes for measuring the activities of Ag^+ or S^{2-} ions, or for being used as matrixes for various salts. For example, heterogeneous mixtures with various salts having low solubility products yield electrodes that are selective for halides (with AgCl, AgBr, AgI as mixed salts) or for heavy metals (with CuS, CdS, PbS as added salts). Due to mixed conduction in these materials, metallic silver can be used as the internal contact, and thus the elaboration of the ISE is simplified.

4.2.3.1.1. Example of a Ag^+ (or S^{2-})-selective electrode

It is constituted by the following electrochemical chain:

$Ag \mid Ag_2S \mid Ag^+$ (sol) $\parallel$ external reference electrode.

As the solubility product of Ag_2S is very low ($Ks_{Ag_2S} = 10^{-50}$), measuring the Ag^+ ions activity is, in theory, possible up to concentrations of about 10^{-20} mol/l^{-1}.

4.2.3.1.2. Example of a Pb^{2+} ion-selective electrode

It is constituted by the following electrochemical chain:

$Ag \mid Ag_2S + PbS \mid Pb^{2+}$ (sol) $\parallel$ external reference electrode.

4.2.3.2. *Selective electrodes for fluoride ions*

The membrane is made of lanthanum fluoride, LaF_3, either pure or doped with europium II fluoride, EuF_2, to increase the ionic conductivity by the F^- ions. The very low solubility product of LaF_3 ($Ks_{LaF_3} = 10^{-29}$) enables the analysis of fluoride ions down to very low concentrations (practical detection limit stated by the electrode manufacturer in the order of 10^{-6} mol/l). This electrode can be used with a

96   Chemical and Biological Microsensors

solid internal reference and it is nearly specific for F^-, as only the OH^- ions interfere in a measurable way.

4.2.3.3. Na^+ ion-selective electrodes

This example was chosen because it shows how the specific structure of the ion-sensitive membrane, the NASICON, can provide the selective character of the electrode. The NASICON, synthesized for the first time in 1976, is a remarkable ionic conductor using Na^+ ions ($\sigma \sim 10^{-3}$ S/cm at room temperature, $Ea \sim 0.35$ eV). It was first used as a solid electrolyte in sodium/sulfur batteries. Subsequently, it was proposed as a sodium ion-sensitive membrane [ENG 88, FAB 88]. The selectivity of this material, with the chemical formula $Na_{1+x}Zr_2Si_xP_{3-x}O_{12}$, arises from its structure consisting of chains of SiO_4 and PO_4 tetrahedra linked by ZrO_6 octahedra, which produce three-dimensional channels inside which the Na^+ cation is mobile (Figure 4.8).

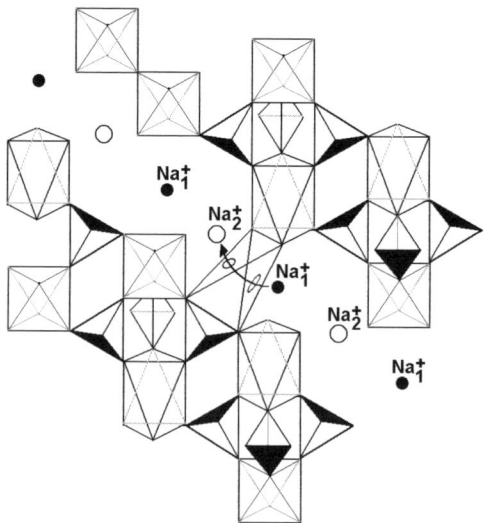

Figure 4.8. *Structure of NASICON*

These channels present bottlenecks comprised of a triangle of oxygen atoms with a size close to that of the Na^+ ions, which provide this material with a high selectivity in the exchange properties with other ions in solution. Thus, the K^+ ion, which is bigger than Na^+, will be blocked very quickly and stopped at the bottlenecks, whereas the Li^+ ion, smaller than Na^+, will have a much-reduced mobility in this type of structure. This is because it will be trapped by the O atoms in the bottleneck, due to its polarizing character. Therefore, these two cations, which are interferents to sodium, will have low potentiometric coefficients [MAU 98]. Let us note that the use of ceramics of the NASICON type, but using conduction provided by Li^+ ions, have been proposed as ion-sensitive membranes for the potentiometric determination of lithium [CRE 97]. The electrodes based on the ceramics of the NASICON type exhibit a higher selectivity to the other alkaline metals and to the protons than that of the electrodes based upon oxide glasses; however, they have the disadvantage of a lower detection limit (10^{-4} mol/l) [MAU 99].

4.2.4. *Polymeric membranes*

The essential element of the ISE based upon charge carriers is the ion-sensitive polymer matrix. From a physical viewpoint, it is a highly viscous liquid phase, non-miscible with water. It is inserted between two aqueous phases, the first containing the analyte and the second being the internal electrolytic solution. It contains various elements, which provide it with its selectivity and conduction properties. It is usually an ionophore (an ionic or molecular recognition entity), which acts as a charge carrier while producing the selective interaction with the species to be detected, associated with a lipophilic salt, which acts as an ion exchanger at the membrane interfaces.

The selectivity of the sensor is thus related to the equilibrium constant of the ion exchange reaction, which occurs between the membrane and the solution studied and involves the target and the interfering species. Therefore, it depends strongly on the ratio of the formation constants of the complexes between these species and the ionophore contained in the membrane. The ionophores, in their non-complexed form, may be neutral or charged. The various equilibriums involved in the analysis of a cation by this type of electrode are schematized in Figure 4.9.

The polymer is considered as an inert matrix, which provides the system with its required physical properties, such as mechanical strength and elasticity. However, polymer-based ISEs present a Nernst-type response only if the polymer phase is doped with ionic impurities. Several types of polymers are used, for example, polyvinyl chloride (PVC; which is the most widely used), polysiloxanes, and polyurethans.

A plasticizer is associated with the polymer, creating a mixture that is often called "plastisol". This mixture is the solvent in which the charge carrier will be dissolved. The mass proportions are about 30% PVC and 70% plasticizer. Such a proportion enables the mechanical properties of the film to be optimized and ensures good mobility for the constituents. The classically used plasticizers are the dioctyl sebacate (DOS), o-nitrophenyl octyl ether (o-NPOE), or dinonyl phtalate (DNP). The selection of the plasticizer is essentially based upon examining its polarity. For example, the selectivity of the analysis of a divalent cation with respect to a monovalent cation is improved if a polar plasticizer, such as o-NPOE, is used rather than a non-polar plasticizer, such as DOS.

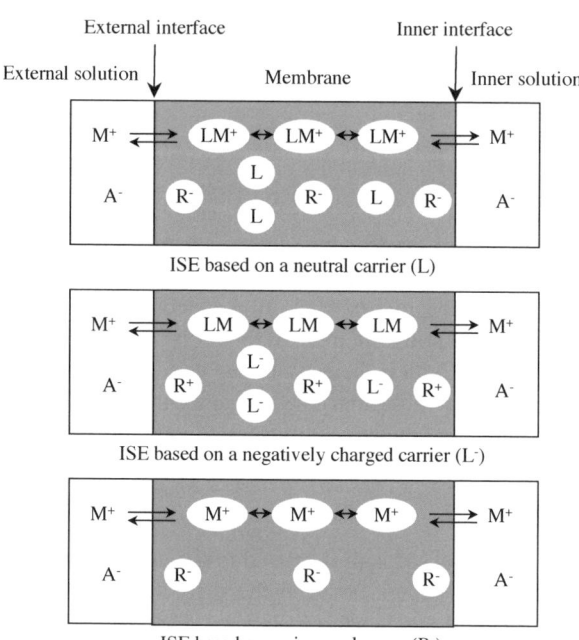

Figure 4.9. *Schematic representation of the equilibriums, involving a cationic analyte (M^+) of the salt MA, between the solution studied, the membrane of an ISE, and its internal electrolytic solution. Top: ISE based upon a neutral carrier and anionic sites R^-; middle: ISE based upon L^- with a negative charge and cationic sites; bottom: a cation (R^+) exchanger*

As already seen previously, the selectivity of the ISE is essentially related to the properties of optical or molecular recognition of the ionophore. Historically, the first polymer membrane ISE was based upon an antibiotic-type ionophore, valinomycin (selective to K^+). However, the development of supramolecular chemistry is indeed the essential origin of the synthesis of a great number of ionophores [UME 00a, UME 00b, UME 02] and, in particular, of the macrocyclic or macroheterobicyclic types for which selectivity depends for a great part on the matching of the sizes of the cavity of the ionophore and of the target to detect.

For a given target species, there are many suitable ionophores. For instance, derivatives of di and triamides (such as cis-tetraisobutylcyclohexane-1,2-dicarboxamide), crown-ethers (12-crown-4...), derivatives of natural carboxylic polyethers (antibiotics, such as ionomycine), or heteroaromatic compounds (such as a bis-1,10-phénanthroline bridged by a hexanediyle chain) can be used as ionophores for Li^+-ISE. Other examples of ionophores are macrocycles (such as calixarenes, selective for Ag^+, Pb^{2+}, Na^+, phosphadithiamacrocycles, selective for ClO_4^-, ARN selective for Cu^{2+}, Fe^{2+}, Cd^{2+}), nonactine, selective for NH_4^+, pyrochatechol, selective for Fe^{3+}, or thioethers groups, selective for Ag^+, phosphates and derivatives, selective for Ca^{2+}. Table 4.2 lists a few examples of membrane compositions for anions, inorganic cations, or organic ions.

4.3. Current developments in potentiometric sensors

The traditional potentiometric sensors are laboratory-specific electrodes, among which is the typical example of the pH electrode. These electrodes are large, fragile, delicate to handle, and rather costly. The internal reference is generally liquid. The ion-sensitive membrane is either a blown bulb-shaped glass with a very thin wall, or a sintered pellet. This type of sensor is not suitable for mass production. Therefore, traditional potentiometric sensors cannot satisfy the current needs for *in situ* measurements, sometimes in aggressive media. Therefore, the current trend is for the development of *all-solid-state* sensors, robust and possibly miniaturized. Miniaturization permits a considerable broadening of the application field (enabling use in confined media that are difficult to access and of small dimensions, for *in vivo* measurements, for example) and a reduction in price, thus making mass production possible.

4.3.1. *All-solid-state sensors*

In this case, all the elements constituting the sensor are solids, the internal reference included. In a selective electrode making use of an ionic conducting membrane, electric transfer is achieved by two types of carriers, electrons in the metallic electrode and ions in the membrane. A reversible mechanism, between both types of charges, is necessary to achieve good operation of the electrode. If the internal reference is liquid, this mechanism is achieved inside the reference (electrode half reaction). For instance, in the case of a Ag/AgCl reference, the charge exchange takes place via the reaction:

$$Ag_{mat} + Cl^- \Leftrightarrow AgCl_{sol} + e^-_{mat}$$

Tables 4.2 to 4.5 give the compositions of a few ISE membranes for dosing various ions, linear response domains, and selectivity coefficients ($log\ K_{i/j}$)

Composition	Linearity, $\log K_{ij}$	Structure of the ionophore (**X**)
CO_3^{2-}-**X** (0.15 M) CO_3TDA(0.01 M) PVC (nd) oNPOE (nd)	$10^{-2.8} - 10^{-7.3}$ $Benz^-$: 2.2 NO_3^-: 0.4 $H_2PO_4^-$: –2.0	$R = C_6H_{13}$
SCN^--**X** (w=3%) PVC (w=32%) oNPOE (w=65%)	$10^{-1} - 10^{-6}$ ClO_4^-: –0.85 I^-:–0.87	
NO_2^--**X** (w=1%) PVC (w=33%) oNPOE (w=66%) $R^1, R^2 = C_2H_4C_6H_5$	$10^{-1} - 10^{-5}$ SCN^-: 0.4 ClO_4^-: –1.0 I^-: –1.6 Br^-, NO_3^-: –2.7	
HPO_4^{2-}-**X** (w=20%) PVC (45%) DBS (w=35%)	$10^{-1} - 10^{-7}$ SCN^-: –2.30 Cl^-: –2.35	$C_{10}H_{21}$
SO_4^{2-}-**X** (w=1%) CTDMA (x=50%) PVC:oNPOE=2:1	$10^{-2} - 10^{-5}$ SCN^-: 2.9 NO_3^-: 1.6 Br^-: 1.1	
Cl^--**X** (w=3%) PVC (w=42%) oNPOE (w=55%)	$10^{-1} - 10^{-6}$ ClO_4^-: 5.3 SCN^-: 4.0 Br^-: 1.5	n =3

Table 4.2. *Inorganic anions [UME 02]*

Composition	Linearity, log K_{ij}	Structure of the ionophore (**X**)
Li$^+$-**X** (w=1%) PVC (w=33%) oNPOE (w=66%)	Nd NH$_4^+$: –0.2 Na$^+$, K$^+$: –0.4 Sr^{2+}, Ba^{2+}: –0.7	cyclohexane-1,2-dicarboxamide derivative R^1, R^2 = (CH$_3$)N(C$_7$H$_{15}$)
Ca^{2+}-**X** (w=1%) PVC (w=33%) oNPOE (w=66%) NaTPB (x=82%)	10^{-1} - 10^{-5} Mn^{2+}, Zn^{2+}: –1.0 Ba^{2+}: –3.1 Sr^{2+}: –3.2	bis(phosphine oxide) with tolyl groups
NH$_4^+$-**X** (w=25%) Nujol (w=50%) Octanol (w=25%) R^1, R^2, R^3, R^4 = CH$_3$	10^{-3} - 10^{-5} K$^+$: –0.4 Rb$^+$: –0.6 Cs$^+$: –1.9	macrocyclic tetralactone with tetrahydrofuran units
Pb^{2+}-**X** (w=1%) PVC (w=33%) oNPOE (w=66%)	3×10^{-3}-4×10^{-6} Cu^{2+}: –1.8 Zn^{2+}: –2.1	dibenzo crown ether
Sm^{3+}-**X** (w=10.2%) PVC (w=28.7%) oNPOE (w=60.4%) KTpClPB (x=6%)	5×10^{-3} – 10^{-7} Cu^{2+}: 1.2 Fe^{3+}: 0.1 Pr^{3+}: –1.5	bis(dithiocarbonate) with ether linkages
Cu^{2+}-**X** (w=2.6%) PVC (w=32%) DOP (w=64%) KTpClPB (x=13%)	Nd K$^+$: high Pb^{2+}: –1.6 Ca^{2+}: –1.7 Cd^{2+}: –2.0	calix[4]arene derivative R^1, R^2, R^3, R^4 = CH$_2$CH$_2$SCH$_2$C(S)N(CH$_3$)$_2$

Table 4.3. *Inorganic cations [UME 00a]*

Composition	Linearity, log $K_{i/}$	Structure of the ionophore (**X**)
AP-**X** (w=0.2%) PVC (w=32.8%) DOP (w=67%)	10^{-1} - 3×10^{-4} $C_6H_{13}NH_3^+$: 0 $(C_2H_5)_2NH_2^+$: –1 NH_4^+: –1	
CA-**X** (w=3%) PVC (w=22.4%) DOP (w=74.6%)	10^{-2} - 10^{-4} malonate: 0 maleate: 0 succinate: –0.1	
AQ-**X** (w=4.9%) PVC (w=39.6%) oNPOE (w=55.5%)	10^{-2} - 10^{-5} $C_{12}H_{25}(CH_3)_3N^+$: 0 $C_8H_{17}NH_3^+$: –1.4 $(C_4H_9)_2NH_2^+$: –1.9 N=5	
G-**X** (w=1.1%) PVC (32.8%) DBP (65.6%) GDTPB (x=37%)	10^{-1} - 6×10^{-4} guanidinium: 0 $(C_2H_5)_4N^+$: 2 pyridinium: 0.3	
NT-**X** (w=3%) PVC (w=22.4%) DOP (w=74.6%)	Nd ATP^{4-}: 0 ADP^{3-}: –0.7	

AP, primary ammonium derivatives; AQ, quaternary ammonium derivatives; CA, carboxylate derivatives; CO$_3$TDA, tetradecylammonium carbonate; CTDMA, tridodecylmethyl-ammonium chloride; DBP, dibutyl phthalate; DBS, dibutyl sebacate; DOP, dioctyl phthalate; G, guanidinium derivatives; GDTPB, guanidinium tetraphenylborate; KTpClPB, potassium tetrakis(4-chlorophenyl)borate; NaTPB, sodium tetraphenylborate; NT, nucleotides; oNPOE, o-nitrophenyl octyl ether; PVC, polyvinyl chloride

Table 4.4. *Organic ions [UME 02b]*

The fixed composition of the internal solution ensures that the reference potential is constant.

In the case of a selective electrode with a solid internal reference, it is more difficult to achieve reversibility and stability for the transition equilibrium between the ionic and electronic conductivities, because the contacts between solids are much more difficult to make than the solid–liquid contacts. Various solutions have been proposed. They depend on the nature of the conductivity of the ion-sensitive membrane.

4.3.1.1. *Mixed conduction membranes*

In this case, the condition of a reversible transition from electronic conductivity of the metallic connection to ionic conduction of the measured solution is made by the membrane itself. In order to maintain a stable electronic equilibrium, i.e. to maintain equality $\tilde{\mu}_{e,met} = \tilde{\mu}_{e,mem}$ (electrochemical potential of the electrons in the connexion metal and in the membrane), the area of the exchange surfaces between the metal and membrane must be important and as well-defined as possible. Several combinations have been tested over many years in order to reach this objective [NIK 85]:

– an intermediate layer between the membrane and the connexion wire can be made by very good mixing of the connexion metal and the membrane material. Selective electrodes with chalcogenide membranes have been elaborated according to the diagram [VLA 80]:

$$M^{2+} \text{ (sol)} / MS_{Z/2} - Ag_2S / MS_{Z/2} - Ag_2S - Ag / Ag$$
$$\text{Membrane} \quad \text{Intermediate layer}$$

The good reproducibility and the stability of these electrodes for the determination of Ag^+, Cu^{2+}, Pb^{2+}, Cd^{2+}, Hg^{2+}, I^-, Br^- and Cl^- ions in solution is undoubtedly due to the very high contact surface between metallic silver and silver chalcogenide. This configuration ensures both a fast electronic exchange and silver saturation;

– a thin metallic film can be used. The techniques of microelectronics, such as RF cathodic sputtering or thermal evaporation, yield good-quality films establishing a very good contact with the ion-sensitive membrane if the surface of the latter has been polished appropriately. Another noteworthy sensor was created for the detection of sulfide ions in neutral media (such as thermal waters) involving a partially crystalline vitreous chalcogenide CuAsSe with a platinum film for internal reference [CAL 01].

4.3.1.2. Membranes with purely ionic conduction

When the membrane is purely ionic, the internal reference can be a mixed conduction material inserted between the metallic connexion and the ion-sensitive membrane. This material exchanges electrons with the metal and common ions with the membrane. For instance, the Na_xWO_y bronzes have been tested in sensors with a Na^+-sensitive NASICON membrane [LEO 94].

The internal reference can also be a purely ionic conductor, called an "ionic bridge". For instance, the $LaF_3/AgF/Ag$ sensor has been tested for the detection of fluoride ions [DEP 94], while a AgCl–NaCl solid solution was used as an ionic bridge between the NASICON and a silver film for the creation of a sodium ion sensor [POI 97]. In this case, the Ag/AgCl redox couple fixes the potential on the metallic connexion side.

4.3.1.3. A particular example: all-solid-state state pH electrodes

Among the specific electrodes, the pH-sensitive glass electrode is probably the most widely used. pH measurement is required in many cases (industrial processes monitoring, medicine, and biology). Therefore, many researchers have investigated the use of solid-state internal references to replace the reference solution of the pH electrode, thus allowing the use of this electrode in more difficult conditions (for instance, at higher temperatures necessary for sterilization of the electrodes in medical use). The following internal references have been chronologically used more or less successfully:

– metallic sodium or its amalgams. This example is recalled purely for historical reasons. The fabrication of a solid internal reference was noted in 1924; however, the high reactivity of the sodium severely damaged the glass membrane and these works were not followed up;

– the use of a metallic alloy containing an alkaline metal (sodium or lithium) according to the composition of the glass membrane used. The activity of the alkaline metal in the alloy is lower than that of the pure metal, and therefore, the problem damage to the glass is reduced significantly;

– use of a mixed conductor: Na_xWO_3 bronze or glass conducting by Na^+ ions containing a redox couple (Fe^{2+}/Fe^{3+}, V^V/V^{IV}, etc).

The steadily growing needs for biological analysis *in vivo* resulted recently in the achievement of a new range of all-solid-state miniaturized pH electrodes involving the properties of iridium oxide [WAN 02]. A new process called "carbonate melting oxidation" enables the growth of a film of iridium oxide *in situ* at the surface of a very thin metallic iridium wire by oxidation in molten carbonate. The resulting electrodes are very stable over long periods, which would enable continuous pH measurements without the necessity of frequent calibrations.

4.3.1.4. *ISE with a polymer membrane and an all-solid-state internal reference*

For numerous applications, there are requirements for all-solid-state sensors able to function under varying environmental, geometrical, and positioning condition, which present less constraints than the traditional sensors. Two main methods have been explored in this respect: ISE with solid contacts based upon an internal reference with a second kind response or based upon a reversible redox couple encapsulated in a membrane phase.

For example, the measuring chain:

Ag/membrane, AgX, RX/ *analyzed* solution of X^-

does constitute an all-solid-state ISE system. The membrane, with a classical composition (PVC and plasticizer), contains the carrier of X^-, RX, where R may be a quaternary ammonium, and the very insoluble salt AgX. The carrier is dissolved in the organic phase, highly concentrated so that the membrane has a high buffering power for X^- and Ag^+ concentrations. Consequently, the silver electrode has a constant potential and acts as an all-solid-state internal reference electrode.

The second way can be schematized by the following measuring chain:

Pt/membrane, redox couple, organic solution, RX/ solution of X^- to be *analyzed*.

In this example, the redox couple (quinonic, metallocenic systems) is dissolved in an organic solvent in the presence of the liquid ion exchanger (typically tetradecylammonium alkyl phosphate) and set in contact with an inert electrode (Pt). Although this type of ISE is not strictly all-solid-state, the fact that the organic phase is not soluble in the aqueous phase provides it with the interesting properties of a solid ISE.

4.3.2. *All-solid-state microsensors*

A microsensor will generally be similar to a macrosensor as far as the measuring chain is concerned, the main difference arising from the disposition of the constituting elements, for which the sensor maker has more liberty. As shown in Figure 4.10, the various constituents are deposited in the form of thin layers (with a width of a few nanometers up to a micrometer) on the same physical base called a substrate. Each element must fulfill specific requirements:

– the substrate must be chemically inert, electrically insulating, and must withstand the temperatures of the thermal treatments that are necessary for the elaboration of each layer. Typical substrates are glass, a ceramic, such as alumina, passivated silicon, or polyimide;

– the metal serves to establish the connexion for picking up the potential of the sensor;

– the internal reference must provide the reversibility of the electrochemical chain. It involves reversible exchanges with the connection and with the membrane. The electrical conductivity of the internal reference material must be sufficient; otherwise the responses would be slowed down. Moreover, the impedance must not be too high in order to prevent the sensor from being sensitive to electric noise and atmospherics;

– the ionic membrane must be selective, insoluble in water, and also impervious.

The microsensor results from the piling up of various layers: sensitive membrane/internal reference/metallic connexion. Two possible configurations for the elaboration of the potentiometric microsensor in a planar geometry are shown in Figure 4.10.

Depending on the process used for elaborating the layers, one will refer to thin films or thick films (thick layers). Both technologies belong to microelectronics. If the first one necessitates sophisticated equipments (thermal evaporation under vacuum, RF cathodic sputtering, laser ablation) and a clean room environment, it will consequently be reserved for mass production because of production costs. The second one, known as serigraphy (screen printing), necessitates more modest equipment (screen printing machine) and is, therefore, adapted to limited production. However, one of the keys of the success of screen-printing is the optimization of suitable ink adapted to the material to be deposited. For instance, screen-printing, as well as the sol-gel process, has been tested in view of the development of Na^+-microsensors making use of NASICON as the ion-sensitive material [CAN 91, FAB 92, CRE 97]. Among the membranes elaborated using thin-film technologies, numerous works concerning chalcogenide membranes [TAI 99, MIL 99, MOU 00, MOU 01] can be cited.

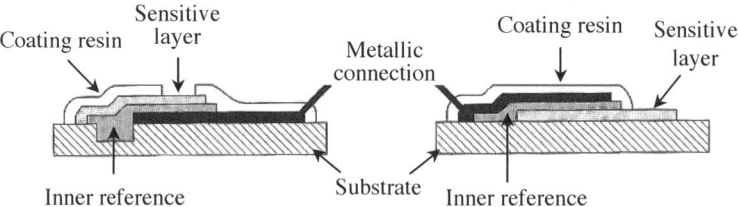

Figure 4.10. *Two typical planar configurations for the elaboration of a potentiometric microsensor*

Potentiometric Sensors   107

Numerous potentiometric microsensors are currently under development. It is not possible to give an exhaustive list, therefore, two examples have been selected to illustrate the new developements. They differ by the nature of their ion-sensitive membrane. In the first case, an inorganic membrane is used made up of a chalcogenide glass, which can detect Cu^{2+} ions. The second microsensor has a polymer membrane, inside which the ionophores are immobilized, which permits the monitoring *in vivo* of pH variations and of K^+, Na^+ and Ca^{2+} concentrations. Before describing them, we will mention the recent research concerning the development of "electronic tongues", which comprise a whole set of sensors having low selectivity, but allowing, thanks to an appropriate mathematical treatment (data fusion), the extraction of pertinent information from complex media [VLA 97, DIN 00].

4.3.2.1. *Selective microelectrode for dosing Cu^{2+} ions in solution [TAI 99]*

The schematic drawing of a microsensor selective to Cu^{2+} ions is given in Figure 4.11. The sensitive part of the sensor, a mixed conduction glass with the composition $(Ge_{28}Sb_{12}Se_{60})_{1-x}Cu_x$ (with x close to 50% mol), the internal reference and/or the collector (Cr) are deposited as thin films by cathodic sputtering.

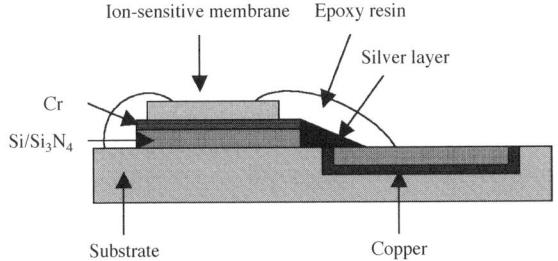

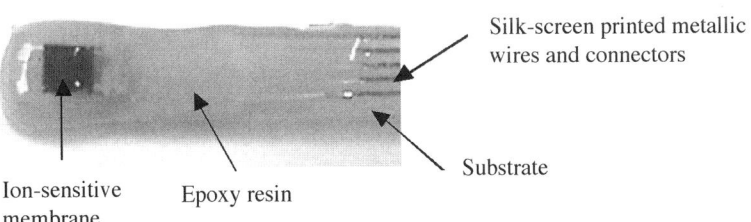

Figure 4.11. *All-solid-state microsensor*

These layers are deposited either on previously nitrided silicon, or on a glass substrate. The detection limit of such a microelectrode is about 10^{-6} mol/l Cu^{2+} ions. Above this limit, quasi-Nernstian behavior is observed, with a slope of 28.5 mV/pCu, which is very close to the theoretical value (29.58 mV at 25°C) for a two-electron transfer at the interface glass membrane/aqueous solution. Table 4.5 illustrates the selectivity of this microelectrode and compares it with other more classical Cu^{2+}-selective electrodes.

	log $K^{pot}_{Cu^{2+},j}$ (concentration of interfering ions (mol/l))			
Interfering ions, j	Amorphous ion-sensitive membranes		Crystallized ion-sensitive materials	
	$(Sb_{12}Ge_{28}Se_{60})_{40}(Cu)_{60}$ (thin film)	Cu-Ag-As-Se	CuS-Ag_2S (polycrystalline)	$Cu_{1.8}Se$ (crystal)
K^+	−5.1 (1)	−6.7 (1)	–	–
Na^+	−5.3 (10^{-1})	−5.4 (1)	–	–
Ca^{2+}	−5.1 (10^{-1})	−5.1 (1)	−5.1 (1)	−2.4 (1)
Ni^{2+}	−4.1 (10^{-1})	−4.5 (1)	−4.6 (1)	−3.5 (1)
Cd^{2+}	−4.1 (10^{-1})	−4.7 (10^{-1})	−4.3 (10^{-1})	−3.7 (10^{-1})
Pb^{2+}	−2.5 (10^{-1})	−4.5 (10^{-1})	−4.1 (10^{-1})	−3.0 (10^{-1})
Mn^{2+}	−3.4 (10^{-1})	−5.2 (1)	−5.1 (1)	−3.0 (1)
Fe^{3+}	−1.3 (10^{-1})	0.6 (10^{-3})	0.5 (10^{-3})	1.1 (10^{-3})

Table 4.5. *Potentiometric coefficients of selectivity of Cu^{2+} ion-sensitive membranes: $(Sb_{12}Ge_{28}Se_{60})_{40}(Cu)_{60}$, Cu-Ag-AsSe, CuS-Ag_2S, $Cu_{1.8}Se$ [TAI 99]*

4.3.2.2. Polymer membrane microsensor [LIN 00]

The schematic drawing of a multiparameter microsensor is given in Figure 4.12. It is meant, for example, for *in vivo* monitoring of H^+, K^+, Na^+ and Ca^{2+} concentrations in the heart muscle.

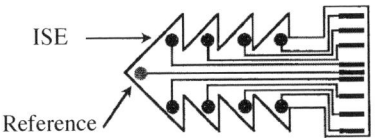

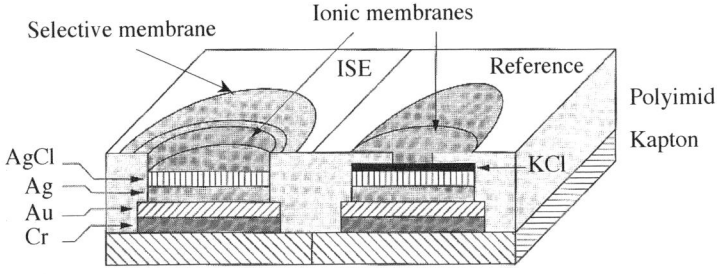

Figure 4.12. *Schematic drawing of a multi-microsensor with a polymer membrane (eight measuring contacts and a reference, dimensions 12 mm x 4 mm x 125 μm) and cross-section*

This type of measurement requires high miniaturization and a flexible substrate. The electrode network of Figure 4.12 is made of eight different measuring sites and one reference contact, deposited on a Kapton film coated with a 100 μm thick insulator (polyimide). Chromium thin films (200 Å), ensuring adhesion, and gold films (2,000 Å), ensuring electric conduction, are successively deposited by evaporation under vacuum, and are covered by a Ag/AgCl layer deposited electrochemically. The suitable membrane, dissolved in a solvent, such as tetrahydrofurane, is, in turn, deposited on the whole lot, in one or two steps and, once the solvent has evaporated, membrane layers about 50 μm thick are commonly obtained. The membrane of the all-solid-state reference electrode can be of the polyurethane type, possibly doped with a cation or anion-exchanging sites (for example, a salt, such as tetraalkylammonium tetraphenylborate).

A particularly interesting point of this microfabrication method is that different types of sensors can be easily integrated on the same platform: for example, a combination of amperometric (lactate sensors) and potentiometric (pH, pK^+) sensors has been developed.

4.4. Bibliography

[BAK 71] BAKER C.T., TRACHTENBERG I., "Ion selective electrochemical sensors – Fe^{3+}, Cu^{2+}", *J. Electrochem. Soc*, vol. 118, p. 571-576, 1971.

[BAK 97] BAKKER E., BÜHLMANN P., PRETSCH E., "Carrier-based ion-selective electrodes and bulk optodes. 1. General characteristics", *Chem. Rev.*, vol. 97, no. 8, p. 3083-3132, 1997.

[BUL 98] BÜHLMANN P., PRETSCH E., BAKKER E., "Carrier-based ion-selective electrodes and bulk optodes. 1. Ionophores for potentiometric and optical sensors", *Chem. Rev.*, vol. 98, no. 4, p. 1593-1687, 1998.

[BAK 99] BAKKER E., BÜHLMANN P., PRETSCH E., "Polymer membrane ion-selective electrodes – what are the limits?", *Electroanalysis*, vol. 11, no. 13, p. 915-933, 1999.

[BAK 02] BAKKER E., TELTING-DIAZ M. "Electrochemical sensors", *Anal. Chem.*, vol. 74, no. 12, p. 2781-2800, 2002.

[BOH 81] BOHNKE C., MALUGANI J.P., SAIDA A., ROBERT G., "Electric-conductivity and ion selectivity of the glasses $AgPO_3$-PbI_2, $AgPO_3$-HgI_2, $AgAsS_2$-PbI_2 and $AgAsS_2$-HgI_2", *Electrochim. Acta*, vol. 26, p. 1137-1142, 1981.

[BYC 95] BYCHKOV E.A., BRUNS M., KLEWE-NEBENIUS H., PFENNIG G., HOFFMANN W., ACHER H.J., "Cu^{2+}-selective thin-films for chemical microsensors based on sputtered copper-arsenic-selenium glass", *Sens. Actuators B*, vol. 24–25, p. 733-736, 1995.

[CAL 01] CALI C., TAILLADES G., PRADEL A., RIBES M., "Determination of sulfur species using a glassy-crystalline chalcogenide membrane", *Sens. Actuators B*, vol. 76, p. 560–564, 2001.

[CAN 91] CANEIRO A., FABRY P., KHIREDDINE H., SIEBERT E., "Performance characteristics of sodium super ionic conductor prepared by the sol-gel route for sodium-ion sensors", *Anal. Chem.*, vol. 63, p. 2550-2557, 1991.

[CHR 97] CHRISTENSEN J.H., CLEMMESEN P., HANSEN G.H., LILTORP K., MORTENSEN J., "Response characteristics and application of chalcogenide glass Cr(VI) selective electrode", *Sens. Actuators B*, vol. 45, p. 239-243, 1997.

[CRE 97] CRETIN M., FABRY P., "Detection and selectivity properties of Li^+-ion-selective electrodes based on NASICON-type ceramics", *Anal. Chim. Acta*, vol. 354, p. 291-299, 1997.

[DEP 94] DÉPORTES C., DUCLOT M., FABRY P., FOULETIER J., HAMMOU A., KLEITZ M., SIEBERT E., SOUQUET J. L., *Electrochimie des Solides*, PUG, 1994.

[DIN 00] DI NATALE C., PAOLESSE R., MACAGNANO A., MANTINI A., D'AMICA A., LEGIN A., LVOVA L., RUDNITSKAYA A., VLASOV YU., "Electronic nose and electronic tongue integration for improved classification of clinical and food samples", *Sens. Actuators B*, vol. 64, p. 15-21, 2000.

[ENG 88] ENGELL J., MORTENSEN S., Ion-sensitive measuring electrode device, Radiometer International Patent WO 84/01829, 1984.

[FAB 88] FABRY P., GROS J.P., MILLION-BRODAZ J.F., KLEITZ M. "NASICON, an ionic conductor for solid state Na^+-selective electrode", *Sens. Actuators*, vol. 15, p. 33-49, 1988.

[FAB 92] FABRY P., HUANG Y.L., CANEIRO A., PATRAT G., "Dip-coating process for preparation of ion-sensitive NASICON thin-films", *Sens. Actuators B*, vol. 6, p. 299-303, 1992.

[GUE 95] GUESSOUS A., PAPET P., SARRADIN J., RIBES M., "Thin-films of chalcogenide glass as sensitive membranes for the detection of mercuric ions in solution", *Sens. Actuators B*, vol. 24–25, p. 296-299, 1995.

[GUE 98] GUESSOUS A., SARRADIN J., PAPET P., ELKACEMI K., BELCADI S., PRADEL A., RIBES M., "Chemical microsensors based on chalcogenide glasses for the detection of cadmium ions in solution", *Sens. Actuators B*, vol. 53, p. 13-18, 1998.

[JAS 74] JASINSKI R., TRACHTENBERG I., RICE G., "A chalcogenide glass electrode sensitive to cupric ions", *J. Electrochem. Soc*, vol. 121, p. 363-370, 1974.

[LEG 99] LEGIN A., RUDNITSKAYA A., VLASOV YU., DI NATALE C., D'AMICO A., "The features of the electronic tongue in comparison with the characteristics of the discrete ion-selective sensors", *Sens. Actuators B*, vol. 58, p. 464-468, 1999.

[LEO 94] LEONHARD V., ILGENSTEIN M., CAMMANN K., KRAUSE J., "NASICON electrode for detecting sodium-ions", *Sens. Actuators B*, vol. 18–19, p. 329-332, 1994.

[LIN 00] LINDNER E., BUCK R. P., "Microfabricated potentiometric electrodes and their *in vivo* applications", *Anal. Chem.*, vol. 72, no. 9, p. 336A-345A, 2000.

[MAU 98] MAUVY F., CRETIN M., SIEBERT E., FABRY P., "Mechanism of selectivity in alkali sensitive membrane based on NASICON", *J. New Mat. For Electrochem. Systems*, vol. 1, p. 71-76, 1998.

[MAU 99] MAUVY F., SIEBERT E., FABRY P., "Reactivity of NASICON with water and interpretation of the detection limit of a NASICON based Na^+ ion selective electrode", *Talanta*, vol. 48, p. 293-303, 1999.

[MID 91] MIDGLEY D., TORRANCE K., *Potentiometric Water Analysis*, John Wiley & Sons Ltd, 1991.

[MIL 99] MILOSHOVA M., BYCHKOV E., TSEGELNIK V., KLEWE-NEBENIUS H., BRUNS M., HOFFMANN W., PAPET P., SARRADIN J., PRADEL A., RIBES M., "Tracer and surface spectroscopy studies of sensitivity mechanism of mercury ion chalcogenide glass sensors", *Sens. Actuators B*, vol. 57, p. 171-178, 1999.

[MOU 00] MOURZINA Y., SCHÖNING M., SCHUBERT J., ZANDER W., LEGIN A., VLASOV Y., LÜTH H., KORDOS P., "A new thin-film Pb microsensor based on chalcogenide glasses", *Sens. Actuators B*, vol. 71, p. 13-18, 2000.

[MOU 01] MOURZINA Y., SCHÖNING M., SCHUBERT J., ZANDER W., LEGIN A., VLASOV Y., LÜTH H., "Copper, cadmium and thallium thin film sensors based on chalcogenide glasses", *Anal. Chim. Acta*, vol. 433, p. 103-110, 2001.

[NIK 85] NIKOLSKII B., MATEROVA E., "Solid contact in membrane ion-selective electrodes", *Ion Selective Electrode Rev.*, vol. 7, p. 3-39, 1985.

[OWE 80] OWEN A.E., "Chalcogenide glasses as ion-selective materials for solid-state electrochemical sensors", *J. Non-Cryst. Solids*, vol. 35–36, p. 999-1004, 1980.

[POI 97] POIGNET N., Elaboration et caractérisation de couches pour microcapteurs électrochimiques à base de NASICON, PhD Thesis, Grenoble Institute of Technology, 1997.

[TAI 99] TAILLADES G., VALLS O., BRATOV A., DOMINGUEZ C., PRADEL A., RIBES M., "ISE and ISFET microsensors based on a sensitive chalcogenide glass for copper ion detection in solution", *Sens. Actuators B*, vol. 59, p. 123-127, 1999.

[TOH 83] TOHGE N., MINAMI T., TANAKA M., "Copper-containing glasses as electrode materials sensitive to cupric ions", *Yogyo-Kyokai-Shi*, vol. 91, p. 32-37, 1983.

[TOH 86] TOHGE N., TANAKA M., "Chalcogenide glass electrodes sensitive to heavy ions", *J. Non-Cryst. Solids*, vol. 80, p. 550-556, 1986.

[UME 00a] UMEZAWA Y., BÜHLMANN P., UMEZAWA K., TOHDA K., AMEMIYA S., "Potentiometric selectivity coefficients of ion-selective electrodes. Part I. Inorganic cations, IUPAC Technical Report", *Pure Appl. Chem.*, vol. 72, no. 10 p.1851-2082, 2000.

[UME 00b] UMEZAWA Y., BÜHLMANN P., UMEZAWA K., HAMADA N., "Potentiometric selectivity coefficients of ion-selective electrodes. Part III. Organic ions, IUPAC Technical Report", *Pure Appl. Chem.*, vol. 74, no. 6, p. 995-1099, 2000.

[UME 02] UMEZAWA Y., UMEZAWA K., BÜHLMANN P., HAMADA N., AOKI H., NAKANISHI J., SATO M., PING XIAO K., NISHIMURA Y., "Potentiometric selectivity coefficients of ion selective electrodes. Part II. Inorganic anions, IUPAC Technical Report", *Pure Appl. Chem.*, vol. 74, no. 6, p. 923-994, 2002.

[VLA 80] VLASOV Y.G., ERMOLENKO YU., ISKHAKOVA O., "Lead containing electrodes on the basis of lead and silver sulfides", *Zh. Anal. Khim.*, vol. 35, p. 15, 1980.

[VLA 85] VLASOV Y.G., BYCHKOV E.A., MEDVEDEV A.M., "Chalcogenide glass electrodes for the determination of copper(II)", *J. Anal. Chem. USSR*, vol. 40, p. 357-363, 1985.

[VLA 86] VLASOV Y.G., BYCHKOV E.A., MEDVEDEV A.M., "Copper ion-selective chalcogenide glass electrodes – Analytical characteristics and sensing mechanism", *Anal. Chim. Acta*, vol. 185, p. 137-158, 1986.

[VLA 87a] VLASOV Y.G., BYCHKOV E.A., "Electrochemical ion-selective sensors based on chalcogenide glasses", *Sens. Actuators*, vol. 12, p. 275-283, 1987.

[VLA 87b] VLASOV Y.G., BYCHKOV E., "Ion-selective chalcogenide glass electrodes", *Ion-Selective Electrode Rev.*, vol. 9, p. 5-93, 1987.

[VLA 88] VLASOV Y.G., BYCHKOV E.A., LEGIN A.V., "Influence of oxidizing-agents on lead-ion-selective chalcogenide glass electrodes – electrochemical studies", *Soviet Electrochemistry*, vol. 24, p. 1087-1092, 1988.

[VLA 89] VLASOV Y.G., MOSKVIN L.N., BYCHKOV E.A., GOLIKOV D.V., "Silver bromide based chalcogenide glassy crystalline ion-selective electrodes", *Analyst*, vol. 114, p. 185-190, 1989.

[VLA 92] VLASOV Y.G., BYCHKOV E.A., LEGIN A.V., "Mechanism studies on lead ion-selective chalcogenide glass sensors", *Sens. Actuators B*, vol. 10, p. 55-60, 1992.

[VLA 94] VLASOV Y.G., BYCHKOV E.A., LEGIN A.V., "Chalcogenide glass chemical sensors – research and analytical applications", *Talanta*, vol. 41, p. 1059-1063, 1994.

[VLA 97] VLASOV Y.G., LEGIN A.V., RUDNITSKAYA A., "Cross-sensitivity evaluation of chemical sensors for electronic tongue: determination of heavy metal ions", *Sens. Actuators B*, vol. 44, p. 532-537, 1997.

[WAN 02] WAN M, YAO S., MADOU M., "A long-term stable iridium oxide pH electrode", *Sens. Actuators B*, vol. 81, p. 313-315, 2002.

Chapter 5

Amperometric Sensors

5.1. Sensors based upon chemically modified electrodes

5.1.1. *Introduction*

5.1.1.1. Position of amperometric sensors among the class of chemical sensors

A chemical sensor can be defined as a measuring device capable of delivering direct information about the chemical composition of its environment. It comprises a physical transducer and a selective chemical element. The term electrochemical sensor is used when the information transfer involves an electrochemical mode (amperometric, potentiometric, or conductimetric).

On July 13, 2009, 12,237 bibliographic references dealing with chemical sensors were inventoried in the database of the ISI (Institute of Scientific Information). This number is somewhat distorted because several papers using the term *sensor* in fact describe detection systems, which are not, strictly speaking, sensors, according to the description given above. The number of publications is steadily growing, with over 80% of them issued during the last decade, which reflects the modern character of this research field. The important production of reviews (about one review article out of four research papers) is also notable.

About 43% of these references concern electrochemical devices. Amperometric sensors cover about 15% of the overall scientific production on chemical sensors. Numerous articles are published in conference proceedings or by professional societies (*NATO, SPIE, ACS Symp. Ser.*, etc.).

Chapter written in French by Alain WALCARIUS, Chantal GONDRAN and Serge COSNIER.

Amperometric sensors can be classified in three categories: gas sensors, sensors detecting chemical species in solution (often based upon the use of chemically modified electrodes, described in the first part of this chapter), and biosensors (discussed in the second part of this chapter).

5.1.1.2. *Differences between amperometric analysis and amperometric sensor*

The difference between the study of the intrinsic electrochemical behavior of a system having an amperometric response and its extension from fundamentals to applications (from the initial amperometric detection of target analytes, and progression to the creation of a sensor device), is poorly defined. These frontiers are not clearly visible in the research articles dealing with the systems with amperometric response liable to yield analytic applications.

The notion of a sensor implies a unique character of its functioning, i.e. for an amperometric sensor, this exclusively implies the amperometric response of the device, whereas amperometric analysis may involve an analytical instrument that requires one or several other operations prior to the delivery of quantitative experimental data [JAN 88]. For instance, the continuous flow analysis devices making use of amperometry or voltamperometry are not considered amperometric sensors, although they sometimes deliver faster and more precise information than the systems with direct response, and despite the fact that a direct (volt)amperometric detection mode is often integrated as part of their detection method.

In cases where the electro-analytic studies of (new) amperometric response systems do not lead directly to the development of a chemical sensor, they may aid the progression in the field, which is likely to be exploited in the future. A clear example is that of the chemically modified carbon paste electrodes, which have almost no practical application (in spite of an abundance of literature on their potentialities towards sensitive and selective electrochemical analysis), whereas laboratory results on these electrodes could be exploited beneficially and transferred to the field of single-use (disposable) sensors via thin film electrodes obtained by screen printing.

5.1.1.3. *Importance of the chemical modification of conventional electrodes*

Conventional electrodes (platinum (Pt), various forms of carbon, mercury (Hg), gold (Au), silver (Ag)) are able to detect a chemical species endowed with redox properties via reduction or oxidation. Amperometry (or voltamperometry) is operated by imposing a potential (or a potential varying with time), and measuring the variations of the current induced by this perturbation. These conventional electrodes have intrinsic sensitivity and selectivity limits. Both characteristics can be considerably improved by chemically modifying their surface.

This concept was discovered in 1975 with the creation of a *chiral electrode* (amino-acid bonded to a graphite surface) [WAT 75] and was extensively developed under the instigation of Murray and his collaborators [MUR 80, MUR 84, MUR 87]. Considerable research efforts during the last 30 years have been generated by the possibility of coupling an electronic transfer reaction with the intrinsic properties of a modifying agent inducing new characteristics for the electrode-solution interface. In particular, modified electrodes have raised much interest in analytical electrochemistry, both for their contribution to the analysis of reactions (characterization of mass transport and charge transfer, reactivity at the electrode-solution interface, characterization of new materials, study of the mechanisms of electrochemical synthesis), as well as for the development of new chemical sensors. The latter aspect has motivated numerous review papers dealing with both amperometric sensors for chemical species and biosensors [ESP 86, WAL 88, WAN 89, DON 89, KAL 90, STO 90, MER 90, WAN 91a, WAN 91b, BAL 91, WRI 92, LAB 92, BUD 92, JOS 93, ARR 94, GIL 94a, KAL 95, GOR 95, NIW 95, UGO 95, GIL 95, COX 96, ALE 96, SEN 96, WAL 96, ALB 97, BON 97, KUL 98, WAL 98, CAS 98, MOU 98, CHU 98, STE 99, WAL 99a, REY 99, MAL 99, LAB 00, WAL 01a, DEM 01]. These reviews illustrate, using many examples, the importance of modified electrodes for elaborating more and more sensitive and selective detection systems; they will be the basis of a large part of this chapter, devoted to amperometric sensors. Particular attention is currently being paid to the elaboration of miniaturized and stable-over-time systems. This constitutes a challenge for the large-scale use of sensitive, stable, fast, selective, reliable, and easy-to-use measurement tools, which are increasingly demanded in the fields of environment, medical analysis, and quality insurance.

The conception of new electrode materials needs multidisciplinary approaches, requiring collaborations with electrochemists in other disciplines, such as polymer chemistry, organic chemistry, solid-state chemistry, biochemistry, surface and interface chemistry, and, as far as applications in the field of sensors is concerned, specialists in engineering, automating, and electronics.

5.1.1.4. Different types of amperometric sensors

For any sensor, the success of a modified electrode as an (volt)amperometric detector is linked to its capacity to identify and/or to quantify the target species in a particular medium. This success will depend on several properties:

– stability (operational and long term);

– reproducibility;

– definition and origin of the response to the target external parameter observable on a large concentration range, and rapidity of the determination;

– absence of interference in the required measuring conditions.

Typically, a sensor based upon the concept of a modified electrode is designed to deliver a response, which would otherwise not have been possible to obtain, to enhance the signal:noise ratio, and to remove the effect of interference.

Amperometric or voltamperometric sensors based upon chemically modified electrodes for the analysis of species in solution can be classified in three main categories based on the analytic principle involved:

– analysis by pre-concentrating, i.e. selective permeability;

– direct electrocatalytic detection;

– indirect amperometric analysis.

The quest for optimal performance in each of these three domains involves creating integrated electrochemical devices that have the greatest sensitivity and the best selectivity towards the target molecule.

The first category implies the presence, at the surface of the electrode, of complexing agents or of molecular recognition structures aimed at selectively preconcentrating a given analyte. The role of the modifying agent is to increase the local concentration of the analyte at the electrode surface, prior to detection itself, in order to perform analyzes in very dilute media. An alternative way is to use a perm-selective membrane to decrease possible interferences.

The second category is based upon the use of electron transfer accelerating agents (chemical mediators or electrode material acting as a solid-state catalyzer). They are selected either for their specific action towards a given species, or for increasing the sensitivity of detection of the target species, or for both properties simultaneously.

The third category concerns amperometric detection of species that are usually undetectable by this means, but which perturb the electrochemical properties of a modified electrode likely to interact preferentially with these species. The optimal composition and final structure of the sensor are conditioned to a great extent by the aimed application, so that every analysis is practically an individual case by itself. The choice and preparation of the "sensitive layer", the electrochemical properties of which will be disrupted by the interactions with the target species, requires a skillful designer; associating this system to a transduction device will permit the quantitative monitoring of these perturbations.

Another way to classify amperometric sensors has been proposed based on the manufacturing stage of the device [GOP 91]:

– working electrodes with no protecting membrane;

– electrodes coated by a membrane;

– electrochemical cells containing a gel electrolyte.

This classification rests on the commercially available amperometric sensor devices (mainly systems derived from the Clark oxygen electrode); it is much more restrictive than the classification based upon the detection modes.

Numerous modifying agents and several electrode structures have been tested, leading on to a multitude of amperometric sensors based upon the abovementioned three main principles. These integrated chemical systems require a arsenal of active functions (complexing agents, catalysts, redox-active polymers or ion exchangers, recognition or barrier structures at the molecular scale, etc.), with different natures (organic, inorganic, organometallic) at several scales (molecular, supramolecular, macromolecular), individual or associated to each other, and organized in different ways (fixed on the surface, deposited as mono or multimolecular films, organized in a three-dimensional network, scattered in a composite matrix, etc.).

The aim of this chapter is not to provide a review of the amperometric sensors described in the literature, but rather to outline the domain via the important principles of amperometric detection with some examples illustrating these principles. Current tendencies will also be evoked.

5.1.2. *Fabrication and characterization*

5.1.2.1. *Different types of modified electrodes*

Based on the final structure of the electrode material, three main families of chemically modified electrodes can be distinguished:

– monomolecular films;

– homogenous multimolecular films;

– microstructured and heterogenous multimolecular materials.

Monomolecular films are essentially constituted by chemical groups either adsorbed or linked by covalent bonds to the surface of a solid conventional electrode. Adsorption, which is often direct and non-specific, may have a physical or chemical nature, and may even be controlled by an electric field (electrosorption). The most famous examples concern the chemisorption of olefin derivatives on platinum [LAN 73] and of molecules containing thiol or sulfur groups on gold [LI 84], or the physisorption of substances containing aromatic centers onto carbon

[BRO 77]. The covalent fixation of molecules at the surface of the electrodes needs the reaction of surface groups (for example, the hydroxyl group on a metal electrode or the carboxylic groups of the carbon electrode); it can also occur via coupling agents, such as the silanization precursors, the acyl chlorides, the amine derivatives, or the dicyclohexylcarbodiimide [SMY 92]. In the latter cases, the primary functions grafted on the electrode surface can serve as anchoring points for more complex molecular systems [MUR 92]. A variant of monomolecular film is the formation of a hydrophobic molecular layer at an electrode surface. The layer can have the form of a Langmuir-Blodgett film or result from the auto-assembly of hydrophobic chains terminating in a chemisorbable entity [MIL 88]. In all these cases, the electroactive species are very close to the electrode, so that charge transfer is not limited by diffusion of the species, and the overall electrochemical behavior is of the "thin layer" type. Because of the monomolecular character of the film, the number of functional groups likely to take part in the electrochemical reaction is limited by the surface of the electrode; it can only be increased by changing to multilayer or multimolecular film systems.

Homogenous multimolecular films are by far the most widely used in devices of the amperometric sensor type. They are in most cases organic polymers (conductors or non-conductors) deposited on a solid electrode, which may or may not be electroactive, and may or may not have reactional groups. The number of newly developed organic films, and even organic–inorganic hybrid films, has increased strongly. The electrodes modified by organic polymer films can be created directly from a pre-formed polymer, soluble in an organic solvent and insoluble in the analyzed electrolytic medium. A given quantity of polymer, dissolved in an appropriate solvent, is then deposited on the surface of a solid electrode, which, once dried, gives rise to the so-called "dip coating" polymeric deposit. A variant with the aim of improving the reproducibility of the deposit, called "spin coating", consists of applying the polymeric solution at the center of a high speed-rotating disk by means of a microsyringe. In the absence of specific bonding groups at the electrode surface, adhesion of the polymer results from adsorption or simply from a precipitation reaction. An additional method of preparing homogenous polymeric films is direct polymerization at the electrode surface from a solution of monomers. The reaction may be either purely chemical or initiated by electron transfer, or totally electrochemical (electropolymerization). The non-electrochemical methods include gaseous phase polymerization (for instance, polyvinyl-ferrocene), thermal treatment of previously adsorbed monomers (for instance, phthalocyanine films), photochemically assisted polymerization, or polymerization of bis(trialkoxy)organosilane species (in which the organic part is derivative from the viologen or from the cobalticinium), which results simultaneously in chemical bonding to the electrode surface and a three-dimensional structure of redox centers interconnected by siloxane bridges. The polymerization induced by electrochemistry involves the generation of reactional intermediates (radical ions, cations or anions), which serve as initiators for the film formation at the surface of the electrode. This approach has mainly been exploited for the formation of polymeric layers based on metallic complexes of bipyridine. Finally, conducting polymer films can be prepared by electrochemical oxidation of monomers, such as pyrrole, aniline, or thiophene. The synthesis process leads to incorporation of counter-ions (dopants) in the bulk of the material during polymerization, which can then be beneficially used as support for the immobilization of reactants (catalytic centers, ligands) concentrated at the electrode-solution interface. In each case, the monomers can be substituted by organic groups or not, resulting in materials that have "to-measure" structures and properties. The addition of particular reactants (for instance, metallic or organometallic catalysts, or macrocyclic ligands) to the medium where the synthesis of polymeric films is undertaken, leads to supported hybrid systems, the diversity of which is only limited by the imagination of the researcher, because the number of attractive reactants for the creation of modified electrodes is enormous. An abundance of literature is available concerning the preparation and the characterization of electrodes modified by organic polymers [ESP 86, SKO 86, WAL 88, DER 89, HEI 90, JOS 93, LEE 96]. In addition, multimolecular films on electrodes can also be prepared starting from inorganic monomers. They often result in rigid, homogenous, three-dimensional structures. Two main classes are distinguished: the metallocyanates [ITA 86] and polyoxometallates [POP 91], which are electroactive, and the metal oxides (mainly of the silicate type) prepared by the sol-gel process [LEV 97, WAL 98, WAN 99 a, WAL 01a, WAL 01b], which are generally electro-inactive. The electrochemical activity of the first type has been exploited for the preferential detection of charge-compensating ions, by taking advantage of the selectivity offered by the three-dimensional rigid network, whereas the second have mainly been used for their ability to immobilize specific reactants (ligands, catalysts, biomolecules, etc.) at an electrode surface. A particular advantage of the sol-gel process is its ability to form hybrid organic-inorganic structures in which exists a non-hydrolyzable covalent bond between the rigid inorganic skeleton and the organic group, thus ensuring a durable immobilization of the reactant.

Among the multitude of film compositions, the most frequently encountered examples are:

– electroactive polymers, also called redox polymers, containing in their bulk electroactive groups, such as derivates of viologen (pyridinium-based), metallocens (ferrocene end cobaltocene), metallic bi-pyridinic complexes, hexacyanoferrate ion, tetracyanoquinodimethane, derivates from quinone among others;

– ion-exchange polymers, such as Nafion® (perfluorosulfonate) or polyvinylsulfonic, polystyrenesulfonic, and polyethylenecarbonic acids, which are used as cation exchangers and polymers presenting quaternary ammonium groups, such as polylysine, protonated polyvinylpyridine or N-methyl-polyvinylpyridine serving as ion exchangers;

– the conducting polymers, such as polypyrrole, polythiophene and polyaniline, which are either used as such or modified by covalent bonding to other chemical groups or structured as interpenetrated polymers with an inorganic matrix prepared by the sol-gel process.

Heterogenous multimolecular films are the last family, in which are found all other systems that are neither monomolecular nor multimolecular homogenous films. They are often conducting composite materials acting as electrodes, or are heterogenous films deposited on a solid electrode. They mainly comprise electrodes modified by entities (usually inorganic), which have the form of particles, or aggregates that can barely or not at all constitute homogenous films. These solid particles are then maintained on the surface of a solid electrode by a macromolecular binder or scattered in a conducting composite matrix. This is the case in electrodes modified by clays, [FIT 90, MAC 98], by zeolites [ROL 90, BED 95, WAL 96, WAL 99a], or silicates and other parent materials [LEV 97, WAL 98, WAN 99a, WAL 01a, WAL 01b]. In this context, the importance of carbon paste as an excellent matrix for incorporating solid, non-conducting, modifying agents should be noted, which has lead to numerous applications in the domain of organic and inorganic species analysis [KAL 95]. However, the direct use of modified carbon paste electrodes in sensors is limited, but these electrodes are the basis for numerous studies that have led to the production of single-use electrodes in the form of disposable sensors.

Electrochemical (linear or cyclic voltamperometry, at motionless or rotating disc electrodes, chronoamperometry or chronocoulometry) methods for the *in situ* characterization of modified electrodes are unparalleled. Cyclic voltamperometry enables fast qualitative interpretation of the reactivity of new systems (for instance, the discrimination between a thin-layer-type behavior and a diffusion-limited response). At the structural level, complementary information can be gained by physical surface analysis methods, such as scanning electron microscopy, X or Auger photoelectron spectroscopy, mass spectroscopy of secondary electrons, and optical (infrared (IR), Raman, ultraviolet (UV)/visible) spectroscopies. The use of spectroscopic surface analysis methods and the electro-analytic approaches for characterizing modified electrodes have been extensively described in the literature [MUR 84].

The various types of chemically modified electrodes that have been briefly described above have been the subject of numerous laboratory studies, including those in the field of amperometric sensors, but they are often of limited use. However, one of the advantages of the chemical sensors is their aptitude to be used, at low cost, as "alert" analysis systems, in comparison with the big laboratory analysis instruments (which have large size, high costs, require calibration procedures) that will be used occasionally for more precise determinations when necessary [EMO 96]. This is the reason why there has recently been an effort to develop the ideas generated in fundamental research to create practical systems using modified electrodes. This effort is mainly concentrated on systems that are miniaturized, autonomous, and usable in the field.

5.1.2.2. *The single-use (disposable) sensors*

The screen-printing technique permits the creation of all-solid-state, flat and mechanically robust sensors. This process involves a sequential deposition of thick films on a solid substrate of a ceramic or organic polymer type. Although several elaboration processes have been described based on the use of different ink compositions and of different substrates, which depend on the intended application, a general frame of successive operations can be given [GAL 95].

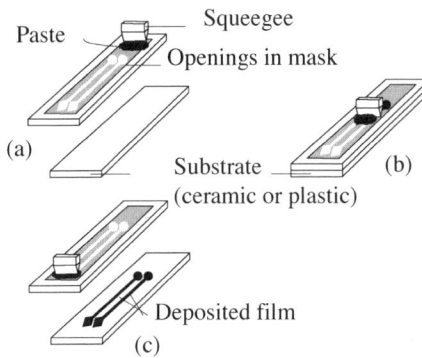

Figure 5.1. *Schematic drawing of the screen-printing process used in the preparation of single-use electrochemical sensors. (a) lined mask and substrate before printing, (b) ink-deposition process, (c) films obtained (shown after mask removal)*

Figure 5.1 schematically illustrates the production of single-use (disposable) sensors by screen-printing. A graphitic ink (dispersion of carbon and of a macromolecular ligand in an organic solvent) is applied to the substrate through a mask covering its surface, which may have been previously modified, if necessary, by the incorporation of an appropriate reactant. Various patterns can be obtained, according to the judicious selection of the mask, and successive deposits can be applied by the use of different inks and different masks, until the desired configuration is obtained (for instance, a carbon-based working electrode and a AgCl/Ag reference electrode on the same substrate). The pieces are then submitted to a thermal treatment to remove the solvent and for "cooking" the patterned

124 Chemical and Biological Microsensors

electrode. An insulating membrane can also be deposited later on some of the patterns to delimit a precise surface of the working electrode on one side of the substrate, and ensure electrical contact on the other side. The flexibility of this process permits numerous variants, fabrication materials, and final structures, and the possibility to integrate the sensor into more complex systems. This type of modified electrode has been reviewed in detail [NAS 98].

A variant of this approach, recently proposed by Girault *et al.* [BAG 99], is illustrated in Figure 5.2. It concerns large-scale fabrication of single-use carbon electrodes by a rotogravure process, which is widely used for different applications in industry. The printing is done by a cylinder covered with metallic copper fitted with engraved cellular structures filled with a graphitic ink that when rotated against a flexible polyester support, deposits regular prints of electrode materials. As in the case of screen-printed electrodes, this microelectrodes network is sickened for the ink(s) solvent(s) to be evaporated. This process is proposed for the mass production of disposable carbon electrodes.

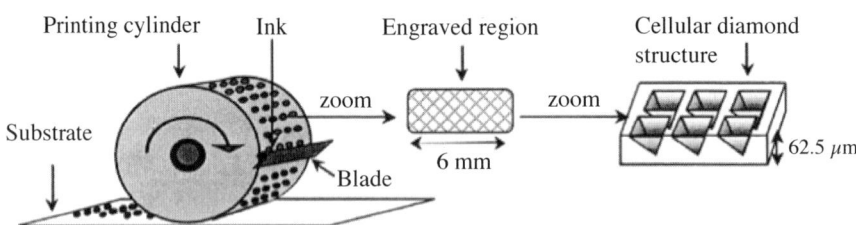

Figure 5.2. *Schematic drawing of the printing process by rotogravure (reproduction with permission from [BAG 99]). a) Rotogravure operation, b) magnification of the engraved areas defining the shape of the electrode, c) magnification of the cellular structure of the engraved part*

5.1.2.3. *Amperometric detectors and decentralized analysis*

Another way of exploiting the fundamental results obtained in the laboratory by means of modified electrodes for optimizing chemical sensors usable *in situ*, is the fabrication of decentralized (at distance) analysis systems. The latter systems are electrochemical-integrated systems capable of functioning at a reasonably long distance from the operator. They can be used for *in vivo* analysis (for instance, lead in blood, glucose in continuous use by a biosensor, which must be bio-compatible), or for the *in situ* monitoring of pollutants (metals, aromatic derivatives) in the environment [WAN 94a, WAN 97].

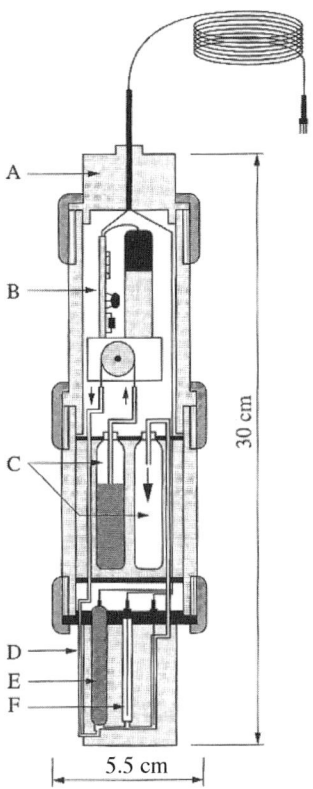

Figure 5.3. *Schematic drawing of a microlab-type submersible system (reproduced with permission, from [WAN 00]). A, cable serving as connection; B, peristaltic micropump; C, tanks (for reactants and refuses; D, sampling tube by microdialysis and electrochemical detector in continuous flux; E, working electrode; F, reference electrode; arrows indicate the flux direction*

An interesting concept in this respect [WAN 00] is called "*lab-on-a-cable*". It is a compact system for continuous flow analysis with amperometric detection (Figure 5.3) integrating the sampling, the possible treatment of the sample, the detection in a hydrodynamic mode, the whole set being controlled at a distance via an electric cable (15 m). Optimization of the arrangement of miniaturized components (micropump, microdialysis, small tanks, microvalves, and microtubings) made possible the development of a complete, compact device. This device, illustrated in Figure 5.3, has been tested for dosing phenol and parent derivatives by means of a tyrosinase-based microsensor [WAN 00], but the concept could easily be enlarged to other applications, because of the flexibility of the design of the prototype [DAN 00].

5.1.2.4. *Advantage of micro and nanotechnologies in the design of amperometric sensors*

Recently, there has been considerable development in the field of "infinitely small" systems in numerous scientific disciplines and, concurrently, of the increasingly localized study of structures at the submicronic scale or even subnanometric scale. This trend also occurred in electrochemistry and was accentuated in this field because of the extraordinary properties of small size electrodes (ultramicroelectrodes).

As far as electrochemical sensors are concerned, miniaturization proves to be an important asset that has allowed the specific behavior of ultramicroelectrodes to be utilized. This has enabled the creation of sufficiently small integrated systems to create *in situ* applications, to prepare networks or modules of electrodes (interconnected or not), to enable the use of extremely low quantities of reactants, to decrease the response times and reduce significantly the diffusional limitations usually encountered in (volt)amperometry. In environmental control, for instance, it is well known that the portableness, the analysis duration, and the automation are very important factors to take into account for the development chemical or biological sensors [SAD 00].

A recent review [STE 99] thoroughly deals with the current state-of-the-art sensors based on electrode networks, including their development, characterization, practical use, and application examples of several types of devices, among which are the amperometric sensors. The fabrication of these arrays of electrodes requires particular techniques, such as microlithography, laser ablation, dry etching, thin films deposits, photolithography, and localized electrodeposition. It often implies the use of silicon-based chips on which are deposited and machined metal or carbon networks, chemically modified or not according to the aimed application [NIW 95]. The major advantage of this type of sensor is to combine, within a small size device, a large number of detectors, the signal processing of which has taken a considerable leap forward with the development of chemometrics. In particular, the intrinsic selectivity of each component to given analytes is not a critical point during their simultaneous analysis. Amperometric sensors based on electrode barrettes are particularly well suited to continuous flow analysis [MAT 93, FIE 94]. Modified by a protective organic membrane, Ir/Hg microelectrode networks have been recommended for the *in situ* determination of heavy metals in natural waters [BEL 98]. Finally, the advantage of localized electrochemical methods for nanometer-level fabrication of metal and semiconductor surfaces must be stressed [STA 98], and is likely to produce new applications in the domain of amperometric sensors in the future.

Another type of microchip has been developed by Girault's group [REY 99, ROS 00], allowing low-cost, mass fabrication of electrochemical analysis microsystems based upon micromachined polymeric supports integrating microelectrodes. These systems can also be used as amperometric sensors, but can also be used in the conception of disposable miniaturized capillary electrophoresis systems with electrochemical detection.

5.1.3. *Fundamental principles and examples of application*

5.1.3.1. *Selective pre-concentrating*

The electro-analytic determinations of a trace species can be considerably improved by a forced accumulation of the electroactive analyte at (or in the vicinity of) the electrode surface, starting from very dilute solutions. This idea of pre-concentrating is not new and is at the origin of the method of analysis by anodic stripping (or cathodic, or potentiometric) [WAN 85] based upon electrochemical accumulation of the analyte as an insoluble form at the electrode surface followed by its electrochemical detection by re-dissolution.

In addition to this potentiostatic accumulation process, which is commonly used at conventional electrodes, resorting to chemically modified electrodes allows the pre-concentrating step to be performed at zero current, by making use of the chemical reactivity of the modifying agent. The analyte can thus be accumulated at the electrode-solution interface either by chemical or physical interaction with a functional group situated at the surface of the electrode, or via a preferential partitioning of the target species in the electrode material with respect to the external solution. The (volt)amperometric detection starts after this zero-current accumulation step. It can be performed in different modes (constant potential amperometry, linear or pulse (normal, differential or square wave voltamperometry)), or even coupled with a method of analysis by redissolution.

The analytical performances of the sensor, in terms of sensitivity, selectivity, and reproducibility, are directly linked to the optimization of various experimental parameters acting on these two successive steps (pre-concentrating and detection). Ion exchange, complexation, preferential recognition, or rejection principles have been exploited for accumulating analytes at the surface of an electrode. In addition to the considerable sensitivity increase, the choice of an appropriate modifying agent may give rise to a better selectivity based on electrostatic interactions (charge selectivity), of preferential molecular recognition (chemical reactivity, stereo-selectivity, specific bonding) or of steric discrimination (size selectivity). To be as beneficial as possible to the process, the pre-concentrating step must be fast and the species must be accumulated in a three-dimensional environment as close as possible to the electrode. This necessitates high accessibility to the reaction sites and fast mass transport. The detection step will be more efficient as the charge transfer reactions will be faster. This condition will be fulfilled if the accumulated electroactive species are either situated in the immediate vicinity of the electrode and undergo fast electron transfer, or if they are likely to be released from their accumulation site and to diffuse quickly towards the electrode surface, or finally, if they are efficiently reached by a charge transfer co-factor (mediator; see section 5.1.3.3) acting as an intermediary between them and the electrode surface. The two most commonly used electrode configurations for elaborating amperometric sensors based on the pre-concentrating-detection scheme [SMY 92] are multimolecular homogenous or heterogenous films (mainly based upon organic polymers) deposited on a solid support on the one hand, and carbon paste electrodes modified by incorporation of a pre-concentrating agent, and the corresponding disposable screen printed electrodes on the other hand.

5.1.3.1.1. Ion exchange

Accumulation of charged electroactive species in the vicinity of the electrode is made possible with the use of materials having ion-exchange properties for elaborating modified electrodes. The thermodynamic driving force inherent to ion-exchange reactions is the basis for numerous electrochemical analysis applications involving pre-concentrating, in spite of the lengthy period required to reach equilibrium during ion-exchange accumulation. This time depends on numerous factors linked to the structure of the ion exchanger and to the electrode configuration; it is determined by the rate of transport of the ionic analytes and the counter-ions from the solution towards the exchange sites. These diffusional limitations depend directly on the size of the analytes and on the porosity of the modifying agent. The faster the mass transport is, the higher the efficiency of the pre-concentrating step and, consequently, the sensitivity of the analysis is. The most commonly used ion-exchanging materials used in modified electrode amperometric sensors are polymeric films with a positive (clays) or a negative (zeolites) charge (see section 5.1.2.1). In addition to their ability to preferentially accumulate a charged analyte on the basis of favorable electrostatic interactions, such materials are also likely to prevent or limit the approach of interfering species (for instance, if they carry a charge opposite to that of the target analyte) to the electrode. The most famous example is probably that of the selective detection of dopamine, which is positively charged at physiological pH values, in the presence of ascorbic acid, by employing a Nafion®-modified electrode, which allows dopamine accumulation while rejecting ascorbic acid. This approach can be extended to the *in vivo* determination of other protonated organic molecules. Heavy metals can be detected by ion-exchange voltamperometry with sensitivity because the ion exchanging material is more hydrophilic.

Beside classical ion exchangers, which can be defined as structures presenting net permanent charges counter-balanced by mobile and exchangeable counter-ions, (semi)conducting polymers as modifying agent of an electrode allow electrochemical control of the ion-exchange reactions (called doping or de-doping, respectively, whether the compensating ion is incorporated into the polymer or rejected during the electrochemical transformation of the latter). Therefore, the accumulation of a target analyte in this type of polymer controlled by an electrochemical switch can be used to inject calibrated quantities, of drugs for instance, in a given environment [MER 90]. The most commonly used polymer in this field is polypyrrole, which is an anion exchanger in its non-modified oxidized form and a cation exchanger in its reduced form in the presence of a polyanion (polystyrene sulfonate) dopant.

The ion-exchange theory is very well described in a detailed book [HEL 62]. It also holds for the pre-concentrating step occurring at the surface of electrodes modified by ion-exchanging materials, with a major reserve linked to the fact that electrochemical analysis by pre-concentrating involves, in most cases, non-equilibrium situations, which are mainly controlled by mass transport. The quantities of accumulated analyte depend on the exchange capacity of the material; the modifying agents used for that contain exchange sites with a typical concentration of the order of 1 mmol/g of material. The selectivity of accumulation is directly linked to the selectivity coefficient associated with the exchange reaction. This parameter has no thermodynamic significance, it is calculated based on the equivalent fractions of the analyte and its counter-ion, both in the exchanger, and in the external solution; it does not take into account the variations of activity in the exchanger. Therefore, it depends on the local conditions, but it may, however, be a good indicator of when an operation is made in well-controlled, known conditions. As a general rule, the selectivity coefficients become more elevated as the charge (density) becomes more important. The competition effects are more marked when the medium is close to saturation; they are lighter when the quantity of matter of the target analyte and the number of the potential interferences are small in comparison with the number of available exchange sites. In the latter case, the limiting factor is the rate of access to the sites, rather than the competition between the various charged species present. The hydrophilic/hydrophobic balance induced at the surface of the modified electrode (for example, distinction between hydrophobic polymers and hydrophilic zeolites) may also induce variations of selectivity, which allow, for instance, differentiation between big hydrophobic ions (marked preference to organic polymers) and small hydrophilic species (affinity towards inorganic exchangers of the aluminosilicate type). Several application examples of ion-exchange voltamperometry are described in the literature [UGO 95, BON 97, and C. Degrand in MOU 98, pp. 74-76].

5.1.3.1.2. Complexation

In comparison to ion exchange, where selectivity is often limited, complexation generally presents more possibilities for accumulating a target analyte by preferential fixation at the surface of the electrode, provided that the complexing group that will be used for the fabrication of the modified electrode is selected judiciously.

The sensitivity and the selectivity of a modified electrode-based sensor depend fundamentally on the formation of a complex between the target analyte and a ligand, which is fixed at the surface of the sensor. The upper limit of sensitivity is mainly ruled by the formation constant of the complex: the higher this constant, the lower the possible detection limit. The liberty in the choice of the ligand theoretically allows exploiting high values of individual formation constants and large differences between the constants associated to the target analyte and the constants involving interferences, which leads to optimization of the detection selectivity.

A multitude of "immobilized ligand-target analyte" couples are described in the literature as operational (volt)amperometric detection systems:

– 2,2'bipyridine–Fe^{2+};

– o-phenanthroline–Cu^{2+};

– polyacrylamidoxime–Ag^{+};

– bathocuproine sulfonate–Cu^{+};

– mercaptans–Hg^{II}, Bi^{III};

– dimethylglyoxime–Ni^{2+}, Co^{2+};

– aminated ligands–Cu^{2+};

– calixarenes–metal cations;

– cyclodextrines–polar neutral molecules;

– tensioactives–charged organic molecules;

– microorganisms, clays, or zeolites–metal cations.

This list is far from being exhaustive and the interested reader can refer to well-documented review articles [WAN 91, SMY 92, KAL 95, GIL 95, CAS 98, MOU 98, LAB 00]. The specific advantage of macrocycles issued from supramolecular chemistry, such as cyclodextrines, calixarenes, or cyclophanes, can be cited for their ability to selectively form inclusion complexes with a target analyte. This type of molecular recognition is beneficially exploited for elaborating chemical sensors [DIA 01], and can easily be coupled to an electron-transfer reaction when the macrocycle is immobilized at an electrode surface [KAI 99]. This gave rise to many applications in the field of amperometric (and potentiometric) sensors [OCO 95, CHU 98] either by making use of their capacity to pre-concentrate an electroactive analyte prior to its electrochemical determination, or by using the modification of the electrochemical behavior of a macrocycle having an electroactive center during accumulation of non-electroactive target analytes.

A key-point of preconcentrating analysis using electrodes modified by complexing agents is the durable immobilization of this reactant at the surface (or in the sensitive microstructure) of the sensor. This immobilization is necessary to avoid reactant loss in solution (which would decrease the sensitivity of the measurement and shorten the lifetime of the sensor) and it makes successive measurements possible after chemical regeneration of the electrode surface between two preconcentrating-detection operations.

Detection can be done at the complexation site or necessitate decomplexation prior to amperometric analysis by itself. The first case is favored by the use of modifying agents of the conducting polymer type, or high-porosity, thin films containing a large density of complexation sites, because they permit a favored charge transfer, an important mobility of the electroactive species, or a proximity allowing charge transfer via electron jumps. Resorting to the second case is necessary when the accumulated species are electrochemically "silent" (for example, complex formation within a non-electronic conducting material for which access to an electronic transfer relay is difficult, if not impossible); for reasons of quantitative analysis and sensor regeneration, this analyte release step must be total and, if possible, quick. In both cases, the composition and structure of the sensitive part of the sensor must be optimized to comprehensively counteract the limitations of charge transfer and the restrictions to mass transport. The current development of hybrid organic-inorganic materials [ECK 01] should enable significant improvements in this respect by the possibility of preparing porous, rigid, and mechanically stable inorganic structures to which specific organic groups are attached by covalent bonds, and which play an increasingly important role in electrochemistry [WAL 01b].

5.1.3.1.3. Adsorption - precipitation

Although less exploited than complexation, the formation of a layer of insoluble matter at the surface of an electrode is also a means of preconcentrating a given analyte in view of its amperometric analysis [KAL 90]. Adsorption on mercury has been widely used for increasing the sensitivity of the analysis by anodic stripping.

Another preconcentrating mode requires the formation of a poorly soluble compound at the surface of the sensor. It involves an electrode modified by a substance, which is likely to precipitate by reaction with the target analyte, thus forming a concentrated electroactive deposit. The electrochemical transformation of the deposit results in an amperometric signal, which can be linked to the analyte concentration; for example, the analysis of iodide ions by means of an electrode modified by cuprous ions. In acidic medium, Cu_2O generates transient Cu^+ species that react with I^- to form CuI, which can be detected electrochemically at the surface of the electrode [LEF 99].

5.1.3.1.4. Selective permeability

Application of a layer having a selective permeability at the surface of an amperometric sensor may cause a significant improvement of its performances. The sensitivity is increased by a favorable distribution of the target molecule within the material in contact with the electrode, whereas the selectivity is much increased by the screening of the species likely to reach the surface of the sensor, preventing the possible interferences from accessing, on the basis of their size, their charge or their polarity. Numerous examples have been described [LEE 96].

Perm-selective membranes (as with any other layer allowing preconcentrating of a given analyte) deposited on the surface of an amperometric sensor also offer the possibility of protecting the surface from being blocked by large molecules (adsorption of macromolecules, formation of biofilms) or of preventing any passivation of electrochemical reactions. They thus increase the lifetime of the sensor. For example, a cellulose acetate membrane constitutes an efficient barrier by size exclusion. It presents a uniform structure of small pores, the size of which is controlled by basic hydrolysis. Such a film permits, among other things, the proteinic matrix effect to be minimized, which is usually an impediment during the detection of hydrogen peroxide (H_2O_2) in a biological medium, and resists the blocking of the electrodes used in routine analysis by anodic stripping of metallic species. The use of cellulose acetate is compatible with the screen-printing technique: it has led to applications in the medical field where, for instance, the covering of disposable carbon-based sensors permits the detection of paracetamol in the presence of physiologic biomolecules [GIL 95].

The ability of ion-exchanging materials to accumulate target species by favorable electrostatic interactions has been mentioned previously. Conversely, coulombic repulsion can be exploited to prevent interfering species from reaching the surface of the electrode. It is based upon the natural tendency of charged films to reject ions having the same charge. The optimal situation will be when the structure of the modifying agent allows rapid incorporation of the ions to be analyzed, while efficiently excluding the impregnation of ions with the opposite charge. It depends on the nature and structure of the film, on the charges of the species (analyte and interferents), on membrane porosity, on the electrolyte concentration and on the pH.

Electropolymerized films, such as polyphenol, polypyrrole, polyaniline or dimethyl-phenylene polyoxide, at the surface of amperometric sensors, have proven to be good molecular filters. The selectivity of their semi-permeable properties depends considerably on their doped state. It is either based upon the Stokes radius of the analytes, or linked to a specific affinity towards interfering substances.

A particular domain where the advantage of semi-permeable membranes is obvious is that of amperometric sensors for dissolved gases, with an operation principle based upon that of the Clark electrode (see Chapter 3). This type of detection is currently applied to the dosage of molecular oxygen and of nitrogen monoxide and involves electrochemical, selective transformation of the analyte, the selectivity being induced by an appropriate choice of membrane separating the measurement medium from the electrode [J. Devynck and F. Bedioui in MOU 98, pp. 72-74]. This type of amperometric sensor is commercially available (mainly for the detection of oxygen) [GOP 91].

A recent addition to the group of perm-selective membranes comprised of polymers with molecular print [SEL 00], which have gradually filtered into the field of chemical analysis [DIC 99], and are used in the production of biomimetic electrochemical sensors [SEL 00]. These polymers are prepared using the target analyte as a directing structural agent, so that after extraction, the material has kept a shape (print) allowing ulterior selective recognition of this analyte, even if it is mixed with other compounds with similar compositions and structures. This approach, although extremely selective, remains poorly exploited in the field of amperometric sensors, because the analyte recognition (and interference rejection) principle is rather long, due to important diffusional limitations. The amperometric analysis being mainly controlled by this mode of mass transport, the sensitivity of the measurement is consequently limited. To overcome this, a solution might be to resort to molecular print inorganic polymers (or organic-inorganic hybrids), which are less affected by such diffusional limitations, as it has been demonstrated for the analysis of dopamine by means of an organic-modified silica film deposited on a carbon electrode [MAK 98].

5.1.3.2. Selectivity and amperometric selectivity coefficients

Evaluating quantitatively the selectivity of an amperometric response sensor requires the notion of amperometric coefficient of selectivity [WAN 94b].

The simplest model consists of considering that the amperometric response of a given analyte, i, under an appropriate applied working voltage, will be significantly more intense if an interfering species that liable to be electroactive at the applied working potential is present.

Assuming an additivity law (similar to that which holds for the Beer-Lambert law of light absorption measurements), one obtains an expression of the total current, i_t, taking into account the analyte concentration, C_i, and the concentrations of the interferents, C_j, according to the relation:

$$i_t = K(C_i + \sum k_{ij}^{amp} C_j) \quad [5.1]$$

where k_{ij}^{amp} is the amperometric selectivity coefficient, i.e. an indication of the preference of the sensor towards the target analyte against the interferents. For a given interfering species, the higher this coefficient is, the more impeding this species will be to the quantitative determination of the analyte. A sensor with ideal selectivity would have selectivity coefficients equal to zero (see Chapter 3).

Beside this purely amperometric selectivity, using chemically modified electrodes may induce other factors affecting the selectivity (and also the sensitivity) of the sensor, which also cause the total measured current to vary. In particular, pre-concentrating analysis (section 5.1.3.1) is very sensitive to that. This is because the selectivity of the measurement is influenced both by the simultaneous electrochemical activities of the analyte and of potential interferents, and by the fact that compounds in the accumulation medium may affect the pre-concentrating process by competing with the analyte for the fixation sites. The first effect will cause the intensity of the signal to increase, while the second will induce a decrease of the signal. The interferences, j, liable to be electroactive at the applied potential, and the interferents, k, liable to limit accumulation of the analyte, i, at the electrode, can then be discriminated. Equation [5.1] becomes:

$$i_t = K(C_i + \sum k_{ij}^{amp} C_j - \sum k_{ik}^{amp} C_k) \quad [5.2]$$

where k_{ik}^{amp} is a pre-concentrating selectivity factor, i.e. an indicator of the aptitude of the interfering species to impede the preferential accumulation of the target analyte at the surface of the sensor. For a given interfering species, the higher this factor is, the stronger the limitation of the pre-concentrating process, and consequently, the lower the analyte detection sensitivity is. For an ideally selective detection, the selectivity factors would be equal to zero. Attempts to correlate these factors to the complexation constants or to the ion-exchange selectivity constants, or to the liquid-liquid extraction distribution coefficients, have been done [LAB 96], but their results remain limited by the fact that these values may fluctuate as a function of the experimental conditions.

5.1.3.3. *Electrocatalytic detection*

One of the major motivations of the electrochemists towards chemical modifications of conventional electrodes arises from the fact that electron transfer reactions depend strongly on the type of electrode used. The electrochemical transformation of many species necessitates high-applied overpotentials, or may not occur at all, in spite of their thermodynamic possibility. This phenomenon can arise from clogging (or other modification of the surface state) of the electrode, but it is due, in most cases, to a kinetic limitation of electron transfer, the origin of which is rarely understood. This is why numerous research efforts have been performed to accelerate electron transfer reactions, by lowering the corresponding activation barrier. This is called electrocatalysis.

The most commonly used method is called "mediated electrocatalysis method". It is based upon a charge transfer mediator that has a reversible (rapid) electrochemical behavior, and serves as a relay (intermediary) between the electrode and the electroactive substrate, which can barely (or not at all) be detected by direct amperometry. This mediator can be introduced in the solution or, which is better, immobilized at the electrode surface, or even as an integral part of the electrode material. In view of applications in sensors, the mediator must be durably integrated to the measuring device. The principle of the method is described by the following equations, which hold in the case of oxidation of a substrate S_{red} catalyzed by a mediator M:

$$S_{Red} - ne^- \rightarrow P_{Ox} \qquad E_S^{o\prime} (>> E_S^o) \qquad [5.3]$$

$$M_{Red} - ne^- \rightarrow M_{Ox} \qquad E_M^{o\prime} (\sim E_M^o) \qquad [5.4]$$

$$S_{Red} + M_{Ox} \rightarrow P_{Ox} + M_{Red} \qquad \text{(possible if } E_S^o < E_M^o) \qquad [5.5]$$

Due to a high activation barrier, direct oxidation of S_{Red} to P_{Ox} necessitates a high overpotential ($E_S^{o\prime} - E_S^o$). In the presence of a suitable mediator, the oxidized component of which is able to react chemically with the substrate ($E_S^o < E_M^o$), the product P_{Ox} can be obtained at a potential $E_M^{o\prime}$ (close to E_M^o because the mediator has been selected among rapid couples), lower than the value that should have been applied, ($E_S^{o\prime}$), had the mediator been absent. The improvement finally obtained is that an amperometric response is now possible at a lower overpotential.

This type of electrocatalysis is applied both for chemical analysis and for the detection of species of biological interest [WRI 92]. It gains particular importance in the domain of the so-called "second-generation" biosensors (see section 5.2 of this chapter) as direct contact between the active redox center of the enzyme and the electrode center is particularly difficult to achieve experimentally. The most commonly used catalysts are mainly organic or organometallic, and hybrid organic-inorganic materials. Among organic mediators, several families exist: (1) phenoxazins, phenothiazins and phenazines, (2) quinones and hydroquinones, (3) tetrathiafulvalen and tetracyanoquino-dimethane, and (4) derivatives of the phenylenediamine type. The organic charge transfer co-factors are ferrocenes and other matallocenes, phtalocyanins, metalloporphyrins, ruthenium oxide-based complexes, and metallic complexes of bipyridine. Transition metal polynuclear hexacyanoferrates, metallic oxides, and polyoxometallates constitute the family of the mostly used inorganic mediators. Finally, it is necessary to mention the advantage of materials prepared by the sol-gel process, which can act as supports for numerous catalysts by keeping them immobilized within a rigid inorganic network, either encapsulated in the structure or covalently bonded in the form of hybrid organic-inorganic nanocomposites [LEV 97, WAL 01b].

The most promising current configurations of electrodes modified by redox mediators, in the frame of amperometric sensors elaboration, are polymeric films deposited on solid electrodes. The charge transfer reactions involving the analyte may occur either at the polymer-solution interface or in a three-dimensional domain in the bulk of the polymer. Depending on the nature of the latter, the overall charge transfer process may be controlled by three factors: the movement of polymeric chains, the movement of electrons, and that of the species through the polymer (analytes and charge-compensating ions). Optimizing the sensor response, therefore, requires the determination of the rate-limiting step, which may be different according to the case. The most efficient systems enable rapid diffusion within the film and a fast charge transfer (for example, an electronic conducting polymer), and they contain a high density of catalytic centers (at least greater than that of the analyte) so that an efficient turnover is ensured.

In addition to electrocatalysis by a chemical mediator, modification (at least partially) of the surface of a conventional electrode by another conducting material is another way of lowering the electron transfer overpotential. In particular, "doping" carbon electrodes with metallic microparticles (platinum, palladium, ruthenium, rhodium) lowers considerably the overpotential for the amperometric detection of H_2O_2 (which is usually produced during the action of oxidase-type enzymes) without catalyzing the electrochemical activity of the majority of the electroactive species contained in biological fluids [WAN 92, WAN 94c, WAN 95a]. This approach was used in the creation of extremely selective sensors without any addition of a mediator, the immobilization of which at the surface of an electrode is still challenging. The reason why these microparticules have catalytic activity is not clearly understood.

5.1.3.4. *Indirect amperometric analyzes*

When species are not electroactive in aqueous solutions for thermodynamic reasons, their direct amperometric detection is not possible. This is the case for most of the alkaline and alkaline-earth cations and ammonium, which are often electrochemically detected by potentiometry (see Chapter 4). Yet their amperometric detection, which is directly proportional to their concentration, would be desirable in the case of the control of low concentration variations, which is not an easy task using potentiometric response sensors (linked to the logarithm of the concentration). This is the reason why three indirect amperometric methods have been developed for detecting these non-electroactive species:

– analysis of the electrochemical response of a modified electrode involving, in its redox transformation, incorporation, or expulsion of an ion;

– electrochemistry at liquid-liquid interfaces;

– analysis of the electrochemical response of electrodes with incorporated zeolites doped by a charge transfer co-factor.

The first method is based upon the fact that the electrochemical transformation of an electroactive film (conducting polymer) [BAR 97, AKH 97] or copper hexacyanoferrate [THO 89] requires incorporating a charge-compensating ion for neutralizing the charge resulting from the electron transfer. So, if the potential of an electrode modified by such a film is kept constant at a given anodic or cathodic value, a current will be produced only when an ion (anion or cation) will come in contact with the electrode surface. This current will be proportional to the solution concentration in anions (or cations) injected at the surface of the detector. Such systems work in continuous flow and some of them have been applied to the detection in ionic chromatography and in capillary electrophoresis.

The second approach, which was achieved by the Girault group [LEE 97], exploits electrochemically-assisted ion transfer at a liquid-liquid interface. This transfer is caused by a potential difference applied between both sides (an aqueous phase and an organic phase) of the interface, and generates a current proportional to the ion concentration in the aqueous phase. Organic gel-aqueous phase micro-interfaces have allowed this concept to adapt to the creation of amperometric sensors usable for routine analysis of metal ions and alkaline cations, and to their detection in ionic chromatography [REY 99].

The third approach is based on the specific electrochemical behavior of electrodes modified by zeolites. It arises from a combination between the ion-exchange capacity and the property of size-selectivity of the zeolites, exploited with benefit in electrochemistry. When an electrode with incorporated zeolites, doped with an electroactive species (for instance Cu^{2+}), is dipped into an aqueous solution containing an electrolyte with a cation too voluminous to penetrate the microporous zeolite network, no electrochemical reduction of Cu^{2+} ions can take place. Conversely, when a sample containing a small cation (with a size smaller to that of the zeolites pores), is injected at the surface of the electrode, an exchange with the Cu^{2+} ions takes place and permits the reduction of these ions at the electrode-solution interface. This induces a faradic current directly proportional to the concentration in injected sample. The response has an amperometric nature, but it is indirect with respect to the injected sample. This analysis mode has been used for detecting alkaline, alkaline-earth, and ammonium cations in continuous flow [WAL 99b]. It can also be applied in the absence of electrolytes and then becomes a detector, which has been applied at the outlet of an ionic chromatograph operating in a chemical-suppression mode [WAL 99c]. This latter aspect can be generalized both to non-electroactive anions and cations, which can be detected by indirect amperometry by resorting to electrodes that have been modified by ion-exchange resins doped by cationic or anionic electroactive mediators, respectively [MAR 01].

5.2. Amperometric biosensors

5.2.1. *Introduction*

The concept of biosensors, which first appeared in 1962 [CLA 62] with the detection of glucose, originates from the association of the catalytic properties of an enzyme to the transduction properties of an electrode.

Biosensors are analytic tools based upon the immobilization of a biologic molecule (coenzyme, vitamin, enzyme, antibody, antigen, oligonucleotide, etc.) at the surface of a transducer (electrode, optical fiber, field effect transistor, microbalance crystal, etc.) [TUR 87].

The molecular recognition properties of the biomolecule are a source of selectivity, or even stereoselectivity, of the biomolecule-target analyte interaction. This interaction is converted to an electric signal by various physical methods. The value of the signal is directly related to the concentration of the target analyte in the solution.

According to the definition of IUPAC, the term biosensor was originally reserved to transducers incorporating a biological molecule having a catalytic activity. The ability of catalyzing a chemical reaction provides the biosensor with the capacity of non-stop determination of the concentration of an analyte. This generic term currently also includes the analytical devices based upon biomolecules involving only complexation phenomena, such as hybridation of simple DNA wisps or antibodies-antigens interactions. Contrary to enzymatic biosensors, the latter cannot perform a non-stop quantitative determination of an analyte, and thus belong to the category of disposable biosensors.

Biosensors have undergone considerable development during the last 20 years and now represent a major component in the field of analytical chemistry. Their main advantage lies in the rapid, specific detection of a target substance in a complex medium, for instance blood plasma. They differ from the classical analysis systems, such as liquid or gas chromatography, by their low original and in-use costs, their compactness and the possibility to use them for direct measurements in the field.

The application fields include medical diagnosis (90% of the applications), sport, national defense, environment, food-processing, and pharmaceutical research, which is rapidly developing. Indeed, the increasing severity of the norms ruling the environmental protection, the food industry, pharmacology, industrial and domestic safety, added to the growing needs of the biomedical sector or the generalization of the automation of industrial processes, explain the considerable development of research in the biosensors field. The significant improvements recorded in the concepts of immobilization of biological macromolecules, in parallel to the technological improvements resulting from transfer from microelectronics technologies, have certainly encouraged this development.

The expansion of the microelectronics technologies thus offered new opportunities for miniaturizing biosensors. The main objectives are implantable biosensors or associated biosensors to gain simultaneous information on several analytes. Microbiosensors can be created by biomolecule immobilization at the surface of a wide variety of transducers, such as the multiplot microelectrodes, interdigitated microelectrodes, or field effect transistors. This microbiosensor production is currently the subject of fierce international competition, with the achievement of biochips for genome sequencing as a perfect illustration.

Among the various possible transduction modes of catalytic or affine biological processes, electrochemical transduction is the most frequently used, in particular, for enzymatic biosensors. Electrochemical biosensors can be separated into at least three families, according to the transducer used [TUR 87, THE 01]. We can then distinguish the potentiometric biosensors, the conductimetric biosensors, and the amperometric biosensors, which are the most widespread:

– potentiometric biosensors are based upon measuring the potential difference between two electrodes, for instance between a sensitive and a reference electrode. The sensitive electrodes are generally specific (ISE) or redox electrodes. The specific electrodes (see Chapter 4) are generally the most used, as for instance pH electrodes, ionic electrodes (Na^+, K^+, NH_4^+, etc.), or else gas electrodes (CO_2, H_2, etc.). The biocatalytic membrane is placed on the specific electrode. The relation between the potential difference and the activity of the detected species is logarithmic for all these sensors. Their detection range is, therefore, wider than that of amperometric transducers, but this is to the detriment of their sensitivity;

– conductimetric or impedancemetric biosensors use generally interdigitated electrodes (see Chapter 10). The conductance of the biocatalytic or bio-affine membrane placed on the electrodes varies under the occurrence of the production of ionic species by an enzyme (e.g. urease) or of a complexation phenomenon (e.g. antigen-antibody). This technique is rarely used because of interference in real samples. Their use necessitates a good knowledge of the ionic composition of the solutions, as these sensors have no intrinsic selectivity;

– amperometric biosensors are based upon the measurement of the current resulting from oxidation or reduction of an electroactive species, the concentration of which is linked to the activity of the biological species. During the measurement, the potential is kept constant via a reference electrode and the current, generated by the oxidation or the reduction of the electroactive species at the surface of the electrode, is measured. This family will be extensively described in this chapter.

Biosensors making use of field effect transistors can be added to these three families. These sensors, called BioFETs, are divided up into two groups according to the type of receptor: if it is an enzyme, they are called ENFETs, if the detection is immunologic, they are called IMFETs. These BioFETs are described in Chapter 6.

Whatever the transduction process used, the biomolecule must be immobilized. Even though this chapter is specifically devoted to amperometric biosensors, we will consider biomolecule immobilization techniques prior to describing the functioning of amperometric biosensors.

5.2.2. *Immobilization of biomolecules*

The main technological barrier in the fabrication of biosensors is the immobilization of biomolecules at the surface of the transducer. Ideally, this immobilization should be totally irreversible and stable in time, without deactivation of the biomolecule, while managing excellent accessibility to this biomolecule and ensuring a certain conformational mobility. Indeed, the major problems limiting the rapid expansion, or even the commercialization of biosensors, lie in the important drop of the biocatalytic or affine activities during the production process of the sensor, and the fragility of the biological macromolecule, which control the sensitivity and the lifetime of the biosensor.

In the following section, the main immobilization techniques of the receptors are presented. Five categories are distinguished:

– adsorption of the receptor on the support is the most simple of the immobilization methods. This adsorption phenomenon is conditioned by the nature of the support, as the affinity between both parties must be as high as possible. The bonds involved are weak, but numerous. Generally, adsorption is a rather soft immobilization method, but it may induce denaturing of the sensitive element. It can be performed directly on the transducer or follow a surface modification meant to increase the affinity between the transducer and the receptor. The Langmuir-Blodgett method, which has long been in existence, is one of the adsorption surface modification methods. Adsorption of tensio-active compounds at the surface of a material produces multilayers similar to the phospholipidic membrane systems. The so-modified surface can be functionalized by including receptors between the layers of tensioactive compounds. The disadvantage of adsorption is the low mechanical strength of the assembly. Indeed, desorption phenomena are regularly observed when environment conditions (pH, ionic strength) change;

– reticulation consists of associating the proteins with polyfunctional agents, such as glutaraldehyde. The reticulation agent possesses two aldehyde groups at its extremities reacting with the amino groups of the envelope proteins. The proteins are then linked together;

– inclusion or trapping of the receptor in a host matrix can be done by a simple mixing of the various components and depositing the mixture onto a suitable support. This method is the most widespread. A large diversity of materials is thus used: inorganic material (clay), natural, or synthetic organic polymer. The other possibility is to trap the sensitive element along with the matrix formation. This alternative solution is possible only with electrogenerated polymers (polypyrrole, polyaniline, polythiophene, etc.) at the surface of the electronic conducting support. Inclusion in a matrix can immobilize a large quantity of sensitive species. Conversely, steric stresses and interactions with the matrix may denature the species. Also diffusional limitations may occur when the receptor is not sufficiently accessible;

– covalent grafting is a chemical immobilization technique consisting of creating a covalent bond between the sensitive element and the support. It is an irreversible phenomenon. This technique may, in certain instances, permit the orientation of the grafted molecules and thus optimization of the recognition probability. Grafting requires the presence of reactive groups and most transducer surfaces do not allow any direct grafting. So, it is often necessary to proceed to prior functionalization of this surface. Many methods are used, such as silanization or deposition of polymers having functional groups, which make grafting possible (CN, COOH, SH, etc.). It is also common to use a reactional intermediate: the chemical coupling agent (carbodiimide, glutaraldehyde);

– molecular assembling by affinity relies upon properties of specific fixation of certain proteins and permits the construction of multilayer assemblies. The most common molecular assemblies are based upon avidin bridges for immobilizing biotinylated proteins. Various techniques are used to form the first avidin layer (electrogenerated biotinylated polymer, Langmuir-Blodgett film, etc.).

Some of these methods (adsorption, grafting, and trapping) can be combined with a reticulation or a co-reticulation procedure. The mechanical stability of the whole protein/membrane assembly is then significantly increased.

Figure 5.4 represents the distribution of the various methods used for enzyme immobilization. It should be noted that immobilization within a matrix consisting of a pre-formed or electrogenerated film is the most commonly used method.

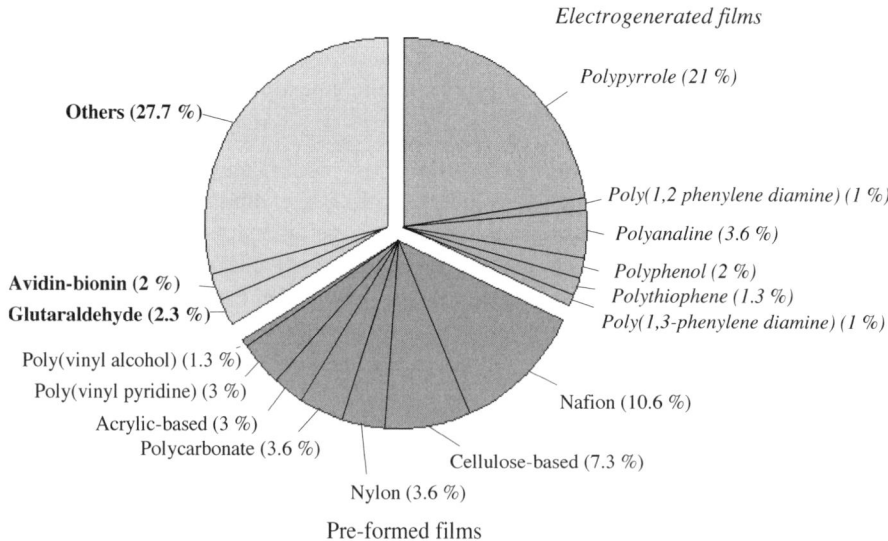

Figure 5.4. *Schematic representation of the distribution (end of the 1990s) of the methods used for immobilizing enzymes at a transducer surface*

5.2.3. *Amperometric biosensors, principle and description*

5.2.3.1. *Principle of an enzymatic biosensor*

Enzymes are biomolecules generally constituted of a coil of proteinic chains containing one or several prosthetic sites. The proteinic chains are macromolecules with a precise sequence of amino acids, which when folded with the other protein subunits making up the enzyme, are described by a quaternary structure.

Many enzymes suitable for the detection of numerous species exist. They are catalysts and some of them are extremely selective, or even stereoselective. These selectivity properties are to be related to the interaction between the active sites and the proteic envelope, which hugs them. Enzymes accelerate by a factor of 10^3 to 10^{16} the reaction, which would occur spontaneously. The substrate binds to the active site (prosthetic site) of the enzyme and forms the enzyme-substrate complex. The active site and the substrate have mutually suitable conformations. By forming the enzyme-substrate complex, which is more stable than the separate entities, the enzyme decreases the activation energy, which corresponds to the difference between the energy of the substrate in its initial state, and the maximum energy reached during the formation of the product. By decreasing the activation energy of the catalyzed reaction, the enzyme accelerates this reaction.

For a given constant quantity of enzyme, the initial rate of a classical enzymatic reaction varies as a function of the substrate concentration, as shown in Figure 5.5.

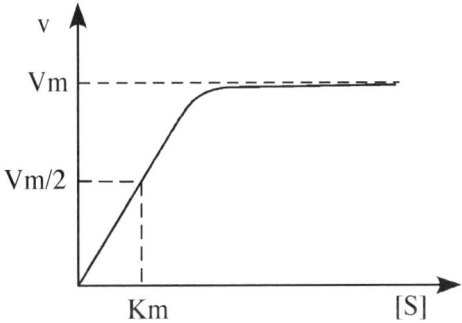

Figure 5.5. *Aspect of the variations of the initial rate of an enzymatic reaction as a function of the substrate concentration*

Michaelis and Menten modeled homogenous enzymatic kinetics in 1913. It takes place in two phases: the enzyme-substrate complex, called the Michaelis complex, is at first quickly formed; it is then slowly transformed into the product, releasing the enzyme, which can enter a new catalytic cycle (see equation [5.6]).

$$E + S \underset{k_{-1}}{\overset{k_1}{\rightleftarrows}} ES \xrightarrow{k_2} E + P \qquad [5.6]$$

In this reaction, E is the enzyme, S the substrate, ES the enzyme-substrate complex, and P the product. The reaction with rate constant k_2 is the limiting step of the process. The following expression can, therefore, be written for the initial rate (v) of the reaction (assuming that the inverse catalytic process is negligible): $v = k_2$ [ES].

Posing that a maximum value of v, noted Vm, is reached when all the catalytic sites are occupied, it can be written that Vm depends only on the initial enzyme concentration e:

$$Vm = k_2\, e \quad \text{with} \quad e = [E] + [ES] \qquad [5.7]$$

Considering that a stationary state is reached, the reaction rate can be written

$$v = \frac{Vm\,[S]}{Km + [S]} \quad \text{with} \quad Km = \frac{k_{-1} + k_2}{k_1} \qquad [5.8]$$

Km, Michaelis-Menten constant, is the substrate concentration for which the reaction rate is equal to $\dfrac{Vm}{2}$ and is expressed in the molarity scale.

Reorganizing the Michaelis-Menten equation gave rise to various types of graphs for the determination of Km and Vm, which characterize an enzyme-substrate system. The two commonly used graphs are those of Lineweaver-Burk and Eadie-Hofstee.

The straight line of Lineweaver-Burk is obtained by reversing [5.8]:

$$\frac{1}{v} = \frac{1}{Vm} + \frac{Km}{Vm\,[S]} \qquad [5.9]$$

Km and Vm can be determined from the plots of the straight line $1/v = f(1/[S])$. The intersection of the straight line with the abscissa yields $-1/Km$ and its intersection with the ordinate gives $1/Vm$ (Figure 5.6).

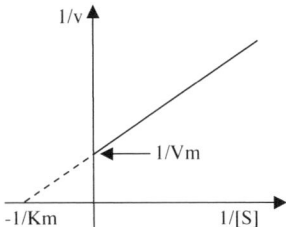

Figure 5.6. *Lineweaver-Burk representation*

The Michaelis-Menten equation, once linearized, leads to the straight line of Eadie-Hofstee:

$$v = V_m - K_m \frac{v}{[S]} \qquad [5.10]$$

Plotting the straight-line $v = f(v/[S])$, yields the values of V_m and K_m (Figure 5.7).

This Michaelis-Menten model holds for free enzymes in solution, identical and uniformly distributed in the reaction medium. In the case of immobilized enzymes, the kinetics of the catalyzed reaction will be different. Several new aspects, with respect to a homogenous kinetics, must be taken into account:

– possible modification of the enzyme conformation,

– steric hindrance limiting accessibility of the substrate to the catalytic site,

– physical interactions between reactants and solid phase,

– occurrence of a distribution coefficient for the substrate,

– diffusional limitations for substrates and products, etc.

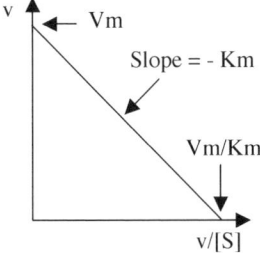

Figure 5.7. *Eadie-Hofstee representation*

Consequently, the characterization of heterogenous catalytic processes is more difficult, and the term "apparent kinetics" is used.

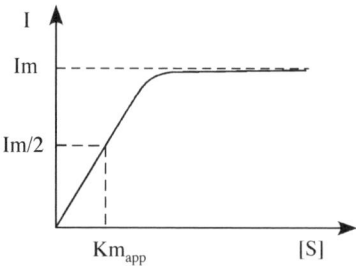

Figure 5.8. *Qualitative representation of the variations of the current as a function of the substrate concentration for an enzymatic amperometric sensor*

In an electro-enzymatic biosensor, the receptor (biomaterial) and the transducer (electrode) are associated. The overall operating process involves both enzymatic kinetics and electrochemical kinetics of the electroactive species (which may be a product or a reactant in the enzymatic reaction). The electroactive species and the detection potential are selected in such a way that the process should not be limited by the electrochemical kinetics (electron transfer and diffusion in the receptor of the biosensor). The overall kinetics is then ruled by the enzymatic reaction. The plots of the response of the biosensor (current intensity) as a function of the enzyme concentration in the substrate look similar to the plots of the enzymatic reaction rate.

It is then possible to use the Michaelis-Menten model for determining the experimental parameters, which characterize the biosensor under study. Thus, the apparent constant of Michaelis-Menten, Km_{app}, will reflect the influence of the biomolecule immobilization method used (possible denaturing of the enzyme, permeability of the biomaterial), whereas the maximum value of the current will depend on the quantity and activity of the enzyme fixed.

Along the first part of the curve, the current delivered by the biosensor is proportional to the substrate concentration in the analyte. This is the concentration range in which the biosensor will be used.

5.2.3.2. Various amperometric biosensors

Amperometry is one of the principal transduction modes of biosensors. The functioning principle of an amperometric biosensor is based upon the determination of the intensity of a current produced by the electrochemical oxidation or reduction of an electroactive species at a given potential. This species may be either the target species itself, such as in the case of the oxidation of guanine bases of an oligonucleotide strand [WAN 77] or a redox probe, the amperometric response of which is perturbed by the molecular recognition phenomenon (hybridation, affine interactions), as when electroactive intercalates are used for the hybridation of DNA strands or when hexacyanoferrate ions are used for the association avidin-biotin [KUR 00]. Yet these approaches remain in the minority in comparison with the amperometric biosensors based upon immobilized enzymes. These biomolecules catalyze the consumption or the production of an electroactive species. The electrochemical detection of this species produces a signal that is directly proportional to the concentration of the target molecules in solution (Figure 5.9).

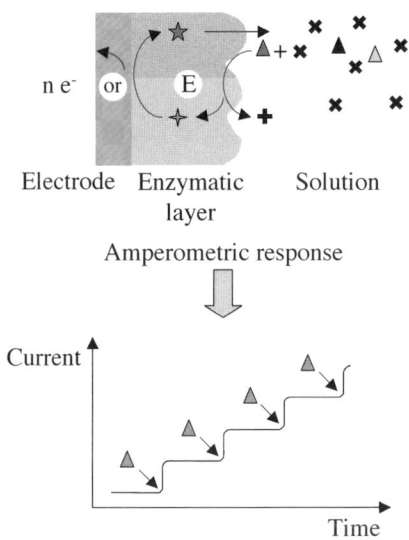

Figure 5.9. *Schematic description of the operating mode of an amperometric biosensor having an enzyme as molecular recognition element*

Among the various families of enzymes, oxidoreductases are practically the only ones able to catalyze the production or consumption of an electroactive species. The most widespread example is the enzymatic oxidation of a substrate in the presence of molecular oxygen, which gets reduced to H_2O_2. A sub-class of oxidoreductases, the dehydrogenases, involve a pyridinic coenzyme nicotinamide adenine dinucleotide (NAD^+) or nicotinamide adenine dinucleotide phosphate ($NADP^+$)

which, once enzymatically reduced to NADH or NADPH, can also be detected via their electro-oxidation.

However, it must be noted that the amperometric biosensors are not limited to the substrates of these enzymes. Indeed, association at the electrode surface of several enzymes having complementary activities extends the range of detectable substrates to other enzyme families, such as lyases, transferases, or hydrolases. For example, cascade reactions catalyzed by choline esterase and choline oxidase or by maltose phosphorylase, mutarotase, and glucose oxidase lead to the production of H_2O_2, detectable by electro-oxidation via the detection of a substrate via an enzymatic reaction involving no electroactive compound:

$$H_2O_2 + acetylcholine \xrightarrow{choline\ oxydase} choline + CH_3COOH$$

$$Choline + H_2O + O_2 \xrightarrow{choline\ oxydase} betain + H_2O_2$$

$$Maltose + phosphate \xrightarrow{maltose\ phosphorylase} \alpha\text{-D-glucose} + \beta\text{-D-glucose-1-phosphate}$$

$$\alpha\text{-D-glucose} \xrightarrow{mutarotase} \beta\text{-D-glucose}$$

$$\beta\text{-D-glucose} + O_2 + H_2O \xrightarrow{glucose\ oxydase} gluconolactone + H_2O_2$$

The amperometric sensors based upon enzymes can be arbitrarily classified in three distinct generations:

– the first generation corresponds to biosensors in which the electroactive species generated or consumed by the enzymatic reaction is directly oxidized or reduced at the surface of the electrode;

– the second generation concerns the use of redox mediators or of electrochemically activable catalysts, which interact with the electroactive species involved in the enzymatic reaction. The detection is then achieved via oxidation or reduction of the redox mediator or of the catalyst;

– the third generation implies an electron transfer between the electrode and the prosthetic center of the enzyme either directly or via redox mediators or electronic conductors.

5.2.3.2.1. First-generation enzymatic biosensors

This operating principle, because of its simplicity, enables the use of numerous methods and common, low price materials for immobilizing the enzyme at the surface of the electrode.

Except for the immobilization of an enzyme monolayer, this working mode generally associates a three-dimensional enzymatic detection with a two-dimensional amperometric detection. This entails a loss of the recognition signal intensity, due to diffusion of part of the enzymatic reaction product towards the solution. Moreover, the current intensity, which, ideally, should reflect the rate of the enzymatic reaction, may be strongly affected by diffusional constraints exerted on the electroactive species during its diffusion towards the electrode. A possible accumulation of the product of the enzymatic reaction in the material may also induce an inhibition of this reaction or even a deactivation of the immobilized enzyme.

α - Principle of the working mode based upon O_2

Owing to economical issues linked to diabetes, the most common biosensor configuration is related to the amperometric determination of glucose via immobilization of a glucose oxidase on an electrode:

$$O_2 + glucose \xrightarrow{glucose\ oxydase} gluconolactone + H_2O_2$$

Historically, the first glucose biosensor elaboration consisted of fixing this enzyme on a Clark electrode, which is an oxygen sensor able to dose this gas by an amperometric method at controlled potential (- 0.7 V). This approach is simple and the measurement lasts less than 30 seconds; moreover, no particular pretreatment of the analyzed sample is necessary prior to the dosage. The electric current response of the biosensor results from oxygen reduction. If the latter is not the limiting factor of the enzymatic reaction, any glucose addition will consume oxygen and consequently cause the current to decrease.

However, the sensitivity of the sensor, in terms of detection limit, is not favored as the detection principle is based upon the disappearance of a signal. Moreover, this sensor configuration is strongly dependant on oxygen concentration fluctuations. This is the reason why the working principle of this sensor was subsequently changed to the detection of enzymatically generated H_2O_2.

β - Principle of the working mode based upon H_2O_2

As H_2O_2 reduction at the surface of most electrodes is concomitant with the reduction of O_2, the amperometric detection of H_2O_2 is generally performed by oxidation:

$$H_2O_2 + 2\ H_2O \rightarrow O_2 + 2\ H_3O^+ + 2\ e^-$$

This method presents the advantage that the dioxygen consumed by the enzymatic reaction is regenerated by the amperometric detection of H_2O_2.

However, the classical electro-oxidation of H_2O_2 at the surface of a platinum electrode requires rather high-applied potential (+ 600 mV versus SCE (saturated calomel electrode)). But, in the case of medical applications, such as the determination of glucose or cholesterol in blood, neurotransmitters (dopamine, glutamate, gamma-aminobutyric acide (GABA)) in cerebral tissues, several endogen or exogen substances, present in biological fluids and directly oxidable at this potential, constitute an important interference to the amperometric quantification of H_2O_2; these substances are ascorbate (vitamin C), urate, paracetamol, adrenaline, noradrenaline, dopamine, L-dihydroxyphenylalanine (L-DOPA), tyrosine, or cysteine.

The biosensor concept always utilizes the specificity or even stereospecificity of the molecular recognition biological reaction; however, it bust be kept in mind that the selectivity of the biosensor relies on the selectivity of the recognition and transduction chain, with the lowest selectivity stage controlling the resulting overall selectivity. Thus, amperometric transduction based upon oxidation of the enzymatically generated electroactive products (H_2O_2, NADH, NADPH, phenols, ortho-diphenols, etc.) often entails a loss of the sensor selectivity.

γ - Elimination of the interfering species

Three main strategies have been developed for H_2O_2 detection in view of overcoming these interferences.

The first strategy consists of inserting an additional membrane between the biomaterial at the center of molecular recognition and the electrode surface, which, via steric hindrance or electrostatic repulsion, would minimize the access of interfering species to the electrode surface without blocking the diffusion of H_2O_2. Using this approach, various types of selective membranes have been used, such as cellulose acetate, PVC, polyesters functionalized by sulfonate groups (Eastman AQ), Nafion®, and electrogenerated polymers (overoxidized polypyrrole, polytyramine, polyphenols, polyphenylenediamine) [GIL 94 b, ZHA 94, MAN 95, MOU 94, LIU 96, BEN 96, SAS 90, YAO 98, MOU 93, EDD 95, SCH 91, PAL 95a, CEN 94, PIH 96, PAL 95b].

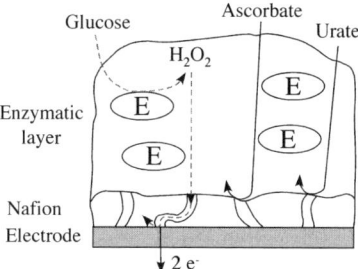

Figure 5.10. *Schematic representation of an amperometric biosensor based upon associated enzyme and Nafion layers*

For instance, a prior deposit of Nafion® (a perfluorated hydrophobic polymer) at the surface of a platinum electrode permits diffusion of species only through its negatively charged channels, thus blocking the permeation of classical anionic interferents, such as ascorbates or urates present in blood (Figure 5.10).

A second strategy is based upon the co-immobilization of several enzymes in view of enzymatically oxidizing the interferents before they can reach the electrode surface. Thus, 4-acetaminophene, well known as paracetamol, is catalytically oxidized to quinone in the presence of O_2 and polyphenol oxidase during its diffusion through a glucose oxidase-containing biomaterial [WAN93]. Similarly, ascorbate undergoes enzymatic oxidation by ascorbate oxidase and catalyzes the reduction of O_2 to H_2O instead of H_2O_2 (Figure 5.11).

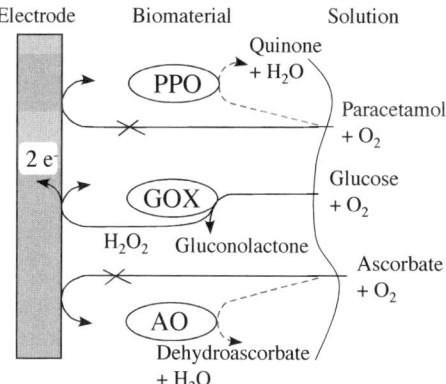

Figure 5.11. *Schematic description of the joint use of polyphenol oxidase (PPO), ascorbate oxidase (AO), and glucose oxidase (GOX) for glucose detection in the presence of interferents*

Finally, the third strategy makes use of enzymatic systems causing an inversion of the transduction mode: the phenomenon of molecular recognition occurs as a reduction (instead of an oxidation) of the electroactive species.

For example, the detection of salicylate in a sample of blood serum used to be based upon oxidation at 0.4 V of the catechol that was enzymatically generated by the salicylate hydroxilase [OLI 99]. Associating polyphenol oxidase and salicylate hydroxilase leads to the formation of quinone in place of catechol via cascade reactions. Consequently, the working mode of the bi-enzymatic sensor is now based upon quinone reduction at – 0.2 V and is no longer affected by interference linked to the oxidation of metabolites present in the analyzed medium [OLI 99, COS 01] (Figure 9). Based upon the same principle, the creation of a bi-enzymatic biosensor alkaline phosphatase-polyphenol oxidase could dose phenylphosphate, not by oxidation at 0.7 V of the phenol generated enzymatically by the alkaline phosphatase, but by the reduction at – 0.2 V of the quinine produced later by the polyphenol oxidase [COS 98].

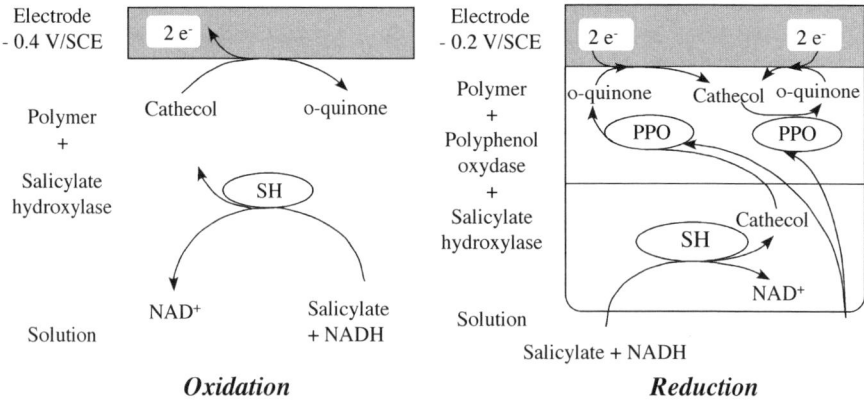

Figure 5.12. *Example of inversion of the transduction mode of a biosensor for dosing salicylate*

δ - Electro-enzymatic amplification

Recycling the substrate of the enzymatic reaction has been largely exploited by immunotests based upon the spectrophotometric detection of dosing hormones, steroids, neuropeptides, antibodies, and viruses. These recycling processes also open the way for phenomena amplifying the amperometric response of enzymatic biosensors.

Under the normal working conditions of biosensors, the enzymatic reaction rate is of the first order to the concentration of the target substrate. Regenerating this substrate by adding a chemical reactor in the analyzed medium causes this substrate to re-enter a catalysis cycle, thus enhancing the amperometric response of the biosensor (Figure 5.13a). For instance, an amperometric biosensor for dosing catechol has been made by immobilization of polyphenol oxidase, which catalyzes the oxidation of catechol to benzoquinone by O_2, which gets reduced to H_2O, the transduction being based upon the amperometric determination of the O_2 consumption.

Adding to the reaction medium some ascorbate, which reacts chemically with benzoquinone and produces dehydroascorbate and catechol, will induce catechol recycling and, therefore, increase O_2 consumption [UCH 93].

Recycling the substrate of the enzymatic reaction, and the resulting amplification of the amperometric signal detection, can also be tackled by an enzymatic reaction adjacent to that of the molecular recognition (Figure 5.13b). For instance, co-immobilizing polyphenol oxidase and glucose dehydrogenase at the surface of a Clark electrode causes the phenol detection signal to be multiplied by a factor of 300. This amplification results from the recycling, in the presence of glucose, of quinine in the form of catechol, another substrate of the polyphenol oxidase, thus lowering the detection limit of phenol down to 0.9 nM [MAK 96].

$$Phenol + O_2 \xrightarrow{polyphenol\ oxidase} benzoquinone + H_2O$$

$$Benzoquinone + glucose \xrightarrow{glucose\ oxidase} catechol + gluconolactone$$

$$Catechol + O_2 \xrightarrow{polyphenol\ oxidase} benzoquinone + H_2O$$

$$O_2 \xrightarrow{Clark\ electrode} H_2O$$

Another example of amplification via a bi-enzymatic configuration has been developed for dosing glutamate [YAO 90], which is a neurotransmitter. In the presence of alanine, the recycling process catalyzed by the glutamate pyruvate transaminase improves the detection limit of the biosensor from 0.2 µM to 0.2 nM (Figure 5.13b).

Regenerating the substrate of the enzymatic reaction by the electrochemical reaction itself can also be envisaged. Thus, catechol can be detected by a biosensor based upon the reduction of benzoquinone enzymatically generated by polyphenol oxidase instead of controlling O_2 concentration (Figure 5.13a). In this case, the catechol formed during the detection of benzoquinone can then undergo several enzymatic oxidation and electrochemical reduction cycles leading to an amplification of the biosensor by a factor of 10 [COS 99]. A similar example concerns the detection of riboflavin or vitamin B_2 by a biosensor based on flavin reductase [COS 97]. The enzymatically reduced riboflavin is regenerated via its detection by electrochemical oxidation (Figure 5.13c). The detection limit is thus pushed down from 0.2 µM for a direct electrochemical detection to 4 nM for the biosensor.

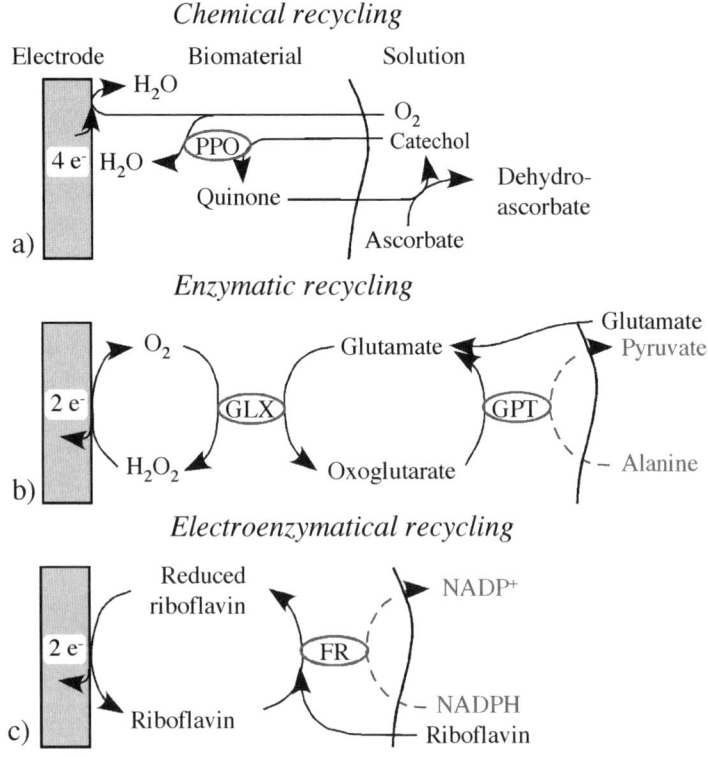

Figure 5.13. *Amplification of the amperometric response of biosensors for dosing catechol, glutamate and riboflavin; immobilized enzymes: a) polyphenol oxidase PPO, b) glutamate oxidase GLX and glutamate pyruvate transaminase GPT, c) flavin reductase FR*

5.2.3.2.2. Second-generation enzymatic biosensors

One of the main problems that might limit the commercialization of biosensors in the biomedical field is the specificity loss due to poor selectivity of the transduction step. The presence of metabolites, easily oxidizable at the detection potential of H_2O_2 or of $NAD(P)H$ (0.5–0.9 V versus SCE) has caused the development of complex structures (multi-enzymatic immobilizations, semipermeable membranes) which affect the sensitivity of the sensor. In this context, improving the sensitivity of the transduction step by redox mediators or catalyst immobilization has led to the second generation of biosensors.

α - Amperometric detection of H_2O_2

To overcome the drawbacks associated with the oxidation of H_2O_2, the detection of this molecule in the presence of O_2 has been performed by electrocatalytic reduction. It has been shown that Prussian blue ($Fe_4[Fe(CN)_6]_3$), once deposited under certain conditions at the surface of an electrode, behaves as a selective electrocatalyst of H_2O_2 at –0.05 V versus SCE [KAR 99]. Consequently, immobilization of enzymes, such as glucose oxidase and glutamate oxidase, at the surface of such a deposit has led to biosensors able to dose glucose and glutamate at –0.1 V and 0 V, respectively, with practically no interference [KAR 95, KAR 00]. This strategy is currently in development with the creation of new mixed deposits similar to Prussian blue, containing copper, cobalt, or nickel. The latter deposits have much increased mechanical and chemical stabilities at pH = 7 in comparison to Prussian blue [WAN 99b].

Another approach concerns incorporation or electrodeposition, in the bulk of the biomaterial, of micro- or nanoparticles of a metal (such as rhodium) that are able to reduce catalytically H_2O_2. These "metallized" biosensors have excellent selectivity for glucose detection in the presence of 0.1 mM paracetamol, urate, tyrosine or salicylate [WAN 94c].

β - Amperometric detection of NADH and of NADPH

Among the oxidoreductases, dehydrogenases represent a large family of enzymes (500) with $NAD(P)^+$ as a coenzyme (Figure 5.14). However, the development of amperometric biosensors based upon dehydrogenases is limited because of $NAD(P)H$ detection difficulties. The reason is that $NAD(P)H$ oxidation is chemically irreversible and requires high overpotentials partially entailing enzymatically inactive forms of $NAD(P)^+$. Moreover, passivation of the electrode surface by $NAD(P)^+$ adsorption is often observed. Consequently, immobilization at the electrode surface of mediators, such as flavins, quinones, phenothiazines, and phenoxazines, which can accept two electrons and one proton, has been extensively studied and applied to the dosage of $NAD(P)H$ [GRU 95, KAR 99b].

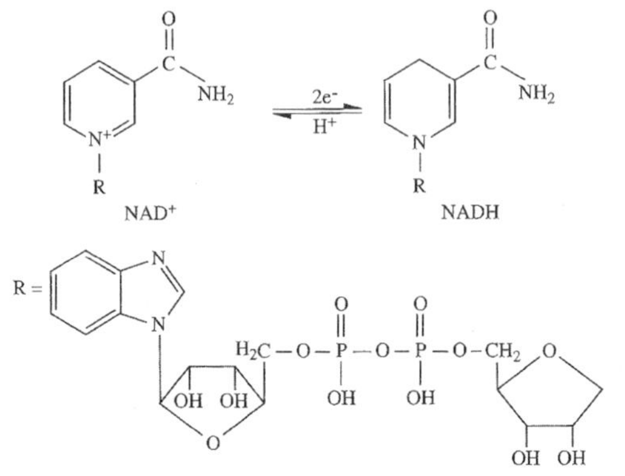

Figure 5.14. *Structure of the pyridinic NAD^+ coenzyme and of its reduced form NADH*

Co-immobilization of dehydrogenases and of these redox mediators by adsorption on graphite, by inclusion into carbon paste or by electropolymerization, has yielded biosensors with an operating mode based upon oxidation of NAD(P)H between –0.2 and 0.2 V [POL 91, KAR 94, WAN 95b]. For example, an amperometric biosensor for dosing lactate at 0 V has been obtained by electropolymerization of methylene blue in a biomaterial containing a lactate dehydrogenase [COS 96] (Figure 5.15).

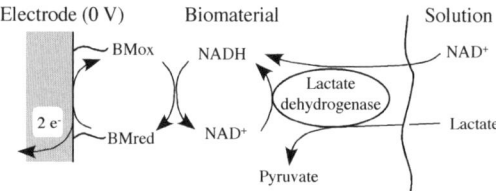

Figure 5.15. *Lactate amperometric biosensor based upon electrocatalytic oxidation of NADH by a film of methylene blue*

5.2.3.2.3. Third-generation enzymatic biosensors

The development of the concept of amperometric biosensors comprised the initiation of direct electron transfer between the electrode surface and the active site of the biomolecule. Any immobilized enzyme would then be connected electrically to the electrode, thus associating simultaneously three-dimensional recognition and three-dimensional transduction. Ideally, this would result in a total information recovery, in the elimination of limitations due to diffusion of the enzymatic reaction product and in an improved selectivity of the biosensor.

The main problem is the accessibility of the active site of the enzyme, which, in most cases, is deeply buried within the proteinic envelope. As the rate of electron transfer decreases exponentially with distance, direct electrode-enzyme transfer is very rarely possible.

Due to the key-role played by glucose in physiological and food processes, initial research efforts focused on the electric connection of glucose oxidase. For that, the planned redox mediators had to meet the following criteria: the measurements must not depend on O_2 partial pressure and must require sufficiently low potentials to avoid interference reactions. For efficient operation, these mediators must react quickly with the reduced enzyme and be chemically stable and non-toxic. The most frequently used redox mediators for connecting the active center of glucose oxidase, flavin adenine dinucleotide (FAD), are derivatives of ferrocene, hexacyanoferrate, quinonic compounds, phenothiazines and phenoxazines, and osmium complexes [HAL 89, KAK 94].

Glucose + glucose oxidase (FAD) $\rightarrow$ *gluconolactone + glucose oxidase (FADH$_2$)*

Glucose oxidase (FADH$_2$) + 2 mediator$_{ox}$ $\rightarrow$ *2 mediator$_{red}$ + 2 H$^+$ + glucose oxidase (FAD)*

2 mediator$_{red}$ $\xrightarrow{electrode}$ *2 mediator$_{ox}$ + 2 e$^-$*

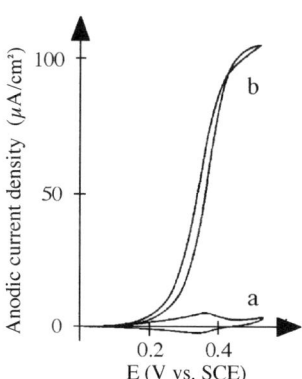

Figure 5.16. *Cyclic voltammetry of a biosensor containing glucose oxidase and ferrocene a) in the absence of glucose, b) in the presence of glucose (30 mM), scan rate: 10 mV/s; from [KUW 95]*

Although the mediators may be dissolved in the analyzed medium or incorporated in the biomaterial via electrostatic interactions, the development of chemical applications, and in particular determinations "*in vivo*", resort preferentially to covalent grafting of the redox mediators. Moreover, this technique overcomes the problems of fluctuations of convection or redox mediator concentration. For instance, an amperometric biosensor for glucose determination at 0.35 V has been fabricated by co-immobilization of aminoferrocene and glucose oxidase by chemical reticulation with glutaraldehyde [KUW 95]. In the absence of glucose, the voltammogram of the biosensor presents the reversible signal of ferrocene. Adding glucose causes a large increase of the oxidation current characteristic of the catalytic oxidation of the enzyme by ferrocene (Figure 5.16).

Although it does not obey the industrial criteria for biosensor fabrication, a sophisticated approach to enzyme connection has been developed by the groups of Heller and of Willner. It consists of chemically modifying the proteinic envelope of the glucose oxidase by redox mediators, in order to ensure direct electron transfer between the glucose oxidase and an electrode [DEG 88, WIL 95].

The difficulty of establishing a simple way to achieve the electric connection of glucose oxidase to the flavinic nucleotides has been circumvented by the joint immobilization of glucose oxidase and peroxidase. Contrary to glucose oxidase, peroxidase has its active site, a Fe porphirin, situated at its periphery, and is, therefore, able to exchange electrons easily (Figure 5.17) [YAN 98, GAS 01]. The principle of glucose detection, based upon the electric connection of peroxidase, is the following:

$$glucose + O_2 \xrightarrow{glucose\ oxydase} gluconolactone + H_2O_2$$

$$peroxidase_{red} + H_2O_2 \rightarrow peroxidase_{ox} + H_2O$$

$$peroxidase_{ox} + 2\ mediator_{red} \rightarrow peroxidase_{red} + 2\ mediator_{ox}$$

$$2\ mediator_{ox} + 2\ e^- \rightarrow 2\ mediator_{red}$$

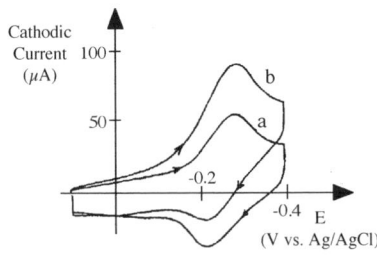

Figure 5.17. *Cyclic voltammetry of a biosensor containing a peroxidase and a film of methylene green a) in the absence of H_2O_2 and b) in the presence of H_2O_2 (1 mM), scanning rate: 60 mV/s; from reference [YAN 98]*

In parallel to the determination of glucose, this concept of electric connection of an enzyme *via* redox mediators has been applied to other enzymes, such as enoate reductase [THA 87], nitrate reductase [WIL 92, COS 97b], diaphorase [COS 96b], laccase [PAL 99], urease [BAR 00] and chymotrypsine [BAR 00], resulting in biosensors for dosing nitrates, NAD^+, $NADP^+$, NADH, urea, and the ethyl acetyltyrosine ester.

Thus, nitrate reductase, which catalyzes the reduction of nitrates to nitrites in the presence of NADPH, which gets oxidized to $NADP^+$, can be connected by 4,4'-bipyridinium or viologens:

$$NO_3^- + NADPH \xrightarrow{\text{nitrate reductase}} NO_2^- + NADP^+$$

Its immobilization in a viologen polymer film and its electrical connection by the viologens lead to an amperometric biosensor for dosing nitrates at –0.7 V (Figure 5.18).

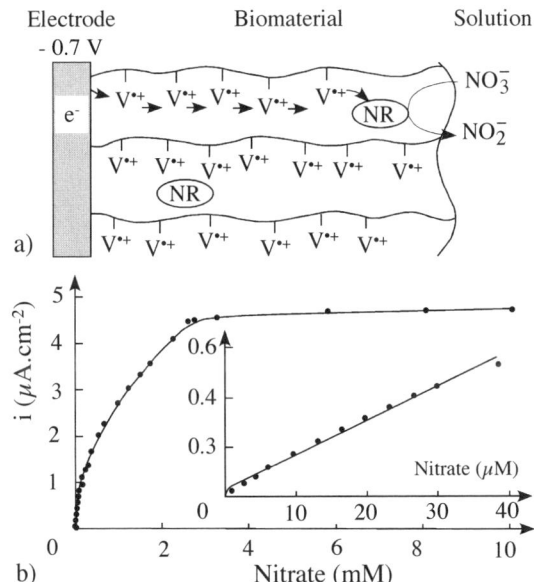

Figure 5.18. *a) Schematic description of the operation principle of a nitrate biosensor containing a nitrate reductase (NR) electrically connected by a film of poly(pyrrole-viologen) (V^{2+}). b) titration curve for nitrates at –0.7 V versus ECS*

Several polymeric structures have been tested for connecting nitrate reductase, which differ from each other by the reticulation degrees of the film, the steric hindrance of the viologen motif, and the length of the viologen-polymeric chain spacing arm.

Significant differences appeared between the biosensors in terms of sensitivity and of detection limit. These differences illustrate the importance of the notion of partial mobility of the redox mediator and of its size for accessing the active site of the enzyme (Figure 5.19).

In addition to the use of redox mediators for fabricating biosensors with an electrically connected enzyme, a much less widespread approach concerns the exploitation of the electronic conduction properties of electrogenerated organic polymers, such as polypyrrole or polyaniline [TAT 92, BAR 98]. In particular, a biosensor based upon the immobilization and connection of a peroxidase in a polypyrrole film proved to be selective and extremely sensitive to H_2O_2 (detection limit: 10 nM). This system has been used for determining glucose at 0.15 V via the additional immobilization of a glucose oxidase [TAT 93].

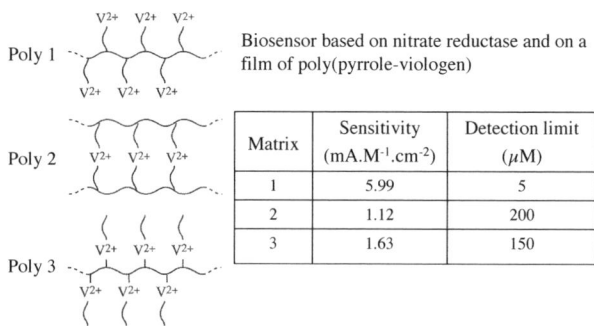

Biosensor based on nitrate reductase and on a film of poly(pyrrole-viologen)

Matrix	Sensitivity ($mA.M^{-1}.cm^{-2}$)	Detection limit (μM)
1	5.99	5
2	1.12	200
3	1.63	150

Figure 5.19. *Influence of the mobility and of the steric hindrance of the viologen V^{2+} upon the efficiency of the electrical connection of nitrate reductase*

5.3. Bibliography

[AKH 97] AKHTAR P., TOO C.O., WALLACE G.G., "Detection of amino acids at conducting electroactive polymer modified electrodes using flow injection analysis .1. Use of macroelectrodes", *Anal. Chim. Acta*, vol. 339, p. 201-209, 1997.

[ALB 97] ALBER K.S., COX J.A., KULESZA P.J., "Solid-state amperometric sensors for gas phase analytes: a review of recent advances", *Electroanalysis*, vol. 9, p. 97-101, 1997.

[ALE 96] ALEGRET S., "Rigid carbon – polymer biocomposites for electrochemical sensing – a review", *Analyst*, vol. 121, p. 1751-1758, 1996.

[ARR 94] ARRIGAN D.W.M., "Voltammetric determination of trace-metals and organics after accumulation at modified electrodes", *Analyst*, vol. 119, p. 1953-1966, 1994.

[BAG 99] BAGEL O., L'HOSTIS E., LAGGER G., OSBORNE M.D., SEDDON B.J., GIRAULT H.H., BRACK D., LOYALL U., SCHAFER H., "Rotograved carbon electrodes for amperometric cadmium and lead determination", *J. Electroanal. Chem.*, vol. 469, p. 189-195, 1999.

[BAL 91] BALDWIN R.P., THOMSEN K.N., "Chemically modified electrodes in liquid-chromatography detection – A Review", *Talanta*, vol. 38, p. 1-16, 1991.

[BAR 97] BARISCI J.N., WALLACE G.G., Clarke A., "Amperometric detection of electroinactive anions using conducting polymer electrodes subsequent to chromatographic separation", *Electroanalysis*, vol. 9, p. 461-467, 1997.

[BAR 98] BARTLETT P.N., BIRKIN P., WANG J., PALMISANO P., DE BENEDETTO G., "An enzyme switch employing direct electrochemical communication between horseradish peroxidase and a poly(aniline) film", *Anal. Chem.*, vol. 70, p. 3685-3694, 1998.

[BAR 00] BARDEA A., KATZ E., WILLNER I., "Biosensors with amperometric detection of enzymatically controlled pH-changes", *Electroanalysis*, vol. 12, p. 731-735, 2000.

[BED 95] BEDIOUI F., "Zeolite-encapsulated and clay-intercalated metal porphyrin, phtalocyanine and schiff-base complexes as models for biomimetic oxidation catalysts – An overview", *Coord. Chem. Rev.*, vol. 144, p. 39-68, 1995.

[BEL 98] BELMONT-HEBERT C., TERCIER M.L., BUFFLE J., FIACCABRINO G.C., DEROOIJ N.F., KOUDELKA-HEP M. "Gel-integrated microelectrode arrays for direct voltammetric measurements of heavy metals in natural waters and other complex media", *Anal. Chem.*, vol. 70, p. 2949-2956, 1998.

[BEN 96] BENMAKROHA Y., CHRISTIE I., DESAI M., VADGAMA P., "Poly(vinyl chloride), polysulfone and sulfonated polyether-ether sulfone composite membranes for glucose and hydrogen peroxide perm-selectivity in amperometric biosensors", *Analyst*, vol. 121, p. 521-526, 1996.

[BON 97] BONTEMPELLI G., COMISSO N., TONIOLO R., SCHIAVON G., "Electroanalytical sensors for nonconducting media based on electrodes supported on perfluorinated ion-exchange membranes", *Electroanalysis*, vol. 9, p. 433-443, 1997.

[BRO 77] BROWN A.P., ANSON F.C., "Molecular anchors for attachment of metal-complexes to graphite electrode surfaces", *J. Electroanal. Chem.*, vol. 83, p. 203-206, 1977.

[BUD 92] BUDNIKOV G.K., LABUDA J., "Chemical modified electrodes as amperometric sensors in electroanalysis", *Uspekhi Khimii*, vol. 61, p. 1491-1514, 1992.

[CAS 98] CASSIDY J.F., DOHERTY A.P., VOS J.G., "Amperometric methods of detection", in D. DIAMOND (ed) *Principles of Chemical and Biological Sensors*, Chem. Anal. Series, vol. 150, Wiley, p. 73-132, 1998.

[CEN 94] CENTONZE D., GUERRIERI A., PALMISANO F., ZAMBONIN P.G., "Electrochemically prepared glucose biosensors – Kinetic and faradaic processes involving ascorbic-acid and role of the electropolymerized film in preventing electrode-fouling", *Fresenius J. Anal. Chem.*, vol. 349, p. 497-501, 1994.

[CHU 98] CHUNG T.D., KIM H., "Electrochemistry of calixarene and its analytical applications", *J. Inclusion Phenom. Mol. Recognit. Chem.*, vol. 32, p. 179-193, 1998.

[CLA 62] CLARK L. C. Jr., LYONS C., "Electrode systems for continuous monitoring in cardiovascular surgery", *Ann. NY Acad. Sci.*, vol. 102, p. 29-45, 1962.

[COS 96] COSNIER S., LE LOUS K., "A new strategy for the construction of amperometric dehydrogenase electrodes based on laponite gel-methylene blue polymer as the host matrix", *J. Electroanal. Chem.*, vol. 406, p. 243-245, 1996.

[COS 96b] COSNIER S., LE LOUS K., "Amperometric detection of pyridine nucleotides via immobilized viologen-accepting pyridine nucleotide oxidoreductase or immobilized diaphorase", *Talanta*, vol. 43, p. 331-337, 1996.

[COS 97] COSNIER S., FONTECAVE M., LIMOSIN D., NIVIÈRE V., "A poly(amphiphilic pyrrole)-flavin reductase electrode for amperometric determination of flavins", *Anal. Chem.*, vol. 69, p. 3095-3099, 1997.

[COS 97b] COSNIER S., GALLAND B., INNOCENT C., "New electropolymerizable amphiphilic viologens for the immobilization and electrical wiring of a nitrate reductase", *J. Electroanal. Chem.*, vol. 433, p. 113-119, 1997.

[COS 98] COSNIER S., GONDRAN C., WATELET J.-C., DE GIOVANI W.F., FURRIEL R.P.M., LEONE F.A., "A bienzyme electrode (alkaline phosphatase polyphenol oxidase) for the amperometric determination of phosphate", *Anal. Chem.*, vol. 70, p. 3952-3956, 1998.

[COS 99] COSNIER S., "Biomolecule immobilization on electrode surfaces by entrapment or attachment to electrochemically polymerized films. A review", *Biosensors & Bioelectronics*, vol. 14, p. 443-456, 1999.

[COS 01] COSNIER S., GONDRAN C., WATELET J.-C., "A polypyrrole-bienzyme electrode (salicylate hydroxylase-polyphenol oxidase) for the interference-free determination of salicylate", *Electroanalysis*, vol. 13, p. 906-910, 2001.

[COX 96] COX J.A., TESS M.E., CUMMINGS T.E., "Electroanalytical methods based on modified electrodes: A review of recent advances", *Rev. Anal. Chem.*, vol. 15, p. 173-223, 1996.

[DAN 00] DANIELE S., BRAGATO C., BALDO M.A., WANG J., LU J., "The use of a remote stripping sensor for the determination of copper and mercury in the Lagoon of Venice", *Analyst*, vol. 125, p. 731-735, 2000.

[DEG 88] DEGANI Y., HELLER A., "Direct electrical communication between chemically modified enzymes and metal electrodes. 2. Methods for bonding electron-transfer relays to glucose-oxidase and d-amino-acid oxidase", *J. Am. Chem. Soc.*, vol. 111, p. 2615-2620, 1988.

[DEM 01] DE MATTOS I.L., GORTON L., "Metal-hexacyanoferrate films: a tool in analytical chemistry", *Quim. Nova*, vol. 24, p. 200-205, 2001.

[DER 89] DERONZIER A., MOUTET J.-C., "Functionalized polypyrroles – new molecular materials for electrocatalysis and related applications", *Acc. Chem. Res.*, vol. 22, p. 249-255, 1989.

[DIA 01] DIAMOND D., NOLAN K., "Calixarenes: designer ligands for chemical sensors", *Anal. Chem.*, vol. 73, p. 22A-29A, 2001.

[DIC 99] DICKERT F.L., HAYDEN O., "Molecular imprinting in chemical sensing", *Trends Anal. Chem.*, vol. 18, p. 192-199, 1999.

[DON 89] DONG S., WANG Y., "The application of chemically modified electrodes in analytical chemistry", *Electroanalysis*, vol. 1, p. 99-106, 1989.

[ECK 01] ECKER H., WARD M. (eds), *Chem. Mater.*, vol. 13(10), p. 3059-3809, 2001, special issue : "From organic-inorganic nanocomposites to functional materials".

[EDD 95] EDDY S., WARRINER K., CHRISTIE I., ASHWORTH D., PURKISS C., VADGAMA P., "The modification of enzyme electrode properties with nonconducting electropolymerized films", *Biosens. Bioelectron.*, vol. 10, p. 831-839, 1995.

[EMO 96] EMONS H., OSTAPCZUK P., "Electroanalysis for the purpose of environmental monitoring and specimen banking: Is there a future?", *Analyst*, vol. 121, p. 1917-1921, 1996.

[ESP 86] ESPENSCHEID M.W., GHATAKROY A.R., MOORE III R.B., PENNER R.M., SZENTIRMAY M.N., MARTIN C.R., "Sensors from polymers modified electrodes", *J. Chem. Soc., Faraday Trans. I*, vol. 82, p.1051-1070, 1986.

[FIE 94] FIELDEN P.R., MCCREEDY T., RUCK N., VAIREANU D.I., "Amperometric arrays in flowing solution analysis", *Analyst*, vol. 119, p. 953-958, 1994.

[FIT 90] FITCH A., "Clay-modified electrodes – A review", *Clays Clay Miner.*, vol. 38, p. 391-400, 1990.

[GAL 95] GALAN-VIDAL C.A., MUNOZ J., DOMINGUEZ C., ALEGRET S., "Chemical sensors, biosensors and thick-film technology", *Trends Anal. Chem.*, vol. 14, p. 225-231, 1995.

[GAS 01] GASPAR S., HABERMÜLLER K., CSÖREGI E., SCHUHMANN W., "Hydrogen peroxide sensitive biosensor based on plant peroxidases entrapped in Os-modified polypyrrole films", *Sens. Actuators B*, vol. 72, p. 63-68, 2001.

[GIL 94a] GILBY J., "Electrochemical sensors a moders success story for an old idea", *Sensor Rev.*, vol. 14, p. 30-32, 1994.

[GIL 94b] GILMARTIN M.A.T., HART J.P., "Novel, reagentless, amperometric biosensor for uric-acid based on a chemically-modified screen-printed carbon electrode coated with cellulose-acetate and uricase", *Analyst*, vol. 119, p. 833-840, 1994.

[GIL 95] GILMARTIN M.A.T., HART J.P., "Sensing with chemically and biologically modified carbon electrodes", *Analyst*, Vol. 120, p. 1029-1045, 1995.

[GOP 91] GÖPEL W., JONES T.A., KLEITZ M., LUNDSTRÖM J., SEIYAMA T. (eds), *Sensors A Comprehensive Survey*, vol. 2. *Chemical and Biological Sensors*, GÖPEL W., HESSE J., ZEMEL J.N. (eds), Wiley-VCH, 1991.

[GOR 95] GORTON L., "Carbon-paste electrodes modified with enzymes, tissues, and cells", *Electroanalysis*, vol. 7, p.23-45, 1995.

[GRU 95] GRÜNDING B., WITTSTOCK G., RÜDEL U., STREHLITZ B., "Mediator-modified electrodes for electrocatalytic oxidation of NADH", *J. Electroanal. Chem.*, vol. 395, p. 143-157, 1995.

[HAL 89] HALE P., INAGAKI T., KARAN H., OKAMOTO Y., SKOTHEIM T., "A new class of amperometric biosensor incorporating a polymer electron-transfer mediator", *J. Am. Chem. Soc.*, vol. 111, p. 3482-3484, 1989.

[HEI 90] HEINZE J., "Electronically conducting polymers", *Topics Curr. Chem.*, vol. 152, p. 1-47, 1990.

[HEL 62] HELLFERRICH F., *Ion Exchange*, Mc Graw-Hill Book Company, Inc., 1962.

[ITA 86] ITAYA K., UCHIDA I., NEFF V.D., "Electrochemically of polynuclear transition-metal cyanides – Prussian blue and its analogs", *Acc. Chem. Res.*, vol. 19, p. 162-168, 1986.

[JAN 88] JANATA J., BEZEGH A., "Chemical sensors", *Anal. Chem.*, vol. 60, p. 62R-74R, 1988.

[JOS 93] JOSOWICZ M., JANATA J., "Electroactive polymers in chemical sensors", Chapter 10 in SCROSATI B. (ed), *Applications of Electro-active Polymers*, Chapman and Hall, p. 310, 1993.

[KAI 99] KAIFER A.E., "Interplay between molecular recognition and redox chemistry", *Acc. Chem. Res.*, vol. 32, p. 62-71, 1999.

[KAK 94] KAKU T., KARAN H., OKAMOTO Y., "Amperometric glucose sensors based on immobilized glucose-oxidase polyquinone system", *Anal. Chem.*, vol. 66, p. 1231-1235, 1994.

[KAL 90] KALCHER K., "Chemically modified carbon paste electrodes in voltammetric analysis", *Electroanalysis*, vol. 2, p. 419-433, 1990.

[KAL 95] KALCHER K., KAUFFMANN J.-M., WANG J., SVANCARA I., VYTRAS K., NEUHOLD C., YANG Z., "Sensors based on carbon-paste in electrochemical analysis – a review with particular emphasis on the period 1990-1993", *Electroanalysis*, vol. 7, p. 5-22, 1995.

[KAR 94] KARYAKIN A., KARYAKINA E., SCHUHMANN W., SCHMIDT H.L., VARFOLOMEYEV S., "New amperometric dehydrogenase electrodes based on electrocatalytic NADH-oxidation at poly(methylene blue)-modified electrodes", *Electroanalysis*, vol. 6, p. 821-829, 1994.

[KAR 95] KARYAKIN A., GITELMACHER O., KARYAKINA E., "Prussian blue based first-generation biosensor – A sensitive amperometric electrode for glucose", *Anal. Chem.*, vol. 67, p. 2419-2423, 1995.

[KAR 99] KARYAKIN A., KARYAKINA E., GORTON L., "On the mechanism of H_2O_2 reduction at Prussian Blue modified electrodes", *Electrochem. Comm.*, vol. 1, p. 78-82, 1999.

[KAR 99b] KARYAKIN A., KARYAKINA E., SCHUHMANN W., SCHMIDT H.-L., "Electropolymerized azines: Part II. In a search of the best electrocatalyst of NADH oxidation", *Electroanalysis*, vol. 11, p. 553-557, 1999.

[KAR 00] KARYAKIN A., KARYAKINA E., GORTON L., "Amperometric biosensor for glutamate using Prussian Blue-based "artificial peroxidase" as a transducer for hydrogen peroxide", *Anal. Chem.*, vol. 72, p. 1720-1723, 2000.

[KUL 98] KULESZA P.J., COX J.A., "Solid-state voltammetry – analytical prospects", *Electroanalysis*, vol. 10, p. 73-80, 1998.

[KUR 00] KURAMITZ H., SUGAWARA K., TANAKA S., "Electrochemical sensing of avidin-biotin interaction using redox markers", *Electroanalysis*, vol. 12, p. 1299-1303, 2000.

[KUW 95] KUWABATA S., OKAMOTO T., KAJIYA Y., YONEYAMA H., "Preparation and amperometric glucose sensitivty of covalently bound glucose-oxidase to (2-aminoethyl)ferrocene on an Au electrode", *Anal. Chem.*, vol. 67, p. 1684-1690, 1995.

[LAB 92] LABUDA J., "Chemically modified electrodes as sensors in analytical chemistry", *Selective Electrode Rev.*, vol. 14, p. 33-86, 1992.

[LAB 96] LABUDA J., VANICKOVA M., "Selectivity of hybrid voltammetric determinations", *Electroanalysis*, vol. 8, p. 343-347, 1996.

[LAB 00] LABUDA J., VANICKOVA M., BUCKOVA M., KORGOVA E., "Development in voltammetric analysis with chemically modified electrodes and biosensors", *Chem. Pap.*, vol. 54, p. 95-103, 2000.

[LAN 73] LANE R.F., HUBBARD A.T., "Electrochemistry of chemisorbed molecules. I. Reactants connected to electrodes through olefinic substituents", *J. Phys. Chem.*, vol. 77, p. 1401-1410, 1973.

[LEE 96] LEECH D., in LYONS M.E.G. (ed.), *Electro-active Polymer Electrochemistry, Part II. Methods and Applications*, Plenum Press, 1996.

[LEE 97] LEE H.J., BEATTIE P.D., SEDDON B.J., OSBORNE M.D., GIRAULT H.H., "Amperometric ion sensors based on laser-patterned composite polymer membranes", *J. Electroanal. Chem.*, vol. 440, p. 73-82, 1997.

[LEF 99] LEFEVRE G., BESSIERE J., WALCARIUS A., "Cuprite-modified electrode for the detection of iodide species", *Sens. Actuators B*, vol. 59, p. 113-117, 1999.

[LEV 97] LEV O., WU Z., BHARATHI S., GLEZER V., MODESTOV A., GUN J., RABINOVICH L., SAMPATH S., "Sol-gel materials in electrochemistry", *Chem. Mater.*, vol. 9, p. 2354-2375, 1997.

[LI 84] LI T.T.T., WEAVER M.J., "Intramolecular electron-transfer at metal-surfaces. 4. Dependence of tunneling probability upon donor-acceptor separation distance", *J. Am. Chem. Soc.*, vol. 106, p. 6107-6108, 1984.

[LIU 96] LIU H.Y., DENG J.Q., "An amperometric glucose sensor based on Eastman-AQ-tetrathiafulvalene modified electrode", *Biosens. Bioelectron.*, vol. 11, p. 103-110, 1996.

[MAC 98] MACHA S.M., FITCH A., "Clays as architectural units at modified-electrodes", *Mikrochim. Acta*, vol. 128, p. 1-18, 1998.

[MAK 98] MAKOTE R., COLLINSON M.M., "Template recognition in inorganic-organic hybrid films prepared by the sol-gel process", *Chem. Mater.*, vol. 10, p. 2440-2445, 1998.

[MAK 96] MAKOWER A., EREMENKO A., STREFFER K., WOLLENBERGER U., SCHELLER F., "Tyrosinase-glucose dehydrogenase substrate-recycling biosensor: A highly-sensitive measurement of phenolic compounds", *J. Chem. Tech. Biotechnol.*, vol. 65, p. 39-44, 1996.

[MAL 99] MALINAUSKAS A., "Electrocatalysis at conducting polymers", *Synth. Met.*, vol. 107, p. 75-83, 1999.

[MAN 95] MANOWITZ P., STOECKER P.W., YACYNYCH A.M., "Galactose biosensors using composite polymers to prevent interferences", *Biosens. Bioelectron.*, vol. 10, p. 359-370, 1995.

[MAR 01] MARIAULLE P., SINAPI F., LAMBERTS L., WALCARIUS A., "Application of electrodes modified with ion-exchange polymers for the amperometric detection of non-redox cations and anions in combination to ion chromatography", *Electrochim. Acta*, vol. 46, p. 3543-3553, 2001.

[MAT 93] MATSUE T., "Electrochemical sensors using microarray electrodes", *Trends Anal. Chem.*, vol. 12, p. 100-108, 1993.

[MER 90] MERZ A., "Chemically modified electrodes", *Top. Curr. Chem.*, vol. 152, p. 49-90, 1990.

[MIL 88] MILLER C.J., MAJDA M., "Microporous aluminium-oxide films at electrodes – Dynamics of ascorbic-acid oxidation mediated by ferricyanide ions bound electrostatically in bilayer assemblies of octadecyltrichlorosilane and an octadecylviologen amphiphile", *Anal. Chem.*, vol. 60, p. 1168-1176, 1988.

[MOU 93] MOUSSY F., HARRISON D.J., O'BRIEN D.W., RAJOTTE R.V., "Performance of subcutaneously implanted needle-type glucose sensors employing a novel trilayer coating", *Anal. Chem.*, vol. 65, p. 2072-2077, 1993.

[MOU 94] MOUSSY F., JAKEWAY S., HARRISSON D.J., RAJOTTE R.V., "In-vitro and in-vivo performance and lifetime of perfluorinated ionomer-coated glucose sensors after high-temperature curing", *Anal. Chem.*, vol. 66, p. 3882-3888, 1994.

[MOU 98] MOUTET J.-C., "Molecular electrode materials: catalysis and analysis", *Actual. Chim.*, vol. 8-9, p. 63-86, 1998.

[MUR 80] MURRAY R.W., "Chemically modified electrodes", *Acc. Chem. Res.*, vol. 13, p. 135-141, 1980.

[MUR 84] MURRAY R.W., "Chemically modified electrodes", in BARD A.J. (ed.), *Electroanalytical Chemistry. A Series of Advances*, vol. 13, Marcel Dekker, New York, p. 191-368, 1984.

[MUR 87] MURRAY R.W., EWING A.G., DURST R.A., "Chemically modified electrodes – molecular design for electroanalysis", *Anal. Chem.*, vol. 59, p. 379A-&, 1987.

[NAS 98] NASCIMENTO V.B., ANGNES L., "Screen-printed electrodes", *Quim. Nova*, vol. 21, p. 614-629, 1998.

[NIW 95] NIWA O., "Electroanalysis with interdigitated array microelectrodes", *Electroanalysis*, vol. 7, p. 606-613, 1995.

[OCO 95] O'CONNOR K.M., ARRIGAN D., SVEHLA G., "Calixarenes in electroanalysis", *Electroanalysis*, vol. 7, p. 205-215, 1995.

[OLI 99] NETO G.O., JUNIOR L. R., KUBOTA L. T., "Electrochemical Biosensors for Salicylate and Its Derivatives", *Electroanalysis*, vol. 11, p. 527-533, 1999.

[PAL 95a] PALMISANO F., GUERRIERI G., QUINTO M., ZAMBONIN P.G., "Electrosynthesized bilayer polymeric membrane for effective elimination of electroactive interferents in amperometric biosensors", *Anal. Chem.*, vol. 67, p. 1005-1009, 1995.

[PAL 95b] PALMISANO F., MALITESTA C., CENTONZE D., ZAMBONIN P.G., "Correlation between permselectivity and chemical-structure of overoxidized polypyrrole membranes used in electroproduced enzyme biosensors", *Anal. Chem.*, vol. 67, p. 2207-2211, 1995.

[PAL 99] PALMORE G., KIM H., "Electro-enzymatic reduction of dioxygen to water in the cathode compartment of a biofuel cell", *J. Electroanal. Chem.*, vol. 464, p. 110-117, 1999.

[PIH 96] PIHEL K., WALKER Q.D., WIGHTMAN R.M., "Overoxidized polypyrrole-coated carbon fiber microelectrodes for dopamine measurements with fast-scan cyclic voltammetry", *Anal. Chem.*, vol. 68, p. 2084-2089, 1996.

[POL 91] POLÀSEK M., "Amperometric glucose sensor based on glucose-dehydrogenase immobilized on a graphite electrode modified with an N, N'-bis(benzophenoxazinyl) derivative of benzene-1,4-dicarboxamide", *Anal. Chim. Acta*, vol. 246, p. 283-292, 1991.

[POP 91] POPE M.T., MULLER A., "Polyoxometalate chemistry – an old field with new dimensions in several disciplines", *Angew. Chem., Int. Ed. Engl.*, vol. 30, p. 34-48, 1991.

[REY 99] REYMOND F., LEE H.J., ROSSIER J.S., TOMASZEWSKI L., FERRIGNO R., PEREIRA C.M., GIRAULT H.H., "Electrochemical sensor research at the Laboratoire d'Electrochimie of the EPFL", *Chimia*, vol. 53, p. 103-108, 1999.

[ROL 90] ROLISON D.R., "Zeolite-modified electrodes and electrode-modified zeolites", *Chem. Rev.*, vol. 90, p. 867-878, 1990.

[ROS 00] ROSSIER J.S., SCHWARZ A., BIANCHI F., REYMOND F., FERRIGNO R., GIRAULT H.H., "Polymer micro-structures: prototyping, low-cost mass fabrication and analytical applications", *Proc. Micro Total Anal. Syst.*, VAN DEN BERG A., OLTHUIS W., BERGVELD P. (eds), Kluwer Academic Publishers, Dordrecht, p. 159-162, 2000.

[SAD 00] SADIK O.A., MULCHANDANI A., *Chemical and Biological Sensors for Environmental Monitoring*, ACS Symp. Ser. 762, MULCHANDANI A., SADIK O.A. (eds), ACS, Washington DC, p. 1-7, 2000.

[SAS 90] SASSO S.V., PIERCE R.J., WALLA R., YACYNYCH A.M., "Electropolymerized 1,2-diaminobenzene as a means to prevent interferences and fouling and to stabilize immobilized enzyme in electrochemical biosensors", *Anal. Chem*, vol. 62, p. 1111-1117, 1990.

[SCH 91] SCHUHMAN W., "Amperometric substrate determination in flow-injection systems with polypyrrole enzyme electrodes", *Sens. Actuators B*, vol. 4, p. 41-49, 1991.

[SEL 00] SELLERGREN B. (ed), *Molecularly Imprinted Polymers*, Elsevier, 2000.

[SEN 96] SENDA M., YAMAMOTO Y., in: VOLKOV A.G., DEAMER D.W. (eds), *Liquid-liquid Interfaces*, CRC Press, 1996.

[SKO 86] SKOTHEIM T.A. (ed), *Handbook of Conducting Polymers*, Vols. 1&2, Marcel Dekker, 1986.

[SMY 92] SMYTH M.R., VOS J.G. (eds), *Analytical Voltammetry*, in SVEHLA G. (ed), *Comprehensive Analytical Chemistry*, vol. 27, Wilson and Wilson's, Elsevier, 1992.

[STA 98] STAIKOV G., LORENZ W.J., "Electrochemistry and Nanotechnology" in LORENZ W.J., PLIETH W. (eds), *Electrochemical Nanotechnology - In situ Local Probe Technique at Electrochemical Interfaces*, Wiley-VCH, p. 13, 1998.

[STE 99] STEFAN R.-I., VAN STADEN J.F., ABOUL-ENEIN H.Y., "Electrochemical sensor arrays", *Crit. Rev. Anal. Chem.*, vol. 29, p. 133-153, 1999.

[STO 90] STOECKER P.W., YACYNYCH A.M., "Chemically modified electrodes as biosensors", *Selective Electrode Rev.*, vol. 12, p. 137-160, 1990.

[TAT 53] TATSUMA T., WATANABE T., WATANABE T., "Electrochemical characterization of polypyrrole bienzyme electrodes with glucose-oxidase and peroxidase", *J. Electroanal. Chem.*, vol. 356, p. 245-253, 1993.

[TAT 92] TATSUMA T., GONDAIRA M., WATANABE T., "Peroxidase-incorporated polypyrrole membrane electrodes", *Anal. Chem.*, vol. 64, p. 1183-1187, 1992.

[THA 87] THANOS I.C.G., SIMON H., "Electro-enzymatic viologen-mediated stereospecific reduction of 2-enoates with free and immobilized enoate reductase on cellulose filters or modified carbon electrodes", *J. Biotechnology*, vol. 6, p. 13-29, 1987.

[THE 01] THÉVENOT D. R., TOTH K., DURST R. A., WILSON G. S., "Electrochemical biosensors: recommended definitions and classification", *Biosensors & Bioelectronics*, vol. 16, p. 121-131, 2001.

[THO 89] THOMSEN K. N., BALDWIN R.P., "Amperometric detection of nonelectroactive cations in flow systems at a cupric hexacyanoferrate electrode", *Anal. Chem.*, vol. 61, p. 2594-2598, 1989.

[TUR 87] TURNER A. P. F., KARUBE I., WILSON G. S. (eds), *Biosensors, Fundamentals and Applications*, Oxford University Press, 1987.

[UCH 93] UCHIYAMA S., HASEBE Y., SHIMIZU H., ISHIHARA H., "Enzyme-based catechol sensor based on the cyclic reaction between catechol and 1,2-benzoquinone, using L-ascorbate and tyrosinase", *Anal. Chim. Acta*, vol. 276, p. 341-345, 1993.

[UGO 95] UGO P., MORETTO L.M., "Ion-exchange voltammetry at polymer-coated electrodes: Principles and analytical prospects", *Electroanalysis*, vol. 7, p. 1105-1113, 1995.

[WAL 88] WALLACE G.G., "Chemically modified electrodes", in EDMONDS T.E. (ed.), *Chemical Sensors*, Blackie, p. 132-154, 1988.

[WAL 96] WALCARIUS A., "Zeolite-modified electrodes: Analytical applications and prospects", *Electroanalysis*, vol. 8, p. 971-986, 1996.

[WAL 98] WALCARIUS A., "Analytical Applications of Silica-Modified Electrodes – a Comprehensive Review", *Electroanalysis*, vol. 10, p. 1217-1235, 1998.

[WAL 99a] WALCARIUS A., "Zeolite-modified electrodes in electroanalytical chemistry", *Anal. Chim. Acta*, vol. 384, p. 1-16, 1999.

[WAL 99b] WALCARIUS A., "Factors affecting the analytical applications of zeolite modified electrodes: indirect detection of nonelectroactive cations", *Anal. Chim. Acta*, vol. 388, p. 79-91, 1999.

[WAL 99c] WALCARIUS A., MARIAULLE P., LOUIS C., LAMBERTS L., "Amperometric detection of nonelectroactive cations in electrolyte-free flow systems at zeolite modified electrodes", *Electroanalysis*, vol. 11, p. 393-400, 1999.

[WAL 01a] WALCARIUS A., "Electroanalysis with pure, chemically modified, and sol-gel-derived silica-based materials", *Electroanalysis*, vol. 13, p. 701-718, 2001.

[WAL 01b] WALCARIUS A., "Electrochemical applications of silica-based organic-inorganic hybrid materials", *Chem. Mater.*, vol. 13, p. 3351-3372, 2001.

[WAN 77] WANG J., CAI X., RIVAS G., SHIRAISHI H., DONTHA N., "Nucleic-acid immobilization, recognition and detection at chronopotentiometric DNA chips", *Biosensors & Bioelectronis*, vol. 12, p. 587-599, 1977.

[WAN 85] WANG J., *Stripping Analysis*, VCH, 1985.

[WAN 89] WANG J., "Voltammetry following nonelectrolytic preconcentration", in BARD A.J. (ed.), *Electroanalytical Chemistry*, vol. 16, Marcel Dekker Inc., p. 1-88, 1989.

[WAN 91a] WANG E.K., JI H.M., HOU W.Y., "The use of chemicaly modified electrodes for liquid-chromatography and flow-injection analysis", *Electroanalysis*, vol. 3, p. 1-11, 1991.

[WAN 91b] WANG J., "Modified electrodes for electrochemical sensors", *Electroanalysis*, vol. 3, p. 255-259, 1991.

[WAN 92] WANG J., NASER N., ANGNES L., WU H., CHEN L., "Metal-dispersed carbon paste electrodes", *Anal. Chem.*, vol. 64, p. 1285-1288, 1992.

[WAN 93] WANG J., NASER N., WOLLEMBERGER U., "Use of tyrosinase for enzymatic elimination of acetaminophen interference in amperometric sensing", *Anal. Chim. Acta*, vol. 281, p. 19-24, 1993.

[WAN 94a] WANG J., "Decentralized electrochemical monitoring of trace metals: from disposable strips to remote electrodes", *Analyst*, vol. 119, p. 763-766, 1994.

[WAN 94b] WANG J., "Selectivity coefficients for amperometric sensors", *Talanta*, vol. 41, p. 857-863, 1994.

[WAN 94c] WANG J., LIU J., CHEN L., LU F., "Highly selective membrane-free, mediator-free glucose biosensor", *Anal. Chem.*, vol. 66, p. 3600-3603, 1994.

[WAN 95a] WANG J., LU F., ANGNES L., LIU J., SAKSLUND H., CHEN Q., PEDRERO M., CHEN L., HAMMERICH O., "Remarkably selective metallized-carbon amperometric biosensors", *Anal. Chim. Acta*, vol. 305, p. 3-7, 1994.

[WAN 95b] WANG J., NASER N., "Amplified biosensing of alcohol based on biocatalytic accumulation of the mediator", *Electroanalysis*, vol. 7, p. 362-364, 1995.

[WAN 97] WANG J., "Remote electrochemical sensors for monitoring inorganic and organic pollutants", *Trends Anal. Chem.*, vol. 16, p. 84-88, 1997.

[WAN 99a] WANG J., "Sol–gel materials for electrochemical biosensors", *Anal. Chim. Acta*, vol. 399, p. 21-27, 1999.

[WAN 99b] WANG J., ZHANG X., PRAKASH M., "Glucose microsensors based on carbon paste enzyme electrodes modified with cupric hexacyanoferrate", *Anal. Chim. Acta*, vol. 395, p. 11-16, 1999.

[WAN 00] WANG J., LU J.M., LY S.Y., VUKI M., TIAN B., ADENIYI W.K., ARMENDARIZ R.A., "Lab-on-a-cable for electrochemical monitoring of phenolic contaminants", *Anal. Chem.*, vol. 72, p. 2659-2663, 2000.

[WAT 75] WATKINS B.F., BEHLING J.R., KARIV E., MILLER L.L., "Chiral electrode", *J. Am. Chem. Soc.*, vol. 97, p. 3549-3550, 1975.

[WIL 92] WILLNER I., KATZ E., LAPIDOT N., "Bioelectrocatalyzed reduction of nitrate utilizing polythiophene bipyridinium enzyme electrodes", *Bioelectrochem. Bioenerg.*, vol. 29, p. 29-45, 1992.

[WIL 95] WILLNER I., KATZ E., WILLNER B., "Electrical contact of redox enzyme layers associated with electrodes: Routes to amperometric biosensors", *Electroanalysis*, vol. 9, p. 965-977, 1997.

[WRI 92] WRING S.A., HART J.P., "Chemically modified, carbon-based electrodes and their application as electrochemical sensors for the analysis of biologically important compounds – a review", *Analyst*, vol. 117, p. 1215-1229, 1992.

[YAN 98] YANG R., RUAN C., DENG J., "A H_2O_2 biosensor based on immobilization of horseradish peroxidase in electropolymerized methylene green film on GCE", *J. Applied Electrochem.*, vol. 28, p. 1269-1275, 1998.

[YAO 90] YAO T., YAMAMOTO H., WASA T., "L-glutamate enzyme electrode involving amplification by substrate recycling", *Anal. Chim. Acta*, vol. 236, p. 437-440, 1990.

[YAO 98] YAO T., TAKASHIMA K., "Amperometric biosensor with a composite membrane of sol-gel derived enzyme film and electrochemically generated poly(1,2-diaminobenzene) film", *Biosensors & Bioelectronics*, vol. 13, p. 67-73, 1998.

[ZHA 94] ZHANG Y., HU Y., WILSON G.S., MOATTI-SIRAT D., POITOUT V., REACH G., "Elimination of the acetaminophen interference in an implantable glucose sensor", *Anal. Chem.*, vol. 66, p 1183-1188, 1994.

Chapter 6

ISFET, BioFET Sensors

6.1. Structure of ISFET sensors

6.1.1. *Introduction*

Due to their design, ISFETs (ion-sensitive field-effect transistors) show decisive advantages for ionic detection compared with classical chemical sensors (ion-specific electrodes or ISE), as presented in the following table. The description of the ISFET and its principle, proposed for the first time in 1970 by P. Bergveld [BER 70], were investigated in detail 2 years later [BER 72].

Characteristics	Ion-specific electrodes (ISE)	Ion-sensitive field-effect transistors (ISFET)
Measuring device input impedance	High >10^{12} Ω	Low 10^4 Ω
Signal treatment	Outside	Integrated
Size	Around 15 cm, diameter 1.5 cm	Quite small: 1.5×2.5 mm
Design	Often with an internal liquid	All solid
Cost	High (low series)	Low (microelectronic collective processes)

Table 6.1. *Compared characteristics of a specific electrode with an ISFET*

Chapter written in French by Nicole JAFFREZIC-RENAULT and Claude MARTELET, translated by Claude MARTELET.

174   Chemical and Biological Microsensors

ISFET sensor structure comes directly from the MOSFET (metal oxide semiconductor field-effect transistor), a classical component of integrated circuits.

6.1.2. MOS (metal-oxide semiconductor) structure

MOS structure consists of a thermal oxide layer (SiO_2 insulator) sandwiched between a p or n doped semiconductor and a metal. In order to ensure the silicon bias versus metal, a rear metallic contact is needed (see Figure 6.1).

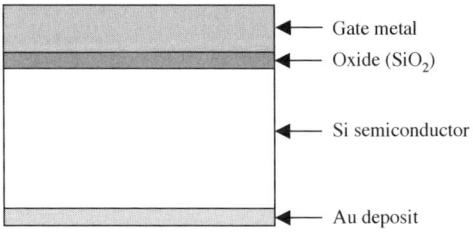

Figure 6.1. *MOS structure*

6.1.2.1. MOS structure at equilibrium

A simplified energy band diagram of a p-type MOS structure with equal Fermi levels is represented in Figure 6.2.

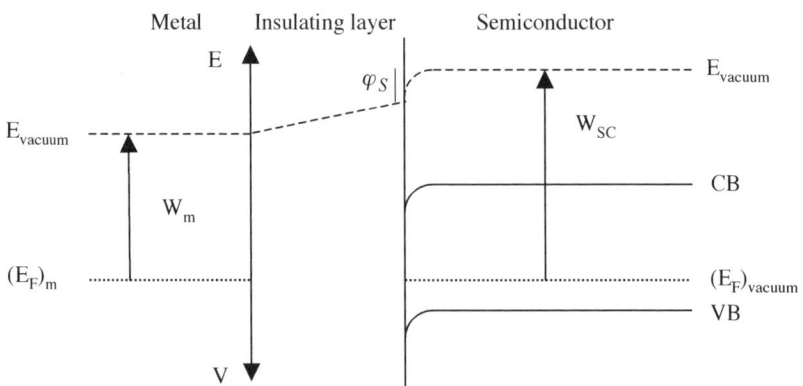

Figure 6.2. *Band diagram of a p-type MOS structure at equilibrium ($W_{SC} > W_m$)*

with:

- $(E_F)_m$: metal Fermi level;
- VB: energetic level of the top of the valence band;
- CB: energetic level of the low semiconductor conduction band;
- ϕ_S: semiconductor surface potential;
- W_{SC}: extraction work of semiconductor;
- E_{vacuum}: vacuum level.

6.1.2.2. *Bias of the MOS structure*

In the case of a negative bias (V_G) of the metal versus semiconductor (see Figure 6.3), main carriers (holes for a p type doping) are attracted to semiconductor surface. The Fermi level near the oxide-semiconductor interface is situated in the valence band, characterizing the situation of accumulation of main carriers. Silicon behaves as a metal.

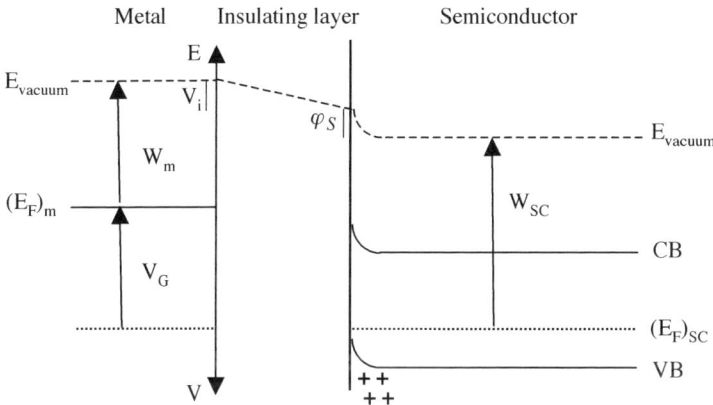

Figure 6.3. *Accumulation situation of a MOS structure*

with:

- W_m: extraction work of metal;
- W_{SC}: extraction work of semiconductor;
- $(E_F)_{SC}$: semiconductor Fermi level;
- ϕ_S: semiconductor surface potential;

- V_i: potential drop in the insulator;
- E_{vacuum}: vacuum level;
- V_G: applied bias (metal/semiconductor).
- V_G can be expressed as:

$$V_G = W_m - W_{SC} - \varphi_S - \frac{(Q_S + Q_F)}{C_i} \qquad [6.1]$$

with:

- Q_S: surface charge of the semiconductor;
- Q_F: oxide fixed charge;
- C_i: oxide capacity.

In case of a positive bias of the metal versus the semiconductor (see Figure 6.4), band bending is inversed. Near the interface, "holes depleted" situation versus the normal situation is established. The semiconductor is in desertion situation for majority carriers (holes).

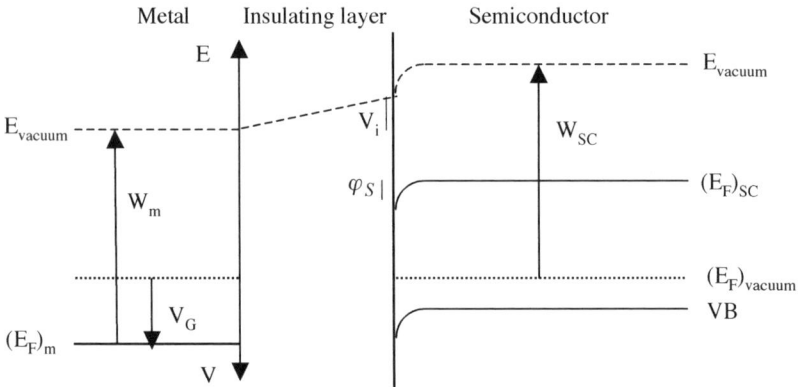

Figure 6.4. *Desertion situation of a MOS structure*

For a higher positive bias (metal versus semiconductor), the semiconductor behavior is then inversed: the density of minor carriers (here electrons) becomes greater than that of major carriers (holes), with major carriers being accumulated at the semiconductor surface, under the insulating layer. For the inversion mode, two cases have to be considered: weak inversion when surface hole density remains greater than that of the electrons at the semiconductor (p type) surface, and high inversion when such a density is lower ($\varphi_S > 2\varphi_0$, φ_0: absolute value of the difference between the Fermi level of the doped silicon and the Fermi level of intrinsic Si).

6.1.3. EOS (electrolyte-oxide semiconductor) structure

6.1.3.1. General considerations

The EOS structure is similar to the MOS structure, the metal role being played by the electrolyte and the reference electrode, generally a saturated calomel electrode (SCE).

The conducting solution allows the bias of semiconducting substrate covered by a thin oxide layer. It forms a blocking structure, preventing all faradic phenomena, and permits a field effect in the semiconductor, as in the case of MOS.

Transport of elementary charges, which was ensured by free electrons of the metal, is done in this case by ionic chemical species of the solution.

Inside the solution, the potential remains constant while at the interface of the insulator solution a low potential drop is generated due to adsorption of charge at the insulator surface. This potential variation can be evaluated using the theory of the Helmoltz double layer.

6.1.3.2. Biasing of the EOS structure

The band structure of the biased EOS in accumulation situation is represented in Figure 6.5, for a p-doped semiconductor.

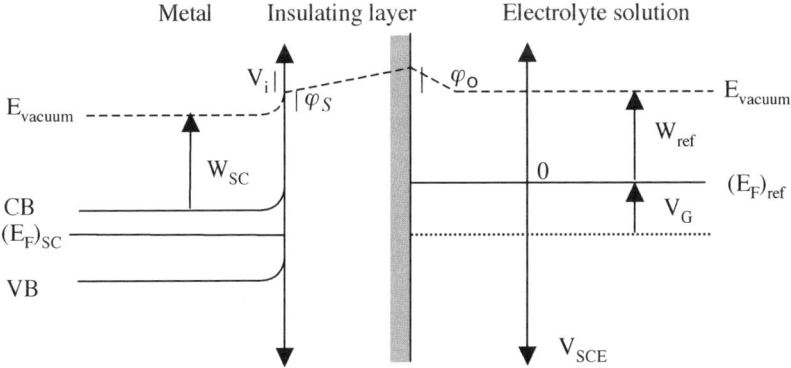

Figure 6.5. *Energetic diagram of an EOS structure biased versus a reference electrode*

with:

- W_{ref}: extraction work of the reference electrode;
- W_{SC}: extraction work of semiconductor;
- CB: energetic level at the bottom of the conduction band;
- VB: energetic level at the top of the valence band;
- $(E_F)_{SC}$: semiconductor Fermi level;
- $(E_F)_{ref}$: reference electrode Fermi level;
- φ_0: potential drop at the surface of the insulator;
- φ_S: semiconductor surface potential;
- E_{vacuum}: vacuum level;
- V_G: applied potential (semiconductor/reference).
- V_G can be expressed as follows:

$$V_G = W_{SC} - W_{ref} + \varphi_0 - \frac{(Q_S + Q_F)}{C_i} - 2\varphi_S \qquad [6.2]$$

with:

- Q_S: semiconductor surface charge;
- Q_F: fixed charge in the oxide.

6.1.4. MOSFET

In this section, the structure and the operating principle of the MOSFET transistor (Figure 6.6) is presented. Inside a p-doped silicon substrate (as is the case of n channel MOSFET) are implanted two n-doping zones composing the drain and the channel on which metallic electrodes are applied. The central zone situated between drain and source is the channel; a thin insulating layer (silica) is laid over the canal and metallization on its surface constitutes the gate electrode, which is the electrode controlling the channel conductivity.

Such a structure is normally blocked; no current can flow through the channel between the source and the drain, as, whatever the potential difference applied between these two parts, as least one of the p-n junctions is reverse biased. The application between gate and source of a positive voltage V_G tends to repel majority holes and to attract, in the part located under the gate, electrons (minority carriers) from the p substrate. When $V_G > V_T$ (V_T threshold voltage), the electrons density becomes greater than the hole density (high inversion situation, $\varphi_S > 2\varphi_0$) thus a n-type channel is formed, ensuring the continuity between drain and source: the flowing of a current I_D between drain and canal is thus possible.

The threshold voltage V_T is expressed by:

$$V_T = W_m - W_{SC} - 2\varphi_b - \frac{(Q_S + Q_F)}{C_i}$$ [6.3]

The current, I_D, depends on voltages V_G and V_D (between drain and source): for $V_D < V_G - V_T$ it is a linear situation and for $V_D < V_G - V_T$ it is a saturation situation.

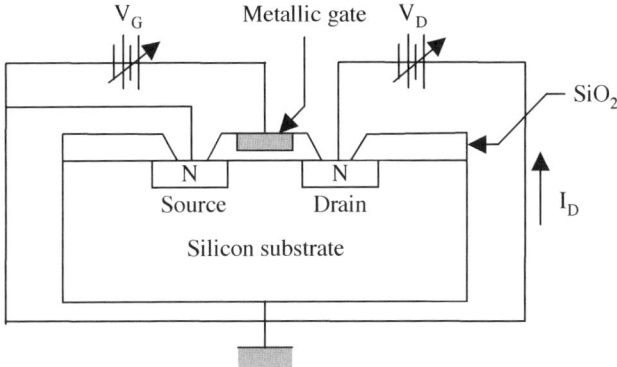

Figure 6.6. *Schematic drawing of a MOSFET*

6.1.5. ISFET

The schematic structure of an ISFET is represented in Figure 6.7. This structure is similar to the MOSFET in which the metal is replaced by a sensitive membrane, the electrolyte and the reference electrode. The threshold voltage can be expressed as follows:

$$V_T = W_{SC} - W_{ref} + \varphi_O - (Q_S + Q_F)/C_i - 2\varphi_b$$ [6.4]

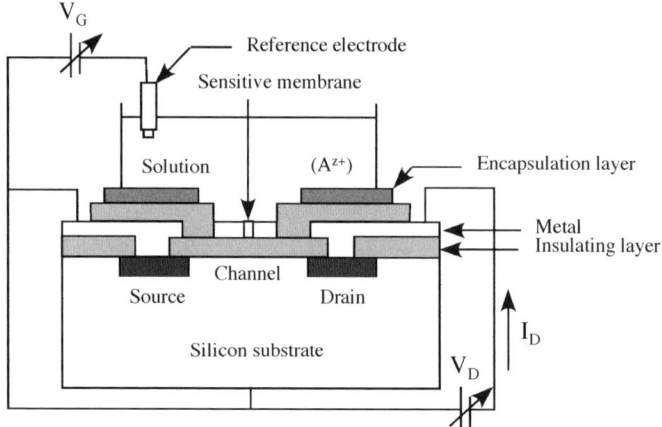

Figure 6.7. *Schematic view of an ISFET*

The threshold voltage V_T thus depends on the chemical properties of the solution:

$$V_T = V_{TO} + \varphi_\circ \qquad [6.5]$$

where V_{TO} only depends on the characteristics of the ISFET component and φ_0 is the voltage difference between the sensitive membrane and the electrolytic solution. As equilibrium exists between the A^{z+} ion to be dosed and its chemical form in the sensitive membrane, φ_0 depends on the A^{z+} activity.

6.2. Techniques used for ISFET fabrication and operation

6.2.1. *ISFET fabrication*

The fabrication process for ISFET production can be divided into two parts. The first deals with component preparation through microtechnologies from the silicon wafer (typically 50 to 75 mm in diameter, increasing in modern processes). The main fabrication steps can be summarized as follows [VLE 88]: fabrication of the insulating box, implantation of the drain and source zones, gate oxide formation, formation of the contacts drain, source, and substrate. A photograph of an ISFET/REFET is shown in Figure 6.8.

ISFET, BioFET Sensors 181

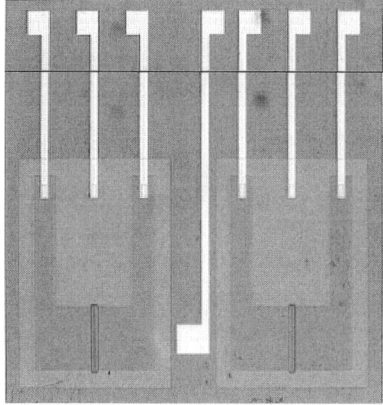

Figure 6.8. *Photograph of an ISFET/REFET structure (documentation LAAS/CNRS)*

The second part of the fabrication process deals with steps occurring after the cutting of the silicon wafer into individual components. It consists of deposition of the sensitive membrane, mounting of the ISFET on the support, electrical connections, and sensor encapsulation. A ready-for-use device is shown in Figure 6.9.

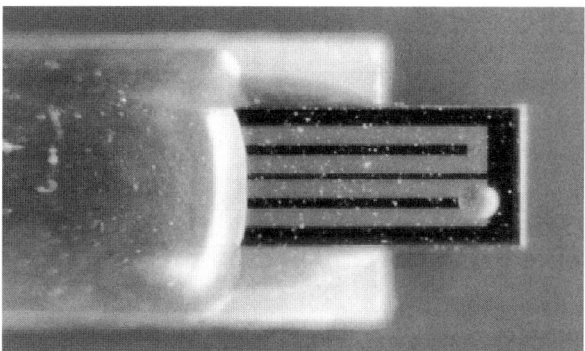

Figure 6.9. *Photograph of an ISFET/REFET device (document from ECL)*

In principle, ISFET manufacturing is identical to that used for producing a field-effect transistor without a metallic gate, but constraints are related to the use in a liquid environment. To overcome such difficulties, specific geometries have been proposed, such as the geometry shown on Figure 6.8, where electrical contacts are moved away from the sensitive zone. Some specific designs allowing, for example,

6.2.2. ISFET Measurement set-up

In the most frequently used method ("source follower" type, see Figure 6.10.), the drain current is kept constant and the voltage V_G is controlled. The voltage variation V_G is equal to the value of the interfacial potential φ_0.

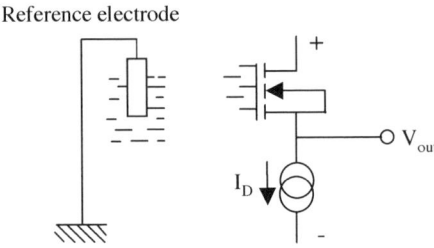

Figure 6.10. *"Source follower" measurement set-up for ISFET*

Miniaturization of the reference electrode has been extensively studied. Various solutions have been proposed: silver wire-silver chloride [JAN 80], integrated reference electrodes [SMI 86], differential measurement set-up with a reference ISFET and a pseudo-metallic reference (Figure 6.11). Recently, new circuits for processing signals from ISFET have been proposed [KOL 06, WAN 08].

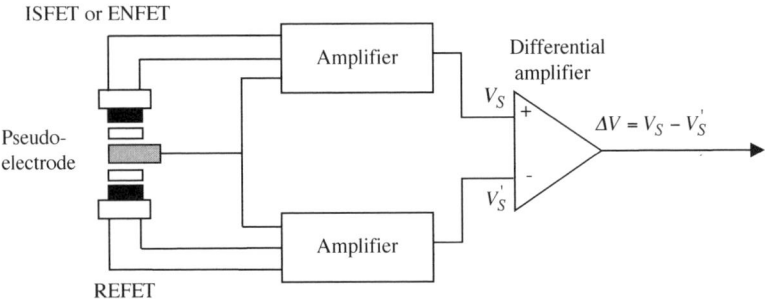

Figure 6.11. *Principle of the differential mode of measurement (reproduced from [PER 89])*

6.3. ISFET membranes

6.3.1. *Detection of H^+ ions*

pH-sensitive membranes consist of thin oxide layers, such as SiO_2, Al_2O_3, Ta_2O_5 or nitrides, such as Si_3N_4. SiO_2 is historically the first membrane used for pH measurement, but it has been relinquished due to its sub-Nernstian response and poor lifetime. All other membranes have Nernstian responses and lifetimes longer than a year.

In the case of oxide membranes such as alumina, silica or nitride, the pH response can be explained through the site-binding model developed by Bousse [BOU 82]. In the basic case where the sensitive membrane consists of a silica-insulating gate of the sensor, H^+ ions inside the solution react with silanol sites Si-OH at the silica surface. Their amphoteric character can be expressed through the two following equilibria:

$$Si-OH_2^+ \leftrightarrow Si-OH + H_S^+ \quad\quad k'_{OH} = \frac{(H_S^+)(Si-OH)}{(Si-OH_2^+)}$$

$$Si-OH \leftrightarrow Si-O^- + H_S^+ \quad\quad k''_{OH} = \frac{(H_S^+)(Si-O^-)}{(Si-OH)}$$

In the case of a solution in which the electrolyte concentration is higher than 0.1 M, it has been shown that:

$$\varphi_0 = 2.303 \frac{kT}{q} \frac{\beta}{(\beta+1)} (pH_{pzc} - pH) \quad\quad [6.6]$$

with the following meaning for the various parameters:

– k: Boltzmann constant;

– T: absolute temperature (K);

– β: sensitivity parameter proportional to the number of silanol sites N_S per unit surface:

$$\beta = \frac{2q^2 N_S}{K_{OH}^{1/2} C_S kT} \quad\quad [6.7]$$

where q represents the elementary charge, with:

$$K_{OH} = \frac{k'_{OH}}{k''_{OH}} \qquad [6.8]$$

and C_s is the double layer capacity at the pH (pH_{pzc}) for which the silica charge surface is zero.

It should be noted that the equation is only valid if:

$$\frac{q\varphi_0}{kT} < \beta$$

and emphasizes the dependence of φ_0 upon the H^+ concentration.

6.3.2. Detection of other ions

6.3.2.1. ISFET with a thick, selective membrane

Such membranes have behaviors close to that of classical selective electrodes, which are conductive, polarizable, and obey Nernst's law. Recognizing sites are distributed along the main part of the membrane thickness.

6.3.2.1.1. Special glasses

Some special glasses have been proposed for the realization of inorganic selective ISFET membranes. The hydrophilic nature of these solid electrolytes is an asset for the membrane adhesion, but such a strategy is limited as only a small number of inorganic materials can be used. ISFETs sensitive for fluoride and sodium have been obtained through the deposition of calcium fluoride films [MOR 88], aluminosilicate glasses [ESA 78], or NASICON on the gate [KLE 87].

6.3.2.1.2. Polymeric membranes

Numerous polymeric membranes containing selective molecules have been used for ion-sensitive ISFETs. Most deal with alkaline metal ion detection (K^+, Na^+), alkaline earth metal (Ca^{2+}) or ammonium and nitrate [BER 88, MAD 89]. Using such a strategy, the detection of other species, such as toxic Pb^{2+} and Cd^{2+} ions [LUT 97] or heavy ions has been undertaken [ALI 00]. The corresponding specific membranes are most often obtained through processes similar to those used for obtaining selective electrode membranes where incorporation of an ionophore into a plasticized polymer, such as PVC or polysiloxane, is needed. Some materials, like sol-gels can avoid plasticization [KIM 97].

The main disadvantages of this type of membranes are:

– poor adhesion to the gate dielectric, thus limiting the life-time;

– progressive leaching of active molecule in the solution;

– distribution of active species along the membrane thickness, thus limiting their accessibility, leading to an increase of the response time.

In order to overcome such problems, several solutions have been proposed. One consists of including a silanization reagent in the membrane, for example polyurethane or silicone-based polymers. Geun Sig *et al.* [GEU 91] have obtained a good adhesion on silicon nitride surfaces. A notable improvement has been brought about by the use of photopolymerizable membranes [KAW 84], which offer a better compatibility with microtechnologies. The latter solution has been tested, for example, in calcium-sensitive ISFETs. In this case the sensitive membrane was chemically bound to the transducer surface and a polyHEMA intermediary membrane, also applied by chemical grafting, enabled a well-defined interface (regarding thermodynamic equilibrium) to be obtained [VAN 91]. The introduction of an intermediate protection layer, such as a thin hydrophobic film of acrylic polymer [VAN 97], also improves the lifetime. Another strategy consists of mechanical blocking of the membrane with a ring of a polyimide mesh, suspended over the gate insulator [BLA 82]. Such a membrane type has also been used for the detection of complex ions, such as salicylate in drugs [TOM 95] or for cocaine analysis [CAM 95].

The use of membranes issued from natural products can give quite good results. Thus, on a selenocyanate selective sensor, lifetimes over 2 months have been obtained by the direct incorporation of the ionophore in natural enamel latex [WAK 86].

6.3.2.1.3. Ion implantation

Although less exploited, the production strategy of selective membranes through ion implantation offers numerous advantages:

– compatibility with ISFET fabrication processes;

– perfect adhesion of the selective membrane;

– superficial treatment without any effect upon the ISFET electric transconductance value;

– quasi-Nernstian response in a concentration range from 10^{-4} M to 10^{-1} M.

An approach consisting of implanting trivalent ions, such as B^{3+}, Al^{3+}, Ga^{3+}, In^{3+}, or Tl^{3+}, which will induce ionic exchange sites for monovalent cations by substitution to tetravalent silicon atoms has been proposed. Quasi-Nernstian responses have thus been obtained for Na^+ and K^+ [PHA 84, WON 89]. This method also allows in-situ generation of aluminosilicate glasses known to be sensitive to sodium ions [SAN 82].

Such a method needs high implantation doses ranging from 10^{15} to 10^{17} ions/cm², which can induce degradation of the dielectric layer of the gate. To reduce such an effect, some researchers have irradiated the sample through an aluminium buffer layer, which is further eliminated [ITO 88]. In addition to these works directed towards alkaline ion detection, there is the possibility to obtain fluoride ion-selective membranes [BAC 97].

6.3.2.2. Modified ISFETs through chemical grafting

For such ISFETs, the response is often sub-Nernstian, however, these devices present numerous advantages:

– chemical continuity with the gate material, thus inducing a potential long lifetime;

– nearly monomolecular thickness of the selective layer allowing the recognition sites to have great accessibility;

– ability for the gate materials (most often oxides) to be chemically grafted using hydroxyl groups at the surface which are present (residual) or created with chemical reagents, such as silanes, for example.

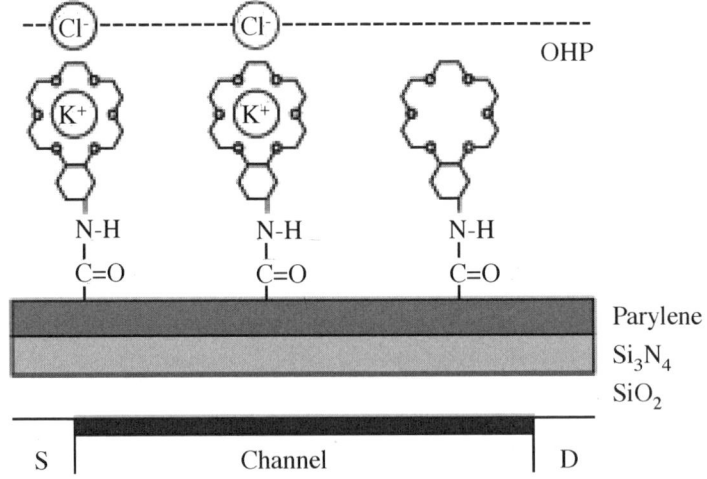

Figure 6.12. *ISFET K^+ obtained by parylene grafting with a crown ether, from [MAT 86]*

This method has been less explored, due to difficulties related to the synthesis of a graftable reagent bearing selective chemical groups. Thus Matsuo *et al.* [MAT 86] have deposited a 100 nm parylene membrane, onto the silicon nitride gate of an ISFET. Potassium ion selectivity, as shown in Figure 6.12, was obtained by direct grafting of the carboxylic ions of parylene by crown ether molecules. To illustrate this type of membrane, the use of bifunctional silanes bearing cyano groups allowing silver ion detection can be cited. In this case, a theoretical approach has been also proposed [PER 90]. Such membranes are also illustrated by aminophosphonate condensation onto a grafted chlorosilane for calcium ion detection [JAF 92] or by grafted gates for the detection of nitrate ions [ROC 92] or potassium ions [ELB 99].

6.3.2.3. *Other processes*

The possibility of depositing thin layers of specific species onto the ISFET gate via vacuum sublimation has been investigated previously [MLI 98]. Thus, ultra-thin layers of thiacalix(4)arene (thickness around 20 nm; Figure 6.13) give a quasi-Nernstian response for Cu^{2+} ions, instead of a marked sub-Nernstian response for higher thickness (around 100 nm) [ALI 01]. In this case, the morphology of the layers favors recognizing site orientation, more so than a thin layer,. This last example underlines the importance of the spatial organization of the selective species on the characteristics of the selective membrane.

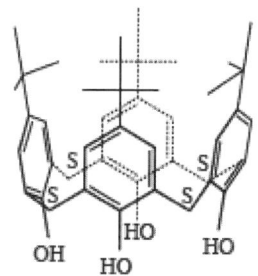

Figure 6.13. *Molecular structure of thiacalix(4)arene showing a possible orientation of the molecule on a surface made hydrophilic by OH groups (from [ALI 01])*

6.4. Detection of molecular species

6.4.1. *Metabolic biosensors*

Metabolic biosensors have the advantage of an auto-regeneration of the recognizing structure. In this case, the recognized substance, substrate S, is transformed into a detectable product on the transducer (L_T means the enzyme fixed onto the transducer surface). The most common case concerns the intermediate formation of a complex L_TS that is decomposed with a rate corresponding to a kinetic constant k_2. At equilibrium, it decomposes as fast as it is produced, according the Michaelis-Menten theory:

$$L_T + S \underset{k_{-1}}{\overset{k_1}{\rightleftarrows}} L_TS \overset{k_2}{\rightarrow} P + L_T$$

with:

$$K_M = \frac{(k_{-1} + k_2)}{k_1} = \left(\frac{V_{max}}{V-1}\right) \cdot [S]$$

V representing the reaction rate: $V = k_2[L_TS]$.

The corresponding biosensor response will depend upon such kinetics, and in this case, the affinity constant, K_M (Michaelis constant), represents the substrate concentration for which the reaction rate reaches half the maximal rate, V_{max}. In reality, this value corresponds to the half saturation of the enzyme, thus only concentrations less than such a value can be exploited for a biosensor-type application. This is because, at saturation, the production of P, and thus the intensity of the signal delivered by the sensor, becomes independent of the substrate concentration (analyte in this case).

Such remarks apply to metabolic biosensors for which the enzyme is readily accessible via the substrate (enzymes fixed under a monolayer state or, of course, in a homogeneous medium) with the reaction product being instantaneously detected at the transducer surface. However, in most of cases, the enzyme is incorporated in a membrane, thus the response is conditioned by diffusion processes of the substrate to the enzyme L_T and by diffusion of P to the transducer surface or to the analyzed solution.

6.4.2. The enzymatic field-effect transistor (ENFET) principle

Due to its principle, ISFET is able to detect any phenomenon, inducing a pH or a charge modification, on its surface.

As reactions catalyzed by enzymes (hydrolase or oxidase types) consume or produce protons, an ISFET could be used to detect substances (substrates) decomposed by the induced pH variation. Such a device is called an ENFET (see Figure 6.14) [KIM 90].

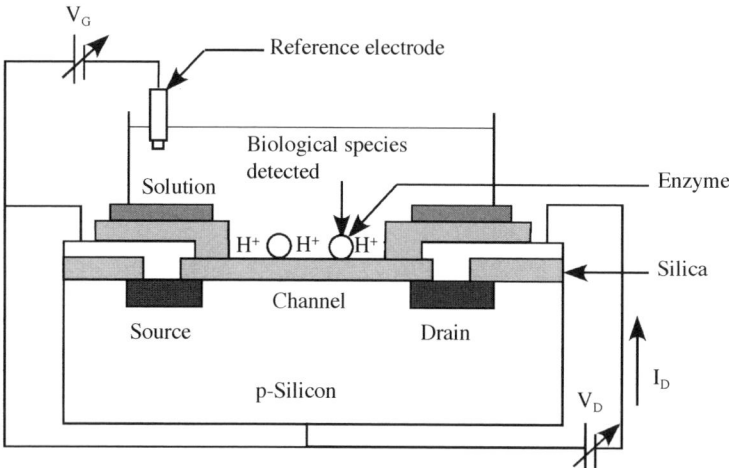

Figure 6.14. *Schematic drawing of an ENFET*

For analytical purposes, species detection using an ENFET is possible through one of the following mechanisms:

– direct detection of reaction products, involving the biocatalyzed degradation of the detected species. Only enzymatic reactions implying pH variations or modification of ionic species concentration can be used;

– inhibition level measurement of an enzymatic reaction by the toxin to be detected. Such an inhibition can be obtained at quite low levels; this strategy can induce high sensibility levels.

6.4.3. *Some ENFET examples*

A great number of research works have been done on this topic in order to detect molecules, such as glucose and urea. Special attention has been paid to these two species, due to their interest in the biomedical field for monitoring diabetes or during renal dialysis.

The following reactions, involving the production or consumption of protons or concentration variations of ionic species, provide an idea of the diversity offered in the design of an ENFET [MAI 02]:

$$\text{Penicillin} \xrightarrow{\text{Penicillinase}} \text{Penicillinoate} + H^+$$

$$\beta\text{ D-Glucose} + O_2 \xrightarrow{\text{Glucose oxydase}} \text{D-glucono }\delta\text{ lactone} + H_2O_2 \xrightarrow{H_2O} \text{D-Gluconate} + H^+$$

$$\text{Urea} + 2 H_2O + H^+ \xrightarrow{\text{Urease}} HCO_3^- + 2 NH_4^+$$

$$\begin{array}{c}\text{N-acetyl-tyrosine}\\ \text{ethyl ester}\end{array} + H_2O \xrightarrow{\alpha\text{-chymotrypsine}} \text{N-acetyl-L tyrosine acid}$$

4-chlorophenol + O_2 + $2 H^+$ $\xrightarrow[\text{tyrosinase}]{PPox}$ 4-chlorocatechol + H_2O

Thus, in the case of urea, it is possible to detect either the concentration variations of protons using a pH-ISFET, or concentration modifications of ammonium using a ISFET selective for such a species [PIJ 01]. For glucose ENFETs, it is also possible to follow concentration variations of oxygenated water by using a peroxidase [SHU 94]. As a two step-reaction is involved, it is also possible to modulate the characteristics through a co-immobilization of glucose oxidase and of gluconolactonase, the latter accelerating the spontaneous hydrolysis step of gluconolactone to gluconic acid [HAN 90].

The review of Kimura and Kuriyama [KIM 90] constitutes a good approach of methods used for ENFET fabrication. In the scope of the present chapter, only some significant examples are discussed. Table 6.2 gives some examples of devices where, in general, the enzymatic membrane is obtained through simultaneous cross-linking, on the ISFET gate, of the enzyme with an inert protein as bovine albumin or by trapping in a photo-cross-linked membrane. To complete the data given in Table 6.2 the interested reader can find more recent applications in the review of Dzyadevych et al. [DZA 06].

The main limitation of ENFETs based on pH variations measurements, is related to the buffer capacity of the tested medium. This buffer capacity is particularly marked in biological media (blood, serum). In order to reduce such effects, it is possible to insert an additional membrane that modulates the diffusion of the involved species through size and charge effects, allowing a better sensor fitting to the tested sample; this strategy has been especially used for urea and glucose ENFETs [SOL 00, DZA 99, SHU 95].

To conclude this short overview of ENFETs, the possibility to produce enzymatic multi-biosensors [HAN 89] can be mentioned as shown on Figure 6.15. Some devices associating ionic and enzymatic detections have also been studied [MUN 97].

Analyzed compound	Enzyme	Linear range of the ENFET/mM	References
Penicillin G (fermentation)	Penicillinase amidase Penicillinase (β lactamase) Penicillinase acylase Penicillinase	0.3–30 0.2–25 0–5 0.05–1	[BRA 89] [CAR 80] [ZHO 93] [SCH 01]
Xanthine Inosine (fish freshness control)	Xanthine oxidase Inosine-monophosphate deshydrogenase	0.02–0;1	[TAM 88] [GOT 88]
L-lactate (food)	L-lactate deshydrogenase	10^{-4}–10	[KHA 01a]
Formaldehyde (environment)	Alcohol oxidase	10–300	[KOR 97]
Triglycerides (clinical, industry)	Lipase	Triacetine 100–400 Tributyline 3–50 Trioleine 0.6–3	[NAK 86]

Table 6.2. *Characteristics of some ENFETs*

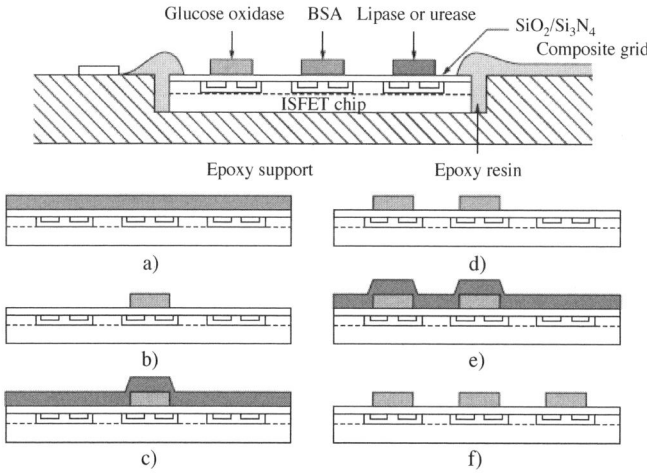

Figure 6.15. *Cross-section of a multi-ENFET showing the fabrication steps: a) deposition of a photo-polymerizable (PVP/BSA) membrane; b) redeveloping and washing; c) deposition of photopolymer containing GOD; d) illumination and developing; e) deposition of the lipase or urease photopolymer; f) revelation of the lipase membrane (from [HAN 89])*

6.4.4. ENFET and inhibition mechanisms

The second application field for the analytical use of ENFET proceeds via inhibition mechanisms, some significant examples can be found in the review by Jaffrezic-Renault [JAF 01]. It concerns the detection of organophosphorous compounds. Such compounds are known to be among the most toxic and are used as pesticides, insecticides, or war agents and can inhibit acylcholinesterases (acetyl-cholinesterase, butyryl-cholinesterase) and acid or alkaline phosphatases. In the case of acetyl-cholinesterase (AchE), the involved reaction can be written as follows:

$$CH_3\text{-}COO(CH_2)_2\text{-}N^+(CH_3)_3 + H_2O \xrightarrow{AchE} CH_3\text{-}COO^- + H^+ + OH\text{-}(CH_2)_2\text{-}N^+(CH_3)_3$$

acetylcholine → acetic acid + choline

Table 6.3 gives an example for paraoxon detection and underlines the lifetime advantages of ENFETs compared with a device based on a classical transducer.

Using inhibition phenomena, among recent possibilities for ENFET sensors, the possibility to quantify glycoalkaloid poisons in food has been reported [KOR 02]. For example, in the case of potatoes, quantitation of the glucose content could lead to a quality criterion of these food products.

Characteristics	Type of enzymatic electrode	
	ISE (butyrylcholine = 6.2×10^{-4} M)	pH-ISFET (butyrylcholine = 5.6×10^{-4} M)
Response time	30 minutes	<5 minutes
Analysis time	60 minutes	<7 minutes
Lifetime	2 days	4 days
Linear regression and correlation coefficient	$y = 1.295 x - 2.777$ (x in µg/l) 0.9876	$y = 1.023 x - 3.684$ (x in µg/l) 0.9927
Linear domain	2.5–40 µg/l	16.6–82 µg/l
Detection limit	1.0 µg/l (ppb)	10 µg/l (ppb)
Exactitude	< 7.1%	< 13.2%

ISE, ion-sensitive electrode; ISFET, ion-sensitive field-effect transistor.

Table 6.3. *Comparison of characteristics of ISE-type enzymatic electrodes and pHFET for paraoxon detection*

6.5. BioFETs

The BioFET acronym should be reserved for qualifying any microsensor based upon a FET (field-effect transistor) transduction and allowing the detection of a biological species or incorporating a biomolecule inside its sensitive membrane. In some cases, the BioFET acronym should also be used to qualify ISFETs applied to the medical field and enabling, for example, a specific ion, such as sodium or potassium, to be monitored in blood. If such a definition of BioFET is accepted, ENFETs devices would occupy a specific position based on their selectivity and biocatalytic potentialities. In fact, due to their specificity and to the importance of ENFETs, the BioFET term is reserved for field-effect devices except for ENFETs.

6.5.1. *Systems based on affinity mechanisms*

For *affinity* biosensors, the interaction between the complexing species and the recognized molecule, which condition the biosensor response, can be written as:

$$L_T + S \underset{k_{-1}}{\overset{k_1}{\rightleftarrows}} L_T S \quad \text{with} \quad K = \frac{[L_T S]}{[L_T].[S]} = \frac{k_1}{k_{-1}}$$

where L_T represents the complexing species, for example, an antibody, fixed onto the transducer surface and S the recognized substance, for example, an antigen, k_1 and k_{-1} being rate constants, respectively, for the direct and reverse reaction. K represents the affinity constant for this equilibrium between the functionalized surface and analyte.

K values vary from 10^3 l/mol for lectins to 10^{15} l/mol for avidin-biotin systems. For a given number of complexing ligand sites L_T, the number of sites $L_T S$ and, in consequence, the signal given by the transducer should be proportional to the analyte concentration. The sensitivity should thus be all the more important, as the value of K is larger. In the case of a FET transducer, this interaction must be accompanied by modification of charge or of electrical capacity in order to be detected.

Antigen/antibodies, DNA/RNA, oligonucleotides/DNA systems should be good candidates for the development of FET transducer-based microbiosensors. In this respect, Blackburn [BLA 87] estimated using a theoretical model of the involved interfaces, that the detection limit of antigens or antibodies should be situated around 10^{-7} and 10^{-1} M, thus justifying the interest of IMFET (immuno field-effect transistors). Such a detection possibility has motivated a lot of discussions, as it seems difficult, on the basis of an immuno-modified FET in static mode, to discriminate the charge of the complex antigen-antibody from the charge of one of the separate constituents, or to consider the various layers as blocking capacities. In addition, this problem is related to the drift problems, to the porous character of the membrane including the immunospecies, which are in the ionic environment of the measuring medium, which screens all dipolar or capacity effects induced by the immunoreaction [BER 91]. However, in such a membrane including proteins, the presence of fixed charges and mobile counter ions can be exploited in a dynamic mode for detection aims. If a strong modification of the ionic concentration is created, a signal perturbation is induced at a pH value different from that corresponding to the isoelectric point of the immobilized protein, which can be related to the fixed charge density in the membrane. This method also enables the determination of the isoelectric point or, for example, its variations during an antibody-antigen interaction [SCH 90].

Another approach of immunodetection using FET devices is more related to the structure of layers of biomolecules and to the understanding of the formation of the immune complex. It consists of analyzing the phenomena involved by using impedance spectroscopy. This method is based upon the determination of the transfer function versus frequency (see Chapter 10). The analysis of such a transfer function allows an estimation of the thickness of the involved proteic layers. This method was first tested on multi-layers, including model proteins such as glucose oxidase; it was then applied to monitoring the formation of the antibody-antigen complex on a dinitrotoluene antibody system [KHA 01b].

Other systems based on affinity mechanisms present similar drawbacks and have led to a small number of works using FET-based transducers. The good example would be the detection of heparin on an ISFET coated with protamine [KER 95].

Some strategies for IMFET elaboration are derived from enzyme-linked immunosorbent assay (ELISA) techniques. In this case the antigen-antibody interaction is detected using an immunoenzymatic test allowing the immunoreaction amplification through an enzymatic label (for example, glucose oxidase (GOD) or urease) able to induce a variation in the pH via biocatalysis, which is detectable on an ISFET, similar to that used for ENFETs. Such a system has been tested in the fields of environmental and biomedical applications for atrazine detection and for an immunoglobulin model [COL 91] or for human IgG determination at levels between 0.1 and 2 mg/ml [SAK 90]. Figure 6 shows the building up of successive layers of immunospecies on an ISFET gate.

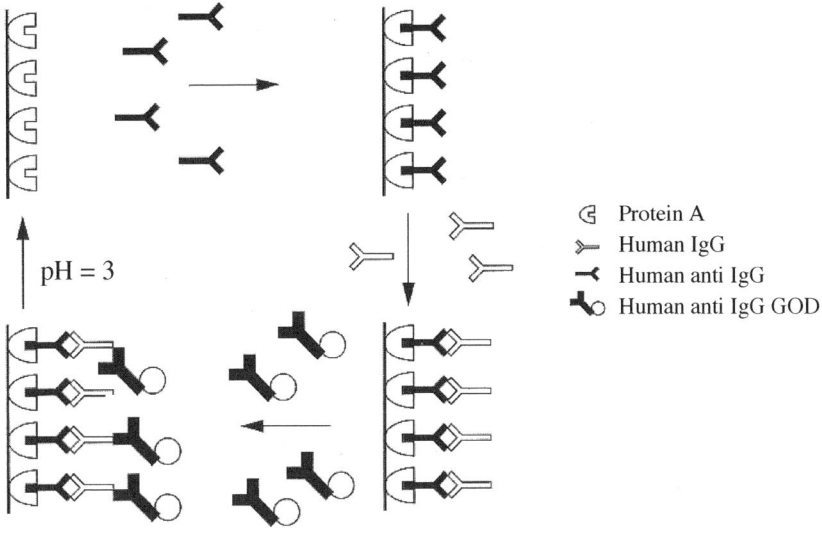

Figure 6.16. *Schematic representation of the sequence of events produced during human immunoglobulin G detection on an ISFET gate (from [COL 91])*

To conclude on this point we will mention the possibility of using protein engineering to build-up the sensing interface of BioFET [ETE 07].

6.5.2. *BioFET based on cells and living organisms*

The most futuristic studies concerning BioFET are well represented by the association of a part or an entire living organism with circuits based on field-effect transistors. On this topic, research using neuronal cells is the most spectacular, even though the neuronal chip still remains a fiction.

These works, linked to the field of bioelectronics, have been continued with neuronal cells of rat brain [VAS 98] and of various mammals brain [OFF 97], and constitute a first step in the development of neuronal biosensors and of neuroelectronic circuits.

Some works, with an analytical aim, are known as microbial FET associating whole cells to ISFET devices. Thus, as early as 1983, microorganisms derived from a plant for water waste treatment have been fixed on a FET in a differential system

196   Chemical and Biological Microsensors

with the view of developing a glucose sensor [HAN 83]. In the same way, an ethanol sensor has been designed using immobilized bacteria producing acetic acid [KAR 88]. The possibilities of using *Gluconobacter oxydans* as a complete cellular receptor have also been explored for developing a xylose FET sensor. This compound is involved, for example, in fermentation processes for conversion to xylitol [RES 96].

A further step for the conception of BioFETs, illustrated in Figure 6.17, was made by the fixation of an entire insect (or of a part) directly on the gate of a field-effect transistor [SCH 99].

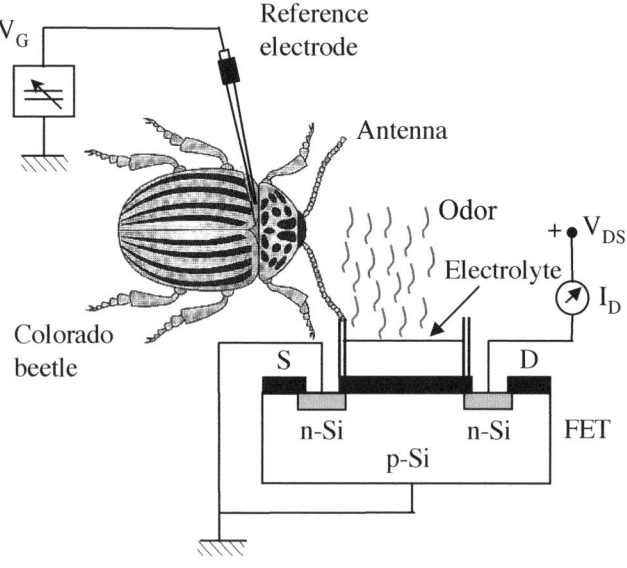

Figure 6.17. *Insect BioFET. Detail of the device showing the connection of the living chemo-receptor with the FET (from [SCH 98])*

Such research constitutes a good approach to a future bioelectronic nose. This concept has been tested on a Colorado beetle for detecting the odor of potato leaves in order to detect insect-attracting aromas (Figure 6.18) [SCH 00]. But, in spite of the high detecting potentialities at a parts per billion level, such devices are limited by their short lifetime.

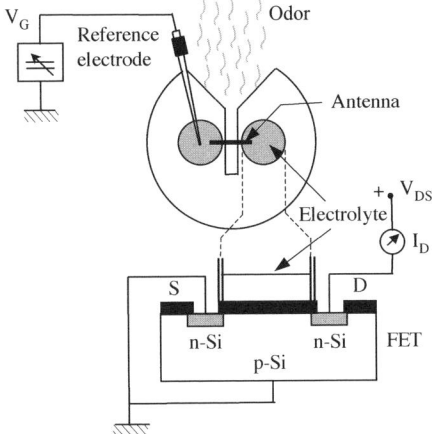

Figure 6.18. *A BioFET for odor detection. Detail showing the device connecting the FET with a reference electrode via an electrolytic solution in connection with the insect antenna (from [SCH 00])*

6.6. Commercial devices

6.6.1. pH ISFETs

Only pH ISFETs have led to a significant commercial development. They are not very present in the traditional physico-chemical analysis in industry or in the laboratory, nevertheless they can be found in the field of applications where glass electrode brittleness constitutes a serious handicap. Laboratory and industry-based examples are the direct control of water or soils, or the analysis of viscous and soft products used in the food industry. ISFETs are also used in the domestic field for water quality control in swimming pools and aquariums. Table 6.4 presents the main manufacturers of ISFETs and lists the specific advantages of the available devices. Currently, around 50 companies in the world manufacture or distribute ISFETs.

Commercialized ISFETs have common characteristics similar to those of glass electrodes with possible pH measurements ranging from 0 to 14 for laboratory devices and from 1 to 13 for pocket systems, with respective reproducibility of ± 0.01 pH unit and ± 0.1 pH unit. Their response time is around 30 seconds. They are operational between 0 and 60°C. A temperature sensor is often integrated, allowing an automatic compensation of temperature effects. The average lifetime is around 3 years with reduced maintenance. The main drawback comes from drift, which, in continuous operation, can be estimate around 0.01 pH unit per hour. Figure 6.19 shows an example of a pocket device with a detail of the sensitive element.

Manufacturers, distributors, websites	Model(s)	Specific advantages
Mettler-Toledo SAS http://fr.mt.com/fr/fr/home.html	InPro 3300	Complies with 3-A sanitary standard, tested in compliance with FDA regulations.
Shindengen Co. http://www.shindengen.co.jp/products_e/ph/index.html	KS 701/KS723	Automatic calibration well-suited for teaching applications
Van London pHoenix Co. http://www.vl-pc.com/	Tuff-Fet™ sensor CyberScan pH 6000 pH FET / ORP / Meter	Perfect for food grade applications The reference can be field rebuildable
Sera France http://www.sera.fr/aquatest.htm	Pocket model	Easy to use for pH measurements in aquariums
RUN Elec http://www.run-elec.fr	pH Marker	Specific technology of encapsulation with silicon nitride
Orion http://www.orion-electrochimie.com	pHuture MMS™	Simultaneous measurement of pH conductivity, oxido/reduction potentials... Sure-Flow® system allows measurements in viscous or contaminated media
IQ Scientific Instruments http://www.phmeters.com/	Broad range from pocket model to laboratory apparatus	Possibility of connection to a pocket computer
Sentron Europe BV http://www.sentron.nl	Broad range of probes with specific geometries	Specific probes for a lot of applications, operating from –5 to 150°C.
Delta Trak http://www.deltatrak.com/isfet-ph-meters.phtml	Probes adapted to solid samples	Automatic compensation of temperature

Table 6.4. *Main ISFETs manufacturers and some commercialized models*

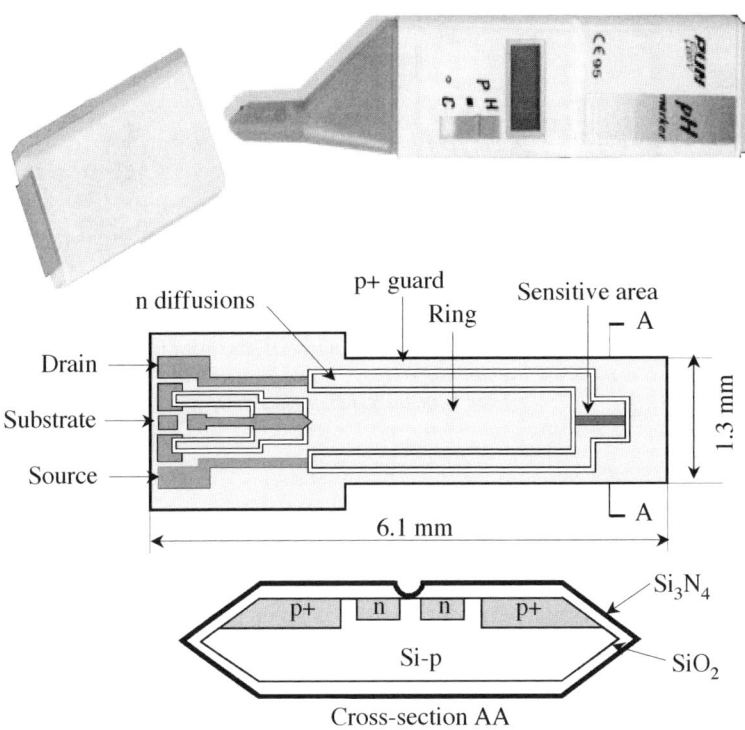

Figure 6.19. *The pocket ISFET meter pH MARKER ^{TM} (French fabrication). Schematic drawing showing the design of the sensitive area with a detail of the cross-section of the substrate and its encapsulation (from RUN Elec® data)*

Some manufacturers can provide pH-FET sensors fitting directly to classical pH meters. It is also possible to find specialized sensors on the market, with specific geometric designs adapted to the analyzed product. Thus, Sentron® has three types of probes on the market: LanceFET™ for solid products, ConeFET™ for viscous materials, and SurFET™ for plane samples. In Figure 6.20 a compact ISFET, with a knife design, allowing meat freshness tests, is shown with its special geometry.

200 Chemical and Biological Microsensors

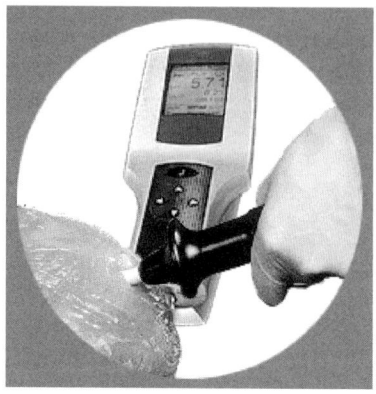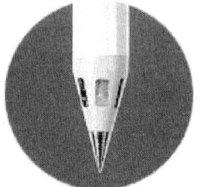

Figure 6.20. *pH measurement on a sample of raw meat with a device using ISFET technology. Detail of the LanceFET™ probe (from Sentron® documentation)*

6.6.2. Multidetection systems

Combining several ISFETs, with gates modified by polymer layers and using a specific signal processing, provides the possibility of developing "electronic tongues" able to discriminate products within a given category. Such devices appear to be quite powerful, especially in the food industry, as illustrated in Figure 6.21 showing unambiguous discrimination between six trademarks of beers.

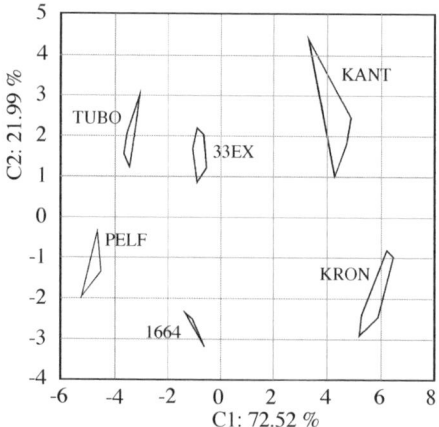

Figure 6.21. *Use of an ISFET-based "electronic tongue" for discrimination of six trademarks of beers from Alpha MOS® data*

Specific advantages of ISFETs (size, low impedance, high signal:noise ratio, fast response, integration possibilities) have not been entirely exploited until now for quantitative multidetection systems. A particularly promising field would be agriculture for soil control. So far, measurements of soil nutriments have been done in Illinois, using a flow analysis system equipped with numerous ISFETs for pH, nitrate, and phosphate. The system remained operational at levels as low as 10^{-5} M [BIR 97] and could include detection of other ions of interest for soil analysis, such as K^+, NH_4^+, Na^+, Ca^{2+}, Mg^{2+}, and Cl^-.

The biomedical field is another area where multidetection systems are required. Some preclinical tests have been performed using a multifunctional ChemFET. Sibbald et al. [SIB 85] have made a simultaneous analysis of pH, K^+, Na^+, and Ca^{2+} in a patient's blood. I-STAT® cartridges, distributed by Abbott, where several parameters can be detected using biosensor technology and advanced microfluidics, serve as a basis for commercial, portable systems.

6.7. Conclusion and perspectives

Since their discovery in the 1970s, ISFET-based devices have lead to a great number of research works in various fields; some have been briefly presented in this chapter. In spite of this academic dynamism, only pH ISFETs have undergone commercial development. Even if the latter have emerged after more than 20 years, during the forthcoming years, ISFET and ENFET-based microsensors should find new areas of application because of the requirements for multidetection systems. Their characteristics (robustness, size, cost, etc) should emphasize their appeal in their potential association with other miniaturized devices as microelectrodes and microconductimeters. Thus, in the medical field, where the need for disposable multisensors is important, as proved by the recent dynamism of DNA biochips, the development of microanalysis modules, integrating ISFETs and allowing simultaneous information on several vital parameters, illustrates one of the privileged sectors of their future development. To conclude this chapter, we will mention work done in the laboratory where the ISFET was born. This approach of a system of integrated sensors for washing processes [LAN 99] gathering domestic and industrial applications is a good illustration of future application of ISFETs in a world more concerned with the problems of quality of life and the environment. An overview of the field of ISFETs over the past 30 years and what may happen in the future has been proposed by the ISFET's inventor [BER 03] and the interested reader can complete the present review with further readings.

Other devices derived from MOSFET structures can serve as a basis for chemical sensors, thus submicron gap suspended-gate FETs, namely SGFET [MOH

08], appear to be quite promising to detect chemical and biologic species with high sensitivity.

The present review was exclusively devoted to silicon-based ISFET and BioFET sensors, but in the future, other FET devices, such as planar field-effect transistor could be used. Thus, a recent paper showing the potential uses of AlGaN planar devices [GUP 08] is quite representative of this trend.

6.8. Bibliography

[ALI 00] BEN ALI M., KALFAT R., SFIHI H., CHOVELON J.M., BEN OUADA H., JAFFREZIC-RENAULT N., "Sensitive cyclodextrin-polysiloxane gel membranes on EIS structure and ISFET for heavy metal ion detection", *Sens. Actuators B*, vol. 62, p. 231-235, 2000.

[ALI 01] BEN ALI M., JAFFREZIC-RENAULT N., MARTELET C., BEN OUADA H., DAVENAS J., CHARBONNIER M., "Characterization of copper ion sensing thiacalix 4 arene films evaporatedon semiconductor substrates", *Mater. Sci. Eng.*, vol. C 14, p. 17-23, 2001.

[BAC 97] BACCAR Z., JAFFREZIC-RENAULT N., MARTELET C., JAFFREZIC H., MAREST G., PLANTIER A., "New fluoride-sensitive membranes prepared through an ion implantation process", *J. Mater. Sci.*, vol. 32, p. 4221-4225, 1997.

[BER 70] BERGVELD P., "Development of an ion-sensitive solid-state device for neurophysiological measurements", *IEEE Trans. Biomed. Eng.*, vol. 17, p. 70–71, 1970.

[BER 72] BERGVELD P., "Development, operation and application of the ion sensitive field effect transistor as a tool for electrophysiology", *IEEE Trans. Biomed. Eng.*, vol. 19, p. 342–351, 1972.

[BER 88] BERGVELD P., SIBBALD A., *Analytical and Biomedical Applications of ISFET*, Elsevier, 1988.

[BER 91] BERGVELD P., "A critical evaluation of direct electrical protein detection methods", *Biosens. Bioelectron.*, vol. 6, p. 55-72, 1991.

[BER 03] BERGVELD P., "Thirty years of ISFETOLOGY. What happened in the past 30 years and may happen in the next 30 years?", *Sens. Actuators B*, vol. 88, p. 1-20, 2003.

[BIR 97] BIRRELL S.J., HUMMEL J.W., "Multi-ISFET sensors for soil nitrate analysis", in *Precision Agriculture '97*, BIOS Scientific Publishers, 1997.

[BLA 82] BLACKBURN G.F., JANATA J., "The suspended mesh ion selective field effect transistor", *J. Electrochem.Soc.*, vol. 129, p. 2580-2584, 1982.

[BLA 87] BLACKBURN G.F., "Chemically sensitive field effect transistors", Chapter 26 in P.F. TURNER, I. KARUBE (eds), *Biosensors Fundamentals and applications*, Oxford Science Publications, 1987.

[BOU 82] BOUSSE L.J., The chemical sensitivity of electrolyte/interface/silicon structures, PhD Thesis, University of Twente, 1982.

[BRA 89] BRAND U., SCHEPER T., SCHÜGERL K., "Penicillin G sensor based on penicillinamidase coupled to a field effect transistor", *Anal. Chim. Acta*, vol. 226, p.87-97, 1989.

[CAM 95] CAMPANELLA L., COLAPICCHIONI C., TOMASSETTI M., BIANCO, DEZZI S., "A new ISFET device for cocaine analysis", *Sens. Actuators B*, vol. 24-25, p. 188-193, 1995.

[CAM 96] CAMPANELLA L., COLAPICCHIONI C., FAVERO G., SAMMARTINO M.P., TOMASETTI, M., "Organophosphorus pesticides (Paraoxon) analysis using solid state sensors", *Sens. Actuators. B*, vol. 33, p. 25-33, 1996.

[CAR 80] CARAS S., JANATA J., "Field effect transistor sensitive to penicilin", *Anal. Chem.*, vol. 52, p.1935-1937, 1980.

[COL 91] COLAPICCHIONI C., BARBARO A., PORCELLI F., GIANNINI I., "Immunoenzymatic assay using CHEMFET devices", *Sens. Actuators B*, vol. 4, p. 245-250, 1991.

[DZA 06] DZADEVYCH S.V., SOLDATKIN A.P., EL'SKAYA A.V., MARTELET C., JAFFREZIC-RENAULT N., "Enzyme biosensors based on ion-selective field-effect transistors", *Anal. Chim. Acta*, vol. 568, p. 248-258, 2006.

[ELB 99] ELBHIRI Z., CHOVELON J.M., JAFFREZIC-RENAULT N., CHEVALIER Y., "Chemically grafted field effect transistors for the detection of potassium ions", *Sens. Actuators B*, vol. 58, p. 491-496, 1999.

[ESA 78] ESASHI M., MATSUO T., "Integrated micro multi ion sensor using field effect semiconductor", *IEEE Trans. Biomed. Eng.*, vol. 25, p. 184-192, 1978.

[ETE 08] ETESHOLA E., KEENER M.T., ELIAS M., SHAPIRO J., BRILLSON L.J., BHUSHAN B., CRAIG LEE S., "Engineering functional protein interfaces for immunologically modified field effect transistor (ImmunoFET) by molecular genetic means", *J. R. Soc. Interface*, vol. 5, p. 123-127, 2008.

[GEU 91] GEUN S.C., DONG L., MEYERHOFF M.E., CANTOR H.C., MIDGLEY A.R., GOLDBERG H.D., BROWN R.B., "Electrochemical performance, biocompatibility and adhesion of new polymer matrices for solid-state ion sensors", *Anal. Chem.*, vol. 63, p. 1666-1672, 1991.

[GOT 88] GOTOH M., TAMIYA E., SHIMIZU I., KARUBE I., "Inosine sensor based on an amorphous silicon field-effect transistor", *Anal. Lett*, vol. 21, p. 1783-1800, 1988.

[GRA 98] GRATTAROLA M., MASSOBRIO G., *Bioelectronics Handbook: MOSFETs, Biosensors, Neurons*, McGraw-Hill, 1998.

[GUP 08] GUPTA S., ELIAS M., XUEJIN W., SHAPIRO J., BRILLSON L., WU L., CRAIG LEE S., "Detection of clinically relevant levels of protein analyte under physiologic buffer using planar field effect transistors" *Biosens. Bioelectron.*, vol. 24, p. 505-511, 2008.

[HAN 83] HANAZATO Y., SHIONO S., "Bioelectrode using two hydrogen ion-sensitive field effect transistors and a platinum wire pseudo reference electrode", *Anal. Chem. Symp. Series*, vol. 17, p. 513-518, 1983.

[HAN 89] HANAZATO Y., NAKAKO M., SHIONO S., MAEDA M., "Integrated multi-biosensors based on an ion-sensitive field-effect transistor using photolithographic techniques", *IEEE Trans Electron Devices*, vol. 36, p. 1303-1310, 1989.

[HAN 90] HANAZATO Y., SHIONO S., MAEDA M., "Response characteristics of the glucosesensitive field effect transistor. Computer simulation of the effect of gluconolactonase coimobilization in a glucose oxidase membrane", *Anal. Chim. Acta*, vol. 231, p. 213-220, 1990.

[ITO 88] ITO T., INAGAKI H., IGARASHI I., "ISFETs with ion-sensitive membranes fabricated by ion implantation", *IEEE Transactions on Electron Devices*, vol. 35, p. 56-64, 1988.

[JAF 92] JAFFREZIC-RENAULT N., CHOVELON J.M., PERROT H., LE PERCHEC P., CHEVALIER Y., "Ion-sensitive field effect transistor sensors with a covalently bonded monolayer membrane: example of calcium detection", *Sens. Actuators B*, vol. 5, p. 67-70, 1992.

[JAF 01] JAFFREZIC-RENAULT N., "New trends in biosensors for organophosphorus pesticides", *Sensors*, vol. 1, p. 60-74, 2001.

[JAN 80] JANATA J., HUBER R.J., "Chemically Sensitive Field Effect Transistors", in H. FREISER (ed.), *Ion Selective Electrodes in Analytical Chemistry*, vol. 2, Plenum Press, p. 107, 1980

[KAR 88] KARUBE I., TAMIYA E., SODE K., YOKOYAMA K., "Application of microbiological sensors in fermentation processes", *Anal. Chim. Acta*, vol. 213, p. 69-77, 1988.

[KAW 84] KAWAKAMI S., AKIYAMA T., UJIHIRA Y., "Potassium ion-sensitive field effect transistor using valinomycin doped photoresist membrane", *Frezenius Z. Anal. Chem.*, vol. 318, p. 349-351, 1984.

[KER 95] VAN KERKHOF J.C., BERGVELD P., SCHASFOORT R.B.M., "The ISFET based heparin sensor with a monolayer of protamine as affinity ligand", *Biosens. Bioelectron.*, vol. 10, p. 269-282, 1995.

[KHA 00] KHARITONOV A.B., ZAYATS M., LICHTENSTEIN A., KATZ E., WILLNER I., "Enzyme monolayer-functionalized field-effect transistors for biosensor applications", *Sens. Actuators B*, vol. 70, p. 222-231, 2000.

[KHA 01a] KHARITONOV A.B., ZAYATS M., ALFONTA L., KATZ E., WILLNER I., "A novel ISFET-based NAD^+-dependent enzyme sensor for lactate", *Sens. Actuators B*, vol. 76, p. 203-210, 2001.

[KHA 01b] KHARITONOV A.B., WASSERMAN J., KATZ E., WILLNER I., "The use of impedance spectroscopy for the characterization of protein-modified ISFET devices: application of the method for the analysis of biorecognition processes", *J. Phys. Chem. B*, vol. 105, p. 4205-4213, 2001.

[KIM 90] KIMURA J., KURIYAMA T., "FET biosensors", *J. Biotechnol.*, vol. 15, p. 239-254, 1990.

[KIM 97] KIMURA K., SUNAGAWA T., YOKOYAMA M., "Applications of sol-gel-derived membranes to neutral carrier-type ion sensitive field effect transistors", *Anal. Chem.*, vol. 69, p. 2379-2383, 1997.

[KLE 87] KLEITZ M., MILLION-BRODAS J.F., FABRY P., "New compounds for ISFETs", *Solid State Ionics*, vol. 22, p. 295-299, 1987.

[KOL 06] KOLL M., "A new preprocessing circuit for ISFET", *Meas. Sci. Rev.* vol. 6, p. 39-43, 2006.

[KOR 97] KORPAN Y., SOLDATKIN A., GONCHAR M., SINIRNY A., GIBSON T., EL'SKAYA A., "A novel enzyme biosensor for formaldehyde based on pH-sensitive field effect transistors", *J. Chem. Tech. Biotechnol.*, vol. 68, p. 209-213, 1997.

[KOR 02] KORPAN Y.I., VOLOTOVSKY V.V., MARTELET C., JAFFREZIC-RENAULT N., NAZARENKO E.A., EL'SKAYA A.V., SOLDATKIN A.P., "A novel enzyme biosensor for steroidal glycoalkaloids detection based on pH-sensitive field effect transistors", *Bioelectrochemistry*, vol. 55, p. 9-11, 2002.

[LAN 99] LANGEREIS G.R., An integrated sensor system for monitoring washing processes, PhD Thesis, University of Twente, 1999,

[LUT 97] LUTENBERG R.J.W., EGBERINK R.J.M., ENGBERSEN J.F.J., REINHOUDT D.N., "Pb^{2+} and Cd^{2+} selective chemically modified field effect transistors based on thioamide functionalized 1,3-alternate calix[4]arenes", *J. Chem. Soc. Perkin Trans.*, vol. 2, p. 1353-1357, 1997.

[MAD 89] MADOU M.J., MORRISON S.R., *Chemical Sensing with Solid State Devices*, Academic Press, 1989.

[MAI 02] MAI ANH T., DZYADEVYCH S.V., SOLDATKIN A.P., DUC CHIEN N., JAFFREZIC-RENAULT N., CHOVELON J.-M., "Development of tyrosinase biosensor based on pHsensitive field-effect transistors for phenols determination in water solutions", *Talanta*, vol. 56, p. 627-634, 2002.

[MAT 86] MATSUO T., NAKAJIMA H., "Parylene-gate ISFET and chemical modification of its surface with crown ether compounds", *Sens. Actuators*, vol. 9, p. 115-123, 1986.

[MLI 98] MLIKA R., BEN OUADA H., JAFFREZIC-RENAULT N., LAMARTINE R., GAMOUDI M., GUILLAUD G., "Study of an ion selective evaporated calixarene film used as sensitive membrane in ISFET sensors", *Sens. Actuators*, vol. 47, p. 43-47, 1998.

[MOH 08] MOHAMMED-BRAHIM T., SALAÜN A.-C., LE BIHAN F., "SGFET as charge sensor: application to chemical and biological species detection", *Sens. Transducers J.*, vol. 90, Special Issue, p. 11-26, 2008.

[MOR 88] Moritz W., Meierhofer I., Muller L., "Fluoride sensitive membrane for ISFETs", *Sens. Actuators*, vol. 15, p. 211-219, 1988.

[MUN 97] MUNOZ J., JIMENEZ C., BRATOV A., BARTROLI J., ALEGRET S., DOMINGEZ C.,"Photosensitive polyurethanes applied to the development of CHEMFET and ENFET devices for biomedical sensing", *Bios. Bioelectron.*, vol. 12, p. 577-585, 1997.

[NAK 86] NAKAKO M., HANAZATO Y., MAEDA M., SHIONO S., "Neutral lipid enzyme electrode based on ISFETs", *Anal. Chim. Acta*, vol. 185, p. 179-185, 1986.

[OFF 97] OFFENHÄUSSER A., SPRÖSSELER C., MATSUZAWA M., KNOLL W., "Field-effect transistor array for monitoring electrical activity from mammalian neurons in culture", *Biosens. Bioelectron.*, vol. 12, p. 819-826, 1997.

[PER 89] PERROT H., JAFFREZIC-RENAULT N., DE ROOIJ N.F., VAN DEN VLEKKERT H.H., "Ionic detection using differential measurement between an ion-sensitive FET and a reference FET", *Sens. Actuators*, vol. 20, p. 293-299, 1989.

[PER 90] PERROT H., JAFFREZIC-RENAULT N., CLECHET P., WLODARSKI W.B., DE ROOIJ N.F., VAN DEN VLEKKERT H.H., "A generalized theory of an Ag^+ sensitive electrolyte-insulatorsemiconductor field-effect transistor with silica surface modified by chemical grafting", *Sens. Actuators B*, vol. 1, p. 380-384, 1990.

[PHA 84] PHAM M.T., HOFFMANN W., "Ion-sensitive membranes fabricated by the ion-beam technique", *Sens. Actuators*, vol. 5, p. 217-228, 1984.

[PIJ 01] PIJANOWSKA D., JAFFREZIC-RENAULT N., MARTELET C., TORBICZ W., "ISFET type urea biosensors for biomedical applications", *Recent Res. Devel. Electroanal. Chem.*, vol. 3, p. 7-20, 2001.

[RES 96] RESHETILOV A.N., DONOVA M.V., DOVBNYA D.V., BORONIN A.M., LEATHERS T.D., GREENE R.V., "FET-microbial sensor for xylose detection based on *Gluconobacter oxydans* cells", *Biosens. Bioelectron.*, vol. 11, p. 401-408, 1996.

[ROC 92] ROCHER V., JAFFREZIC-RENAULT N., PERROT H., CHEVALIER Y., LE PERCHEC P., "Nitrate-sensitive field effect transistor with silica gate insulator modified by chemical grafting", *Anal. Chim. Acta*, vol. 256, p. 251, 1992.

[SAN 82] SANADA Y., AKIYAMA H., UJIHIRA Y., NIKI E., "Preparation of Na^+-selective electrodes by ion-implantation of lithium and silicon into single-crystal alumina wafer and its application to the production of ISFET", *Frezenius' Z. Anal. Chem.*, vol. 312, p. 526-529, 1982.

[SAK 90] SAKAI H., KANEKI N., HARA H., ITO K., "Availability and development of an enzyme immunosensor based on an ISFET for human immunoglobulins", *Anal. chim. Acta*, vol. 220, p. 189-193, 1990.

[SCH 90] SCHASFOORT R.B.M., KOOYMAN R.P.H., BERGVELD P., GREVE J., "A new approach to immunoFET operation", *Biosensors Bioelectron.*, vol. 5, p. 103–124, 1990.

[SCH 98] SCHÖNING M.J., SCHÜTZ S., SCHROTH P., WEIßBECKER B., STEFFEN A., KORDOS P., LÜTH H., HUMMEL H. E., "A BioFET on the basis of intact insect antennae", *Sens. Actuators B*, vol. 47, p. 234-237, 1998.

[SCH 99] SCHROTH P., SCHÖNING M.J., SCHÜTZ S., MALKOC Ü., STEFFEN A., MARSO M., HUMMEL H.E., KORDOS P., LÜTH H., "Coupling of insect antennae to field-effect transistors for biochemical sensing", *Electrochim. Acta*, vol. 44, p. 3821-3826, 1999.

[SCH 00] SCHÜTZ S., SCHÖNING M.J., SCHROTH P., WEIßBECKER B., KORDOS P., LÜTH H., HUMMEL H.E., "An insect-based BioFET as a bioelectronic nose", *Sens. Actuators B*, vol. 65, p. 291-295, 2000.

[SCH 01] SCHÖNING M.J., POGHOSSIAN A., SCHROTH P., SIMONIS A., LÜTH H., "An ISFET-based penicillin sensor with high sensitivity, low detection limit and long lifetime", *Sens. Actuators B*, vol. 76, p. 519-526, 2001.

[SHU 94] SHUL'GA A.A., KUDELA-HEP M., DE ROOIJ N.F., NETCHIPOROUK L.I.,"Glucose sensitive enzyme field effect transistor using potassium ferricyanide as an oxidizing substrate", *Anal. Chem.*, vol. 66, p. 205-210, 1994.

[SHU 95] SHUL'GA A., KOUDELKA-HEP M., DE ROOIJ N., NETCHIPOROUK L., "Glucose sensitive ENFET using potassium ferricyanide as an oxidizing subtrate: the effect of an additional lysozyme membrane", *Sens. Actuators B*, vol. 24-25, p. 117-120, 1995.

[SHU 96] SHUL'GA A.A., NETCHIPOROUK L.I., SANDROVSKY A.K., ABALOV A.A., FROLOV O.S., KONONENKO YU.G., MAUPAS H., MARTELET C., "Operation of an ISFET with non-insulated substrate directly exposed to the solution", *Sens. Actuators B*, vol. 30, p. 101-105, 1996.

[SIB 85] SIBBALD A., COVINGTON A.K., CARTER R.F., "Online patient-monitoring system for the simultaneous analysis of blood K^+, Ca^{2+}, Na^+ and pH using a quadruple-function ChemFET integrated-circuit sensor", *Med. Biol. Eng. Comput.*, vol. 23, p. 329-338, 1985.

[SMI 86] SMITH R., SCOTT D.C., "An Integrated Sensor for Electrochemical Measurements", *IEEE Trans. Biomed. Eng.*, vol. 33, p. 83-90, 1986.

[SOL 00] SOLDATKIN A.P., VOLOTOVSKY V., EL'SKAYA A.V., JAFFREZIC-RENAULT N., MARTELET C., "Improvement of urease based biosensor characteristics using additional layers of charged polymers", *Anal. Chim. Acta*, vol. 403, p. 25-29, 2000.

[TAM 88] TAMIYA E., SEKI A., KARUBE I., GOTOH M. SHIMIZU I., "Hypoxanthine sensor based on an amorphous silicon field-effect transistor", *Anal. Chim. Acta*, vol. 217, p. 301-305, 1988.

[TOM 95] TOMASSETTI SU.Y., SAMMARTINO M.P., CRESCENTINI G., CAMPANELLA L., "A new salicylate ISFET for the determination of salicylic and acetylsalicylic acid in drugs", *J. Pharm. Biomed. Anal.*, vol. 13, p. 449-457, 1995.

[VAN 91] VAN DEN BERG A., GRISEL A., VERNEY-NORBERG E., "An ISFET-based calcium sensor using a photopolymerized polysiloxane membrane", *Sens. Actuators. B*, vol. 4, p. 235-238, 1991.

[VAN 97] VANIFATOVA N.G., ISAKOVA N.V., KOLYCHEVA N.V., MYASOEDOV B.F., NAD' V.YU., OTMAKHOVA O.A., PETRUKHIN O.M., PLATE N.A., SPIVAKOV B.YU., TAL'ROZE R.V., "Ion-selective field effect transistors with plasticized membranes: nitrate ion sensor", *J. Anal. Chem.*, vol. 52, p. 52-55, 1997, translation of *Zhurnal of Analitichesko Khimii*, vol. 52 p. 62-65, 1997.

[VAS 98] VASSANELLI S., FROMHERZ P., "Transistor records of excitable neurons from rat brain", *Appl. Phys. A*, vol. 66, p. 459-463, 1998.

[VLE 88] VAN DEN VLEKKERT H.H., DE ROOIJ N.F., "Design, fabrication and characterization of pH-sensitive ISFETs", *Analusis*, vol. 16(2), p. 110, 1988

[WAK 86] WAKIDA S., TANAKA T., KAWAHARA A., YAMANE M., HIIRO K., "Methods for the preparation and characterization of a selenocyanate ion-sensitive field effect transistor with Urushi as the membrane matrix", *Analyst*, vol. 111, p. 795-797, 1986.

[WAN 08] WANG C., ZHAO Y., "A novel current-mode readout circuit for ISFET sensor", *APCCAS, IEEE Asia Pacific Conference on Circuits and Systems* 2008, p. 407-410, 2008.

[WON 89] WONG H.S., HU Y., WHITE M.H., "The ion sensitivity of boron implanted silicon nitride chemical sensors", *J. Electrochem. Soc.*, vol. 136, p. 2968-2972, 1989.

[ZHO 93] ZHONG LI-CHAN., LI GAO-XIANG., "Biosensor based on ISFET for penicillin determination", *Sens. Actuators B*, vol. 13-14, p. 570-571, 1993.

Chapter 7

Biosensors and Chemical Sensors Based Upon Guided Optics

7.1. Introduction

Any complex measuring and controlling system comprises sensors. These sensors ensure acquisition of external or internal parameters (to the system) and, therefore, the monitoring and servo-control of this system. Metrologically, the potentialities of optics are important if the medium is sufficiently transparent [AND 91]. Optical fibers and integrated optics provide interesting possibilities.

Optical fibers sensors appeared in 1980 and in addition to classical optical sensors, present their own characteristics, with some advantages: immunity to electromagnetic fields, electrical insensitivity, possibility of a broad bandwidth, low weight, small dimensions, and very versatile geometrical configurations, high sensitivity, and the ability to function at high temperatures and in aggressive or explosive media. The sensors based on integrated optics appeared a few years later.

In optical telecommunications for transmitting signals over long distances, efforts were made to manufacture optical fibers that were insensitive to phenomena liable to cause transmission perturbations. Thus, a thorough knowledge of the sensitivity of fibers to environing parameters has been gained and is used for the fabrication of sensors, which, therefore, arose as a consequence of the research into optical telecommunications.

Chapter written in French by Jean-Pierre GOURE and Loïc BLUM.

Numerous integrated optics or optical fiber sensors for physical, chemical, or biological parameters have been studied or achieved since 1980. All the references cannot be cited, but some synthesis documents are proposed in the bibliography [PIT 85, GOU 90, DAK 89, FER 92, WOL 91, JON 87]. The possibilities offered are wide and the instrumentation needs are huge: robotics, automated production, aeronautics, the car industry, etc. Evolution towards compact systems can only be achieved by using specific fiber components: couplers, etc. [GOU 89, GOU 01] or integrated optics.

Integrated optics allows the reproduction of the quasi-totality of the classical optical components in a few micrometers-thick multilayer structures. It offers the advantage of confining light to the two directions perpendicular to propagation, thus achieving a light microguide similar to an optical fiber. Physical sensors (for instance, proximeters) or chemical sensors [VAL 90] could thus be fabricated. The supports used are silicon, glass, and lithium niobate.

Finally, the industrial devices necessitate linking a set of sensors to a single measuring station. Systems able to interrogate several sensors via only one fiber appeared recently.

Presently, the optical fiber or integrated optical microsensor market has not yet been fully explored. However, in a few particular niche markets, industrial developments are in progress.

7.2. Definitions

7.2.1. *Luminous wave*

An optical wave is an electromagnetic wave comprising an electrical field, E, and a magnetic field, H, which are perpendicular to each other and to the z propagation direction (direction of the speed C in vacuum or v in an isotropic medium). In fact, a scalar representation, rather than a vectorial representation, is used. The optical vibration corresponds to the electric field. When the wave is not polarized, it can generally be written as:

$$S = a\, e^{-\alpha z}\, e^{i(\omega t - \varphi)} \qquad [7.1]$$

where φ is the phase [$\varphi = (2\pi/\lambda)$], n is the refractive index ($n = C/v$), α the absorption coefficient, z the coordinate along the direction of the wave propagation, a is an amplitude term, $\omega = 2\pi/T$ is the pulsation, $\lambda = vT$ the wavelength ($\lambda o = CT$ in vacuum). The wave intensity is given by the product: $I = a^2 e^{-2\alpha z}$. This quantity is proportional to the energy received by a detector or by human eye.

In the case of natural light, the electric field perpendicular to the luminous ray is not restricted in direction. Conversely, after crossing certain media (polarizers) the electric field may have a privileged direction (linear polarization) or have its extremity rotating round the propagation axis (elliptic vibration). If the medium is birefringent, two waves are created, the ordinary wave with the n_o index and the extraordinary wave, with n_e index along the propagation direction with respect to that of the optical axis of the medium. Some media, such as quartz, become birefringent when they are submitted to a mechanical stress.

7.2.2. Optical fibers

An optical fiber is made of a transparent cylindrical core with a n_c index surrounded by an optical transparent sheath with an index $n_s < n_c$ (see Figure 7.1).

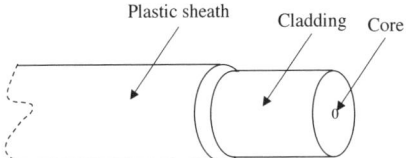

Figure 7.1. *Principle of step index optical fibers*

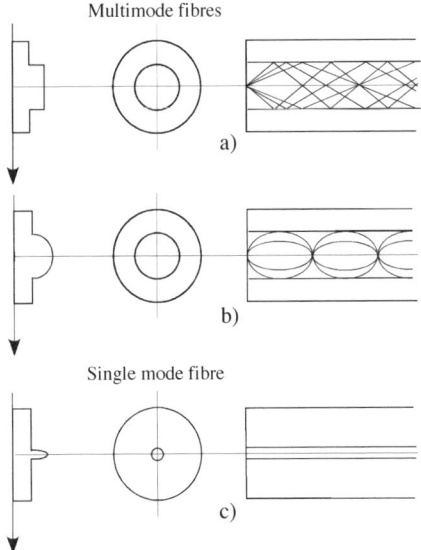

Figure 7.2. *Classical optical fibers. a) Step index multimode fiber, b) graded index fiber, c) single-mode fiber*

The optical fibers can be classified into three broad categories (Figure 7.2):

– step *index multimode fibers*. The core index is constant, and a luminous ray entering the fiber has straight-line propagation inside it. It is reflected each time it reaches the core-sheath interface, according to the Snell-Descartes laws;

– *graded index* multimode *fibers*. The core index varies as a function of the distance from the fiber axis, and no straight-line propagation of the light ray occurs. They carry several modes of the electromagnetic field;

– *single-mode fibers*. The core diameter is very small as the possible diameters lie between 3 and 10 µm. Only one mode is transmitted.

– *special fibers:* polarization-maintaining single-mode fibers or rare-earth-doped fibers (see Figure 7.3).

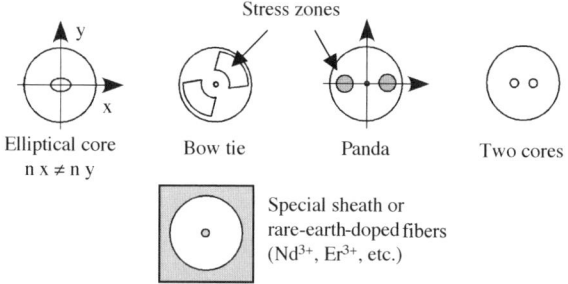

Figure 7.3. *Special fibers*

7.2.3. Planar guides

Integrated optics allows the elaboration of optical sensors based on the same physical principles as those described in the preceding section with the fibers using the same modulations (except for the phenomena involving long distances as in the case of continuous sensitivity fibers). In the multilayer structures of integrated optics, complex optical components, such as interferometers, have been achieved. The main substrates used are silicon and glass, but studies have also been undertaken on lithium niobate. One of the advantages of integrated optics, with technologies derived from those of microelectronics, is to enable the fabrication of a high number of components (collective fabrication) on the same plate, which decreases the production costs. The simultaneous use of fibers and integrated optics is certainly one of the methods in development; however, a solution for connecting guides to fibers is required. Another advantage of integrated optics is to achieve a first treatment of the signal on the substrate.

7.3. Principles of optical microsensors

7.3.1. *Definition*

A sensor is a system, which delivers a signal under the action of the physical parameter to be measured. The signal contains information on this physical parameter and transmits this signal directly or indirectly. In an optical sensor, the transmitted signal is a luminous wave. An optical fiber(s) sensor is a device comprising one or several fibers which enable information to be gathered that is representative of the measured parameters, without any other energy supply than that of the observed phenomena and/or of the light waves circulating in the fibers (definition NF C 93-800, September 1991). This definition precludes assemblies comprising a conventional sensor and an optical fiber transmission system. So, an optical fiber sensor is called "extrinsic" when the characteristics of light are modified by the measured parameter outside the optical fiber(s). It is not a true fiber sensor. The schematic drawing of an optical sensor is presented in Figure 7.4.

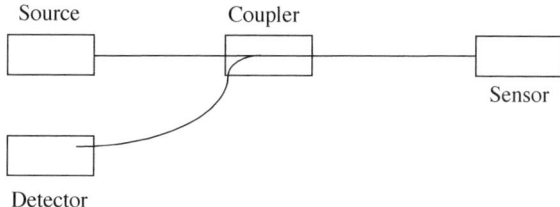

Figure 7.4. *Sensor with intensity modulation. The intensity variation may be due to absorption within the cell or to fluorescence (extrinsic sensor)*

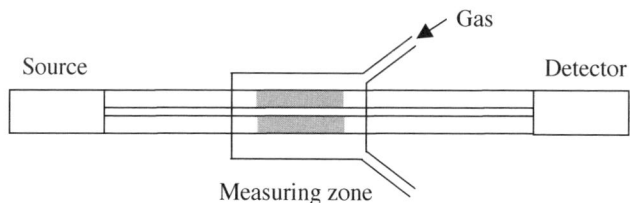

Figure 7.5. *Sensor with intensity modulation. The intensity variation is due to the modified propagation in the measuring zone where the optical sheath is replaced with a specific sheath that absorbs the gas to be detected (evanescent wave). Intrinsic sensor*

An optical fiber sensor will be called "intrinsic" when the sensitive element is constituted by one or several optical fibers having one or several characteristics of light transmission, reflection, or emission, which depend on the measured parameter. A sensor is punctual when the physical or chemical effect is measured at the extremity of the fiber. If the measurement involves a certain length, the sensor is called "continuous" (and the measurement is spread). The measurement is called distributed if it is localized on various discrete points of the optical fiber. It is called integrated if the sensor response arises from the whole ensemble of the effects over the optical fiber. One of the outstanding problems is to discriminate the parameters by a differential method, that complicate the apparatus.

7.3.2. Modulation of the optical signal

Generally speaking, optical technologies are distinguishable from the others by two essential aspects: sensitivity (photons can be counted) and the spatial, spectral, time and orientational resolutions, which enable the exploration of the space-time-frequency. The remarkable improvements of photodetectors (photomultipliers, photodiode barrettes, charge transfer devices) are exploited.

The measurement techniques used are generally based upon light-matter interaction or on fundamental phenomena, such as interference, diffraction, etc. In the case of interactions with no frequency change, the measured quantities are either optical parameters (absorption coefficient, wavelength, reflectance, refractivity index), or modifications of the properties of light beams (diffusion, birefringence, etc.), or spatial parameters (distance, diameter, etc.). For example, applications utilizing measurements of width and particle size using lasers. Absorption spectroscopy systems are widely commercialized.

Strong matter-light interaction is a wide field: the domains of Raman and of molecular emission spectroscopies. The field of fluorescence imaging in fluid mechanics is growing. Moreover, due to the very high sensitivity of the physical phototechniques, numerous biological determination devices have been commercialized. The molecular probes have the potential to be used as environmental probes. Substances of this type can be used in a medium to be characterized or placed at one extremity of a fiber (optodes).

Modulation of	Physical phenomena
Intensity	Absorption
	Diffusion
	Modification of propagation
	Fluorescence
Phase	Variation of length
	Variation of index
Polarization	Rotation
	Birefringence
Frequency	Raman
	Doppler

Table 7.1. *Physical principles of optical sensors*

In order to convert optical guides to produce a sensor (fibers or guides in integrated optics), we can provoke various modulations on an optical wave:

– intensity modulation,

– phase modulation,

– wavelength modulation,

– polarization modulation, and use other various physical phenomena.

7.3.2.1. Intensity-modulation sensors

Intensity modulation is the simplest method. It necessitates a source, a propagation medium, a measuring zone (cuvette, etc.), and a detector. The intensity modulation sensors, which began to be developed many years ago, offer the advantages of simplicity, safety, and low cost. They comprise a source, a detector, a fiber, and possibly a coupler (Figures 7.4 and 7.5). However, to obtain good stability, the problems of source intensity variation must be corrected. Modulation by losses due to core-sheath interface, fluorescence, or luminescence is mainly used.

In the case of losses at the core-sheath interface, the sensor is made of a core coated with a layer that absorbs the target gas and is placed in the analyzed atmosphere, or with a specific sheath, which is sensitive to the analyzed gas. The resulting variation of the index of the sheath modifies the light propagation. This device enables measurement of the concentration of ammonia (NH_3), methane (CH_4), hydrogen (H_2), and other gases, such as benzene, acetone, etc. Propagation modifications caused by surface plasmons can also be used.

Numerous extrinsic sensors based on fluorescence extinction exist for chemistry: a first fiber carries the excitating light to an indicator, and a second fiber recaptures the emitted fluorescence. In the case of an oxygen sensor, the indicator is dibutylperylen, the excitating light is blue, and the emitted light is green. In the case of the pH detector, the indicator is fluoresceinamine; the source is a white lamp with an emission filter at 520 μm.

7.3.2.2. Phase-modulation sensors

Phase modulation, $\varphi = (2\pi n z)/\lambda$, is based upon the principles of interferometry. Two coherent waves E_1 and E_2 are superimposed on a given point, the resulting wave is represented by par $E_1 + E_2$ and the corresponding intensity by $I = 2 I_0 [I + \cos(\varphi(t))]$ where $\varphi(t)$ is the phase difference between wave 2 and wave 1. The intensity is thus modulated. This necessitates the use of an interferometer (Figure 7.6).

The optical fiber interferometric sensors based on phase modulation are very sensitive and have good dynamics. Hydrophones, magnetometers, accelerometers, stress gauges, thermometers have thus been fabricated. The phase variation may for example be due to a temperature shift, which causes the refractory index and the dimensions of a silicon monomodal fiber to change. A phase variation of about 17 fringes per meter and degree is obtained; this sensor is thus extremely temperature-sensitive.

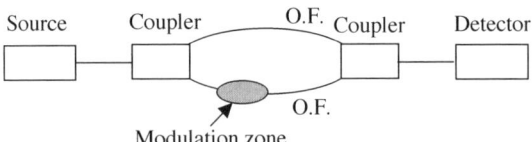

Figure 7.6. *Mac Zenhder interferometer. The phase $\varphi = (2\pi/\lambda_0) n L$ can be modified by the variation of a parameter which changes either the length, L, of the modulation zone or the value of the index, n, in this zone*

A chemical device sensitive to hydrogen is obtained by coating an optical fiber with a palladium layer by cathodic pulverization. The fiber is one of the arms of an interferometer. In the presence of hydrogen, palladium hydride is formed. This product causes axial and longitudinal stresses that modify the phase of the wave. The effects of stresses (traction, pressure) and curvature may induce the presence of two indexes (ordinary and extraordinary), which result in two different propagation speeds, thus creating a phase difference given by $\Delta\varphi = (n_0-n_1) 2\pi L/\lambda_0$.

In polarimetric assemblies, the two arms of the interferometer are constituted from the two polarized modes of a given monomodal fiber, and the phase difference is deduced from the resulting polarization state.

7.3.2.3. *Polarization-modulation sensors*

Polarization modulation is a characteristic of birefringence changes and, therefore, of variations of wave ellipticity. In polarimetric assemblies, we measure the phase difference deduced from the polarization state of two polarized modes of a given arm. In certain assemblies, we measure the polarization rotation as a function of the influence quantity. The optical polarization properties of a fiber can be influenced by the action of an electric or magnetic field. Two main effects illustrate this phenomenon: when an electric field is applied to an optical medium, the material becomes linearly birefringent. The phase retardation introduced is proportional to the electric field (Pockels effect) or to the square of the field (Kerr effect).

When a fiber is strongly illuminated by a coherent light, we observe a granularity at the fiber output due to intermodal interferences (different light paths), which leads to the presence of a speckle figure in remote field. The effects of stresses imposed to the fiber (vibrations, temperature, etc.) induce a modification of the propagation constants contained in the phase terms. This phenomenon can be used for localizing an imposed stress (presence sensor), but it may also be a source of noise.

7.3.2.4. *Wavelength-modification sensors*

Wavelength modulation corresponds to variations of the spectral position or of spectral width. An apparatus using wavelength modulation thus comprises a source, spectrometer elements (prisms, networks, and filters), a detector, and a signal processing system. An example of wavelength modulation is given by a temperature measuring system. Light pulses from a laser are coupled in an optical fiber. Part of the diffused light at a point, z, in the core of the fiber corresponds to the numeric aperture and propagates towards the photodetector (retrodiffusion). This diffused light is not monochromatic. It comprises three rays according to the diffusion process to which the molecule in the core of the fiber is submitted: Stokes and anti-Stokes rays with an intensity ratio depending on the temperature (Raman effect). We thus obtain a measurement, as a function of point z where diffusion occurs.

7.3.2.5. *Other interrogation techniques*

Among the new interrogation and measurement techniques used, let us also mention the Bragg networks [KAS 99], the techniques of reflectometry, and coherence modulation.

7.3.3. Techniques

7.3.3.1. Interferometry

Phase variation may be observed by usual interferometric techniques. In the case of gas sensors, we use two wave interferometers: Mach Zehnder and Michelson. The device (Figure 7.6) starting from an initial wave, produces two coherent waves, S_1 and S_2, represented by

$$S_1 = S_{01} e^{i\omega t} \quad \text{and} \quad S_2 = S_{02} e^{[i\omega t - \varphi(t)]}$$

where $\varphi(t)$ is the phase difference between wave S_1 and wave S_2. The resulting wave at a spatial point r is given by $S = S_1 + S_2$ with the corresponding intensity $I = S\,S^*$ where the symbol * represents the conjugated complex number:

$$I = S_1 S_1^* + S_2 S_2^* + S_1 S_2^* + S_1^* S_2$$

$$I = I_1 + I_2 + I_1 I_2 \,(e^{i\varphi(t)} + e^{-i\varphi(t)})$$

In the case where $I_1 = I_2 = I_0$

$$I = 2 I_0 [\,1 + \cos(\varphi(t))\,] \qquad [7.2]$$

The intensity variation δI is linked to the phase variation $\delta\varphi$ by $\delta I = 2 I_0 \sin \varphi(t)\,\delta\varphi(t)$. This shows that the maximum sensitivity will be obtained for $\sin \varphi(t) = 1$, i.e. when the two arms of the interferometer are in quadrature.

7.3.3.2. Spectrometry

Spectrometry is a whole set of techniques that enable the use of spectral variations of a transducer. Due to the use of fibers and guides, the range covered extends from 0.25 µm to 5 µm depending on the material used.

7.3.3.3. Absorption

The measurement by absorption is based upon Beer-Lambert's law in which the absorbance, $\alpha(\lambda)$, of a light beam, with an initial intensity, I_0, decreased to a value, I, after crossing a given medium over a distance L (in cm), is proportional to the concentration, C, of the measured species:

$$\alpha(\lambda) = \log(I_0 / I) = \varepsilon(\lambda)\,L\,C \qquad [7.3]$$

where $\varepsilon(\lambda)$ is the molar absorption coefficient. We can perform a measurement at one wavelength, at the absorption peak, or at two wavelengths, one of which serves as internal reference, or a multiwavelength measurement with a network and a photodiode array.

7.3.3.4. *Reflectance*

Light reflection can be operated on reflecting surfaces (weak absorption) or on absorbing species having a concentration C. Such a diffuse absorption follows Kubelka-Munk's law:

$$I(R) = \varepsilon(\lambda)\,C/S = (1-R)^2/2R \qquad [7.4]$$

where S is the diffusion coefficient and R the reflectance.

7.3.3.5. *Luminescence*

When a substance is lit by light radiation, the latter may be transformed into thermal energy or light emission may occur with a less energetic wavelength, thus shifted to the red with respect to the excitating wavelength. In fluorescence, the lifetimes are of the order of 1 ns to 10 ms, whereas in phosphorescence they are longer than 100 ms. The fluorescence intensity of the fluorophore (sensitive element) is measured:

$$F = k\,P_0\,\varepsilon\,\theta\,C \qquad [7.5]$$

where P_0 is the incident power, k a geometric coefficient of proportionality, θ the fluorescence quantic yield, C the concentration. In other cases, one measures the increase or decrease of the fluorescence (quenching) in the presence of ionic species different from the fluorophore:

$$F_0/F = 1 + K_{sv}\,C \qquad [7.6]$$

where C is the concentration and K_{sv} is the Stern-Volmer constant. The energy transfer between donor and fluorescent receiver can also be determined by measuring the decrease of the donor emission and the receiver regeneration.

7.3.4. *Refractometry*

Light reaching the interface of two transparent media with different indexes may be partially reflected and partially transmitted. By use of either a stripped sheath, or a specific sheath (absorbing the measured gas), the variation of the refractive index of the environing medium modifies the variation of light intensity in the wave-guide, because of the refractive index variation linked to the gas concentration (Figure 7.5).

A plasmon mode obtained at the surface of a multimodal fiber can also be used. The transducer is made by depositing a metallic coating over a small length of the bare core of a multimodal optical fiber and by immersing the metallized part into the measured medium. The refractive index variations of the latter entail modifications of the plasmon propagation. A similar device can be made using integrated optics.

7.4. Optical fiber biosensors

A biosensor intimately combines two major elements, the receptor and the transducer. The bioreceptor is a biological element immobilized on a solid support and is able to specifically recognize a target species present in a complex medium. The transducer is an electronic device that detects physico-chemical modifications, caused by the bioreceptor in contact with its target substance and which converts them into a measurable and interpretable electric signal [BLU 91].

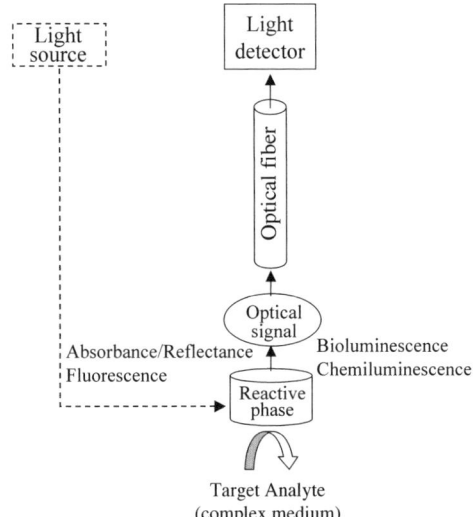

Figure 7.7. *Principle of an optical fiber sensor. The target substance is specifically recognized by a biological element, which is generally an enzyme, present in the reactive phase. The specific recognition causes emission or modification of a light signal that is transmitted by an optical fiber or a bundle of fibers to a light detector, which is usually a photomultiplier tube*

An optical fiber biosensor is the association of a reactive phase and a transduction system composed of an optical fiber or a bundle of optical fibers coupled to a light detector. The reactive phase, constituted by an immobilized biological element, sometimes associated with one or several co-reactants, themselves immobilized, acts to specifically recognize the target substance present in a complex medium. The biospecific recognition then gives rise to an optical signal, which will be transmitted, to the light detector via an optical fiber. Most of these biosensors are based upon fluorescence, absorbance, reflectance or bioluminescence and chemiluminescence measurements. When fluorescence or absorbance/reflectance measurements are exploited, it is necessary to integrate a light source to the device (Figure 7.7).

7.4.1. *Configurations of optical fiber biosensors*

The reactive phase is placed at the tip of the fiber or the fiber bundle. The reactant can be immobilized by adsorption at (or covalent bonding to) a synthetic membrane, or confined behind a semi-permeable membrane.

When a light source is required, light is led to the reactive phase either by a bifurcated optical fiber device or by a unique fiber (or a unique bundle) which also transports the optical signal coming from the reactive phase. In the first case, one tip of the fiber bundle has been divided into two parts. One is coupled to the light source and the other one to the detector. At the common extremity, the bundle is generally distributed at random and is in contact with the reactive phase. When a given fiber transports light from the source to the reactive phase, and then from the reactive phase to the detector, a separator directs the emergent light towards the detector. Finally, when light is emitted directly by the reactive phase (in the case of bioluminescent and chemiluminescent reactions), the fiber or fiber bundle is connected directly to the light detector (Figure 7.8).

7.4.2. *Chemical sensors integrated in optical fiber sensors*

Numerous optical fiber chemical sensors, based for the most part upon measurements of fluorescence or fluorescence extinction, upon absorbance and/or reflectance, have been described [WOL 91]. Among these sensors, the pH, O_2 or NH_3 sensors are the main devices constituting the basis of the transduction system of certain optical fiber sensors.

7.4.2.1. *pH sensors*

For these sensors, the variations of the absorbance of phenol red as a function of pH have been exploited. Once immobilized, the colored indicator, has an acid dissociation constant, pK_a, of 7.6 instead of 7.9 in solution. The basic form of the indicator absorbs at 558 nm whereas, at 600 nm, no form is absorbing. The ratio of the reflected intensity at 558 nm to the reflected intensity at 600 nm is a function of the pH, which can be measured with a precision of 0.01 unity of pH between 7 and 7.4.

The basic and acid forms of certain fluorophores have different properties. The acid and basic forms of 4-methylumbelliferone emit at different wavelengths when they are excited by the same wavelength. The ratio of the measured intensities at these two emitted wavelengths is proportional to the pH. For another fluorophore, the 8-hydroxy-1,3,6-pyrenetrisulfonate (HPTS), the basic form is excited selectively at 470 nm, while the acid form is excited at 405 nm. In this case, the value of the ratio of the fluorescence intensities resulting from excitation at 405 and 470 nm is now correlated to the pH. The response range of these sensors extends only over 1 to 2 pH units around the value of the pK_a of the indicator used.

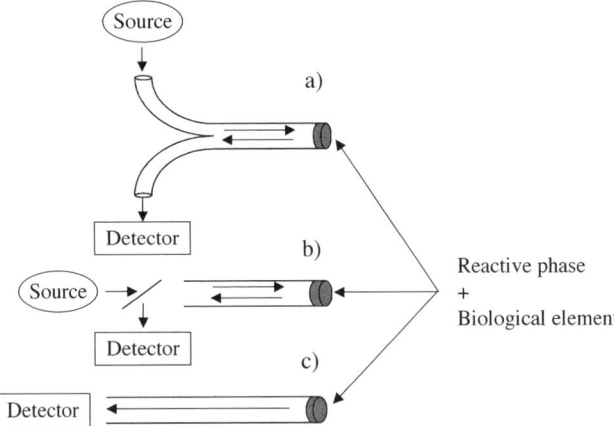

Figure 7.8. *Configurations of an optical fiber biosensor depending on the detection type used. a) Bundle of bifurcated optical fibers; (b) the same fiber or the same fiber bundle transports light from the luminous source to the reactive phase and then from the latter to the light detector; c) the light directly emitted by the reactive phase is transported to the detector by the optical fiber or by a bundle of optical fibers. Configurations a) and b) are used for optical fiber biosensors based upon fluorescence or absorbance measurement; configuration c) is used for optical fiber biosensors based upon bioluminescence or chemiluminescence measurements*

7.4.2.2. Oxygen sensors

Oxygen sensors are essentially based upon fluorescence extinction measurements. The fluorophores used are pyrene butyric acid, perylen dibutyrate, and 9,10-diphenylanthracene. The fluorescence intensity decreases as the inverse of the partial oxygen pressure.

7.4.2.3. NH_3 sensors

NH_3 are in fact pH sensors modified by establishing contact with a microtank containing an ammonium salt. As the wall of this microtank is only permeable to gases, ammonia gas present in the sample diffuses into the internal solution, modifying its pH. At equilibrium, the measured pH of the internal solution is proportional to the ammonia partial pressure in the sample.

7.4.3. Optical fiber enzymatic biosensors

These biosensors involve an enzymatic reaction and detect one of the substrates or products of this reaction, either directly if it has optical properties (fluorescence or absorbance), or indirectly by an optical fiber chemical sensor (pH, O_2, NH_3) if it has no optically measurable property. The enzyme is immobilized on a membrane placed at the extremity of a fiber or a fiber bundle, or confined at the extremity of the biosensor.

7.4.3.1. Indirect detection by chemical sensor

The main optical fiber biosensors developed from chemical sensors are presented in Table 7.2 in which the detected species, the enzymes involved, and the types of chemical sensor are listed.

7.4.3.2. Direct detection by fluorescence

The fluorescence properties of NADH (the reduced form of nicotinamide adenine dinucleotide) have been exploited to develop optical fiber biosensors associated with an immobilized specific dehydrogenase. This is because NADH is the co-enzyme of many dehydrogenases (equation 7.7) and its rate of formation or consumption along an enzymatic reaction can be monitored by fluorimetry and correlated to the reaction co-substrate concentration.

$$\text{Substrate (oxidized form)} + \text{NADH} + H^+ \rightarrow \text{Product (reduced)} + NAD^+ \quad [7.7]$$

A few sensors with this working principle have been described for measuring glucose, lactate, pyruvate, and ethanol. In general, the sensitivity of these sensors is poor and the detection limit is around tens of micromoles per liter to about one millimolar.

The intrinsic fluorescence of glucose oxidase (GOD) and lactate monooxygenase (LMO), which is due to the FAD (flavine adenine dinucleotide) and the FMN (flavine mononucleotide) active groups, respectively, has been used for optimizing optical fiber sensors for glucose and lactate. During these reactions catalyzed by these flavoenzymes, the specific substrate oxidation is coupled with the reduction of the flavinic co-enzyme, which is then re-oxidized by molecular oxygen. The oxidized and reduced forms of these two flavoenzymes (GOD and LMO) have different excitation and emission spectra. The enzymatic oxidation reaction can thus be followed by measuring the fluorescence intensity of the reduced form. These two optical fiber biosensors, which are original concepts, yet merely determine substrate concentrations in a narrow range (1.5-2 mM for glucose and 0.5-1 mM for lactate) with an extremely long response time in steady state (2 to 30 minutes for glucose and 7.5 to 25 minutes for lactate), to which a regeneration step, of a 1 to 8 minutes duration must be added, between two measurements.

Target substance	Enzyme	Enzymatic reaction	Chemical sensor
Ascorbate	Ascorbate oxidase	L-ascorbate + 1/2 O_2 → 1/2 H_2O + Dehydroascorbate	O_2
Cholesterol	Cholesterol oxidase	Cholesterol + O_2 → Cholestenone + H_2O_2	O_2
Ethanol	Alcohol oxidase	Ethanol + O_2 → Acetaldehyde + H_2O_2	O_2
Glucose	Glucose oxidase	Glucose + O_2 + H_2O → Gluconate + H^+ + H_2O_2	O_2, pH
Lactate	Lactate monooxygenase	L-lactate + O_2 → Acetate + CO_2 + H_2O	O_2
Penicillin	Penicillinase	Penicillin + H_2O → Penicilloate + H^+	pH
Urea	Urease	Urea + H_2O → CO_2 + 2 NH_3	pH, NH_3

Table 7.2. *Examples of optical fiber biosensors developed from chemical sensors*

7.4.3.3. *Direct detection by absorbance*

It is possible to conceive an optical fiber enzymatic biosensor by following the formation of the reaction product of the immobilized enzyme, using visible spectrophotometry. Such an optical fiber biosensor for glucose has been described. Its principle is that of the spectrophotometric determination of oxygen peroxide, formed during enzymatic oxidation of glucose, via an auxiliary reaction catalyzed by peroxidase in the presence of a chromogen substrate. The glucose oxidase is immobilized on a nylon membrane placed at the end of a bifurcated bundle. The latter is placed in a small measure cell containing peroxidase and the chromogen reactant in solution. The formation of the colored product is then followed by means of a fiber bundle coupled to a spectrophotometer, and the relation between absorbance and glucose concentration is linear in the 10 µM to 1 mM range.

A glucose sensor combining a bifurcated fiber bundle and an enzymatic electrode has also been described. A redox mediator, with its oxidized form having absorption properties, is used in place of oxygen in the reaction catalyzed by glucose oxidase. The reduced enzyme re-oxidation is carried along, in the catalytic process, with the reduction of the chemical mediator, which is later electrochemically regenerated in its oxidized form. The consumption rate of the mediator in its oxidized form is then followed by spectrophotometry, thanks to the optical fiber bundle. The response time is about 4 min and the linearity range for glucose determination extends from 1 mM to 12 mM.

7.4.4. Biosensors with non-catalytic bioreceptors (affinity biosensors)

7.4.4.1. Lectin biosensors

The first optical fiber biosensor of this type has been developed for the determination of glucose. Its principle rests on the competition between glucose and fluorescein-labeled dextran (FITC-dextran) for binding on a lectin (concanavalin A (Con-A)) immobilized on the inner wall of a semi-permeable membrane [SCH 82]. This lectin has affinity towards free or polymerized glucose. The tube-shaped semi-permeable membrane is fitted at the extremity of an optical fiber and, in the absence of glucose, the dextran gets bound on the Con-A, because it cannot diffuse through the walls of the semi-permeable membrane. The immobilized lectin is out of the illumination cone of the optical fiber, and the fluorescence is then minimum (Figure 7.9).

When glucose is present, it diffuses through the semi-permeable membrane and can interact with Con-A in place of the FITC-dextran. The latter, which is no longer bound to the lectin, is then inside the volume illuminated by the fiber, and the resulting increased fluorescence can be correlated to the glucose concentration in the sample, in the 0.5 to 2 g/l range.

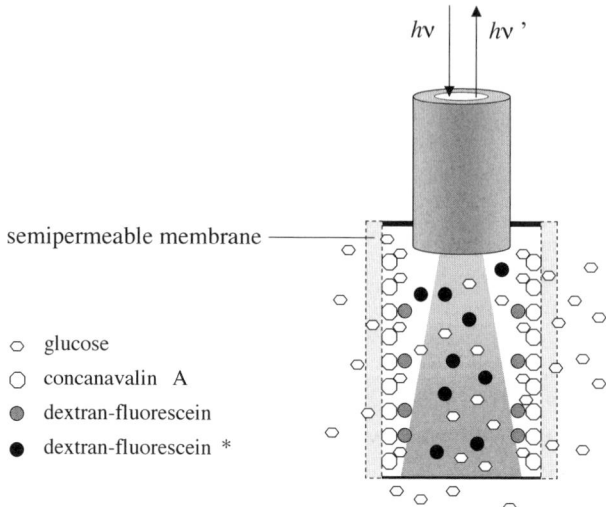

Figure 7.9. *Principle of the optical fiber biosensor for glucose, based upon the competition between fluorescein-labeled dextran and glucose for binding to immobilized concanavalin A*

7.4.4.2. DNA biosensors

A feasibility study has been reported the simultaneous determination of several DNA sequences [FER 96]. A bundle of optical fibers was assembled in which a different oligonucleotide probe was immobilized at the extremity of each fiber. Hybridization of complementary fluorescein-labeled oligonucleotides was detected by the measurement of the fluorescence intensity by means of a CCD (charged coupled device) camera (Figure 7.10).

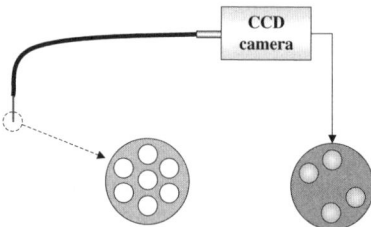

Figure 7.10. *Optical fiber biosensor for the detection of nucleotidic sequences. Nucleic probes are immobilized at the sensitive extremity of each fiber (Ø = 200 µm), the other extremities are connected to a CCD camera, which visualizes the fibers on which hybridization of fluorescein-labeled complementary strands has occurred*

7.4.5. Chemiluminescence and bioluminescence detection sensors

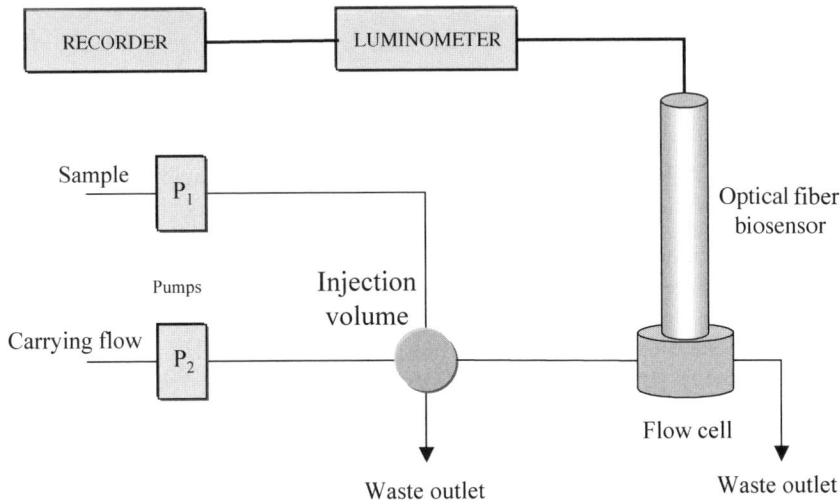

Figure 7.11. *Schematic drawing of an optical fiber biosensor inserted in a flow injection analysis system for bioluminescence or chemiluminescence measurements. The bioreceptor may be the firefly luciferase for the determination of ATP, the bi-enzymatic bacterial system oxidoreductase-firefly luciferase for the determination of NADH or NADPH, or horseradish peroxidase, which catalyzes the chemiluminescence reaction of luminol in the presence of hydrogen peroxide*

These biosensors exploit light emission phenomenon, which occurs during certain chemical (chemiluminescence) or biochemical (bioluminescence) reactions [BLU 97]. The devices based upon this principled have mostly been adapted to continuous flow systems for measurements by FIA (flow injection analysis), thus making possible a complete or partial automation of the analysis processes (Figure 7.11).

7.4.5.1. Chemiluminescent biosensors

The biosensors with chemiluminescence detection have been developed for measuring hydrogen peroxide (H_2O_2) and involve luminophores, which are oxidized in the presence of H_2O_2 by a chemiluminescent reaction [BLU 88, DEG 00, FRE 78]. The main luminophore is luminol (5-amino-2,3-dihydro-1,4-phthalazinedione). Its oxidation can be catalyzed by horseradish peroxidase (Figure 7.12). These chemiluminescent reactions become more interesting when they are coupled to reactions catalyzed by oxidases, such as those used in amperometric detection sensors, and which generate H_2O_2 [BER 94, BLU 93, LAP 96, TSA 00a].

7.4.5.2. *Bioluminescent biosensors*

In bioluminescence detection biosensors, two types of light-emitting reactions are involved:

– the firefly bioluminescence reaction, catalyzed by a luciferase, which enables direct determination of adenosine triphosphate (ATP) in the presence of luciferin, which is a another substrate specific in this reaction (Figure 7.12). Luciferin [D-(-)-2-(6'-hydroxy-2'-benzothiazolyl)-Δ2-thiazoline-4-carboxylic acid] is synthesized naturally by fireflies, but nowadays it can be produced by chemical reaction;

– the bioluminescent system oxidoreductase-luciferase of some marine bacteria, which permits the determination of NADH or NADPH (Figure 7.12).

For such a system, the detection limits for ATP and NADH are 20 pM and 0.3 nM, respectively [BLU 88, BLU 89, GAU 90a]. Moreover, by co-immobilizing specific dehydrogenases with the bacterial bioluminescent system, it is possible to measure the concentration of sorbitol, ethanol, or oxaloacetate with a detection limit in the order of nanomoles per liter [GAU 90b, MIC 96, MIC 97]. Similarly, by coupling the firefly bioluminescence reaction with enzymatic reactions involving ATP, it has been possible to develop optical biosensors for ADP and adenosine monophosphate (AMP) [MIC 98a, MIC 98b].

(a)
$$2\ H_2O_2 + luminol + OH^- \xrightarrow{peroxidase} 3\text{-aminophthalate} + N_2 + 3\ H_2O + h\nu$$
$$(\lambda_{max} = 430\ nm)$$

(b)
$$ATP + luciferin + O_2 \xrightarrow{firefly\ luciferase} AMP + PPi + oxyluciferin + CO_2 + h\nu$$
$$(\lambda_{max} = 560\ nm)$$

(c)
$$NAD(P)H + H^+ + FMN \xrightarrow{oxidoreductase} NAD(P)^+ + FMNH_2$$

$$FMNH_2 + R\text{-}CHO + O_2 \xrightarrow{bacterial\ luciferase} FMN + R\text{-}COOH + H_2O + h\nu$$
$$(\lambda_{max} = 490\ nm)$$

Figure 7.12. *Light-emitting enzymatic reactions. a) Chemiluminescence of luminol catalyzed by peroxidase; b) reaction of bioluminescence of the firefly; c) reaction of bacterial bioluminescence; R-CHO is an aliphatic aldehyde possessing 8 to16 carbon atoms*

7.4.5.3. Electrochemiluminescent sensors

Another possibility for activating the chemiluminescence of luminol in the presence of H_2O_2 is luminol electro-oxidation under an applied potential of about +400 to +500 mV *vs* a Ag/AgCl. This possibility was recently exploited for developing electro-optical biosensors for detecting glucose, cholesterol, lactate, and choline [MAR 99, MAR 00, MAR 01, TSA 00b]. These devices have been miniaturized, particularly by the use of screen-printed carbon electrodes [LEC 00, LEC 01].

7.5. Perspectives and conclusions

Unlike enzymatic electrodes, optical fiber biosensors have not yet given rise to commercialized devices. However, the development of optical biosensors has been undertaken more recently, and it is very likely in the near future that some industrial achievements will appear. Optical biosensors have many advantages over other sensors; they are not sensitive to electric or electrochemical interferences and enable remote measurements. But, as is the case with the other types of biosensors, the key element is the sensitive layer that integrates biological compounds, which, by nature, is fragile. This is the main difficulty, and true improvements will only be made once the fundamental parameters controlling the stability (or the instability) of biocatalysts will be fully understood.

7.6. Bibliography

[AND 91] ANDRE J.C., GOURE J.P., "Capteurs et instrumentation optique pour l'industrie", *Spectra 2000*, vol. 160, p. 39-47, 1991.

[BER 94] BERGER A., BLUM L.J., "Enhancement of the response of a lactate oxidase/peroxidase-based fibreoptic sensor by compartmentalization of the enzyme layer", *Enzyme Microb. Technol.*, vol. 16, p. 979-984, 1994.

[BLU 88] BLUM L.J., GAUTIER S.M., COULET P.R., "Luminescence fibre-optic biosensor", *Anal. Lett.*, vol. 21, p. 717-726, 1988.

[BLU 89] BLUM L.J., GAUTIER S.M., COULET P.R., "Continuous flow bioluminescent assay of NADH using a fibre-optic sensor", *Anal. Chim. Acta*, vol. 226, p. 331-336, 1989.

[BLU 91] BLUM L.J., COULET P.R., *Biosensor Principles and Applications*, Marcel Dekker, 1991.

[BLU 93] BLUM L.J., "Chemiluminescent flow injection analysis of glucose in drinks with a bi-enzyme fibre optic biosensor", *Enzyme Microb. Technol.*, vol. 15, p. 407-411, 1993.

[BLU 97] BLUM L.J., *Bio- and Chemi-luminescent Sensors*, World Scientific, 1997.

[DAK 89] DAKIN J.P., CULSHAW B., *Optical Fibre Sensors*, vol. I and II, Artech House, 1988 and 1989.

[DEG 00] DEGIULI A., BLUM L.J., "Flow injection chemiluminescence detection of chlorophenols with a fibre optic biosensor", *J. Med. Biochem.*, vol. 4, p. 32-42, 2000.

[DIG 99] DIGONNET M.F.J., *Rare Earth Doped Fibre Lasers and Amplifiers*, Marcel Dekker, 1999.

[FER 92] FERDINAND P., *Capteurs à fibres optiques et réseaux associés*, Lavoisier, 1992.

[FRE 78] FREEMAN T.M., SEITZ W.R., "Chemiluminescence fibre optic probe for hydrogen peroxide based on the luminol reaction", *Anal. Chem.*, vol. 50, p. 1242-1246, 1978.

[GAU 90a] GAUTIER S.M., BLUM L.J., COULET P.R., "Multifunction fibre-optic sensor for the selected bioluminescent flow determination of ATP or NADH", *Anal. Chim. Acta*, vol. 235, p. 243-253, 1990.

[GAU 90b] GAUTIER S.M., BLUM L.J., COULET P.R., "Fibre-optic biosensor based onluminescence and immobilized enzymes: microdetermination of sorbitol, ethanol and oxaloacetate", *J. Biolumin. Chemilum.*, vol. 5, p. 57-63, 1990.

[GOU 89] GOURE J.P., VERRIER I., "Linear and non linear optical fibre device", *J. Physics D: Appl. Phys.*, vol. 22, p. 1791-1805, 1989.

[GOU 90] GOURE J.P., "Capteurs optiques—capteurs à fibre optique", *Entropie*, vol. 155, p. 18- 27, 1990.

[GOU 01] GOURE J.P., VERRIER I., *Optical Fibre Devices, Series in Optics and Optoelectronics*, Institute of Physics Publishers, 2001.

[HLA 94] HLAVAY J., GUILBAULT G. G., "Determination of sulphite by use of a fibre-optic biosensor based on a chemiluminescent reaction", *Anal. Chim. Acta*, vol. 299, p. 91-96, 1994.

[JON 87] JONES B.E., *Current Advances in Sensors*, Adam Hilger, 1987.

[KAS 99] KASHYAP R.K.J., *Fibre Bragg Gratings*, Academic Press, 1999.

[LAP 96] LAPP H., SPOHN U., JANASEK D., "An enzymatic chemiluminescence optrode for choline detection under flow injection conditions", *Anal. Lett.*, vol. 29, p. 1-17, 1996.

[LEC 00] LECA B., BLUM L.J., "Luminol electrochemiluminescence with screen-printed electrodes for low-cost disposable oxidase-based optical sensors", *Analyst*, vol. 125, p. 789- 791, 2000.

[LEC 01] LECA B., VERDIER A. M., BLUM L.J., "Screen-printed electrodes as disposable or reusable optical devices for luminol electrochemiluminescence", *Sens. Actuators B*, vol. 74, p. 190-193, 2001.

[MAR 99] MARQUETTE C. A., BLUM L.J., "Luminol electrochemiluminescence-based fibre optic biosensors for flow injection analysis of glucose and lactate in natural samples", *Anal. Chim. Acta*, vol. 381, p. 1-10, 1999.

[MAR 00] MARQUETTE C.A., RAVAUD S., BLUM L.J., "Luminol electro-chemiluminescencebased biosensor for total cholesterol determination in natural samples", *Anal. Lett.*, vol. 33, p. 1779-1796, 2000.

[MAR 01] MARQUETTE C.A., LECA B.D., BLUM L.J., "Electrogenerated chemiluminescence of luminol for oxidase-based fibreoptic biosensors", *Luminescence*, vol. 16, p. 159-165, 2001.

[MIC 96] MICHEL P.E., GAUTIER S.M., BLUM L.J., "Effect of compartmentalization of the sensing layer on the sensitivity of a multienzyme-based bioluminescent sensor for Llactate", *Anal. Lett.*, vol. 29, p. 1139-1155, 1996.

[MIC 97] MICHEL P.E., GAUTIER S.M., BLUM L.J., "A high-performance bioluminescent trienzymatic sensor for D-sorbitol based on a novel approach of the sensing layer design", *Enzyme Microb. Technol.*, vol. 21, p. 108-116, 1997.

[MIC 98a] MICHEL P.E., GAUTIER S.M., BLUM L.J., "A transient enzymatic inhibition as an efficient tool for the discriminating bioluminescent analysis of three adenylic nucleotides with a fibreoptic sensor based on a compartmentalised tri-enzymatic sensing layer", *Anal. Chim. Acta*, vol. 360, p. 89-99, 1998.

[MIC 98b] MICHEL P.E., GAUTIER S.M., BLUM L.J., "Luciferin incorporation in the structure of acrylic microspheres with subsequent confinement in a polymeric film: a new method to develop a controlled release-based biosensor for ATP, ADP and AMP", *Talanta*, vol. 47, p. 167-181, 1998.

[PIT 85] PITT G.D., EXTANCE P., NEAT R.C., BATCHELDER D.N., JONES, R.E., BARNETT J.A., PRATT R.H., "Optical fibre sensors", *IEE Proc.*, vol. 132, p. 214-248, 1985.

[SCH 82] SCHULTZ, J.S., MANSOURI, S., GOLDSTEIN, I. J., "Affinity sensor: a new technique for developing implantable sensors for glucose and other metabolites", *Diabetes Care*, vol. 5, p. 245-253, 1982.

[TSA 00a] TSAFACK V.C., MARQUETTE C.A., PIZZOLATO F., BLUM L.J., "Chemiluminescent choline biosensor using histidine-modified peroxidase immobilized on metal-chelate substituted beads and choline oxidase immobilized on anion-exchanger beads coentrapped in a photocrosslinkable polymer", *Biosensors Bioelectronics*, vol. 15, p. 125-133, 2000.

[TSA 00b] TSAFACK V.C., MARQUETTE C.A., PIZZOLATO F., BLUM L.J., "An electrochemiluminescence-based fibre optic biosensor for choline flow injection analysis", *Analyst*, vol. 125, p. 151-155, 2000.

[VAL 90] VALETTE S., RENARD J.P., JADOT J.P., GIDON P., ERBEIA C., "Silicon-based integrated optics technology for optical sensor applications", *Sens. Actuators A*, vol. 23(1-3), p. 1087-1091, 1990.

[WOL 91] WOLFBEIS O.S., *Fibre Optic Chemical Sensors and Biosensors*, vol. I and II, CRC Press, 1991.

Chapter 8

Sensors and Voltammetric Probes for *In situ* Monitoring of Trace Elements in Aquatic Media

8.1. Introduction

This chapter describes the characteristics of sensors and voltammetric probes (i.e. the whole measuring instrument) for the continuous *in situ* monitoring of the concentrations and eco-chemical properties of trace elements, in particular metals. These sensors and probes must comply with numerous criteria concerning the analyzed element, as well as the measuring conditions (for instance, automated determinations in deep water in the presence of particles in suspension, or in toxic water). In view of obtaining devices for reliable routine measurement, these sensors must be developed from the beginning of their conception, with the whole set of criteria taken into account. The main analytical and technical criteria that must be fulfilled are:

– high sensitivity, precision and reliability,

– ability to determine the overall concentration, but also speciation of the elements (see below),

– continuous and simultaneous analysis of several elements; this is because the environmental interpretation of data concerning a given element is generally linked to that of numerous other elements,

– miniaturization,

Chapter written in French by Marie-Louise TERCIER-WAEBER and Jacques BUFFLE.

– low cost and low energy consumption for allowing *in situ*, automated and long-term measurements,

– robustness, easy use, and maintenance on the ground,

– high storage capacity, fast storage, data transmission from the probe to (one of) the coastal interpretation center(s).

These criteria, taken as a whole, imply new analytical and technical concepts, because even if some laboratory instruments have the desired sensitivity or selectivity, none of them complies with all the criteria nor is adaptable to automated measurements in difficult ground conditions. The ability of a system to determine not only the total concentration of the analyzed element, but also its speciation, is a particularly restrictive criterion. This necessary speciation is linked to the fact that the trace elements emitted in the environment by human activities are distributed among various types of complexes formed with simple inorganic ligands, organic biopolymers, and/or inorganic colloids, such as clays. Only certain species, the proportions of which may vary in time and space, depending on the general physicochemical conditions of the medium, play an important role in a given biogeochemical process (for instance, eco-toxicity or diffusion of an element in a soil or a lake).

So, in order to understand the effects of trace metals upon an aquatic ecosystem, the measuring analytic technique must also yield information on their most important chemical forms (in other words their speciation) and their evolution with time. This information is an absolutely necessity for interpreting the environmental impact (bio-accumulation, eco-toxicity, transport, sedimentation), and for a better understanding of the relation between the anthropologic emissions and their long-term impact.

Few techniques are likely to comply with all the above-cited criteria; the voltammetric techniques belong to them. They have been widely used for the laboratory determination of trace metals in natural water samples during the last thirty years. A few submersible voltammetric probes have been developed recently. The aim of this chapter is to provide an up-to-date review of the current developments in this field, with a focus on the technology. The theoretical principles are very briefly detailed at the beginning of the chapter (for more details, see references [GAL 94, BAR 82]). The speciation aspects are not dealt with in detail; but the interested reader can refer to [HEY 66, BUF 88, BUF 00], which describe in detail all the aspects of *in situ* voltammetric measurements.

8.2. Basic principles of the voltammetric techniques and of their applications to analysis of water

8.2.1. Components and principles

8.2.1.1. Voltammetric measurement

The voltammetric techniques are based upon the electrochemical oxidation or reduction, at a controlled potential, E, of the analyzed element. Schematically, the current, i, resulting from this reaction, is proportional to the concentration of the analyzed component, and the potential at which this reaction occurs is linked to the nature of the electroactive species. The basic units of a voltammetric analyzer are (Figure 8.1):

– a cell, generally including i) a working electrode (at which the reaction studied occurs), ii) a reference electrode, which controls the potential of the working electrode, and iii) an auxiliary electrode, which ensures the circulation of the current, with the working electrode,

– an electronic circuit, called potentiostat, so that the working electrode potential and the current are recorded.

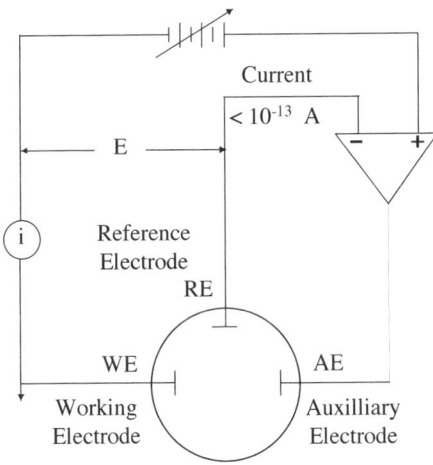

Figure 8.1. *Schematic representation of the electronic circuit of a potentiostat. E = imposed potential; i = measured current*

The working and reference electrodes may have different natures and geometries. Their advantages and drawbacks are discussed in sections 8.4.1 and 8.4.2. The main characteristic of a reference electrode is that its potential remains constant, which implies that the current circulating through it is virtually zero ($<10^{-13}$ A). For this reason, the majority of the current circulates between the working and the auxiliary electrodes. The voltammetric curves are obtained by varying the potential, E, as a function of time, and measuring the current, I, as a function of E (Table 8.1). The variation of E may be linear or modulated, which gives rise to different techniques with different characteristics (see section 8.3.2). In all cases, it is important to note that the current is an electron flux thus depending of dynamic processes in solution (in particular, the rates of diffusion and of the chemical reactions coupled with the redox process): therefore, i is always a function of E and t. The graph of function $i = f(E, t)$ is called voltammetric curve or voltammogram.

The solution contains the solvent (water), an electrolyte (which ensures the current in the solution via ion transport) and the component(s) under analysis, which undergoes the redox reaction at the surface of the working electrode. In the absence of electroactive species, the full curve of Figure 8.2 a, called a polarization curve, has been obtained at a mercury working electrode (its exact form depending on the voltammetric technique used). When electroactive species are present, additional curves (Figure 8.2a, dots), with wave or peak shapes, according to the voltammetric technique, are observed. The key characteristics of Figure 8.2 are discussed hereafter.

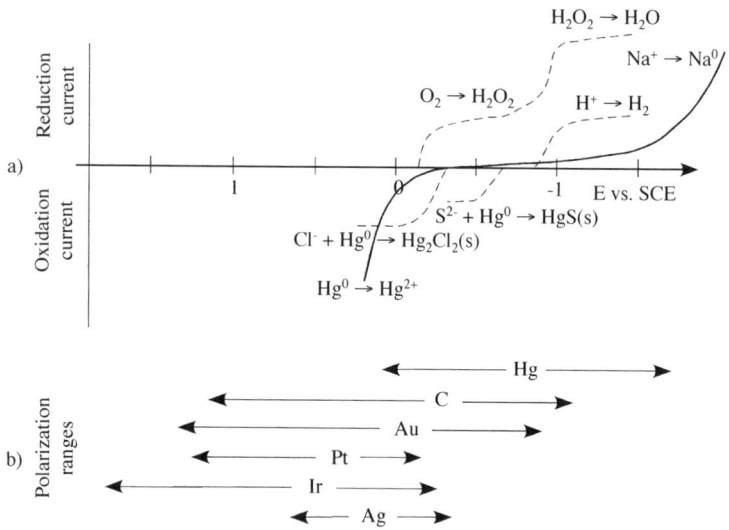

Figure 8.2. *a) Schematic curves of O_2, $S(-II)$, Cl^-, H^+ and Na^+ at a mercury electrode. b) Polarization range of Hg, C, Au, Pt, Ir, Ag electrodes (adapted from [BUF 00])*

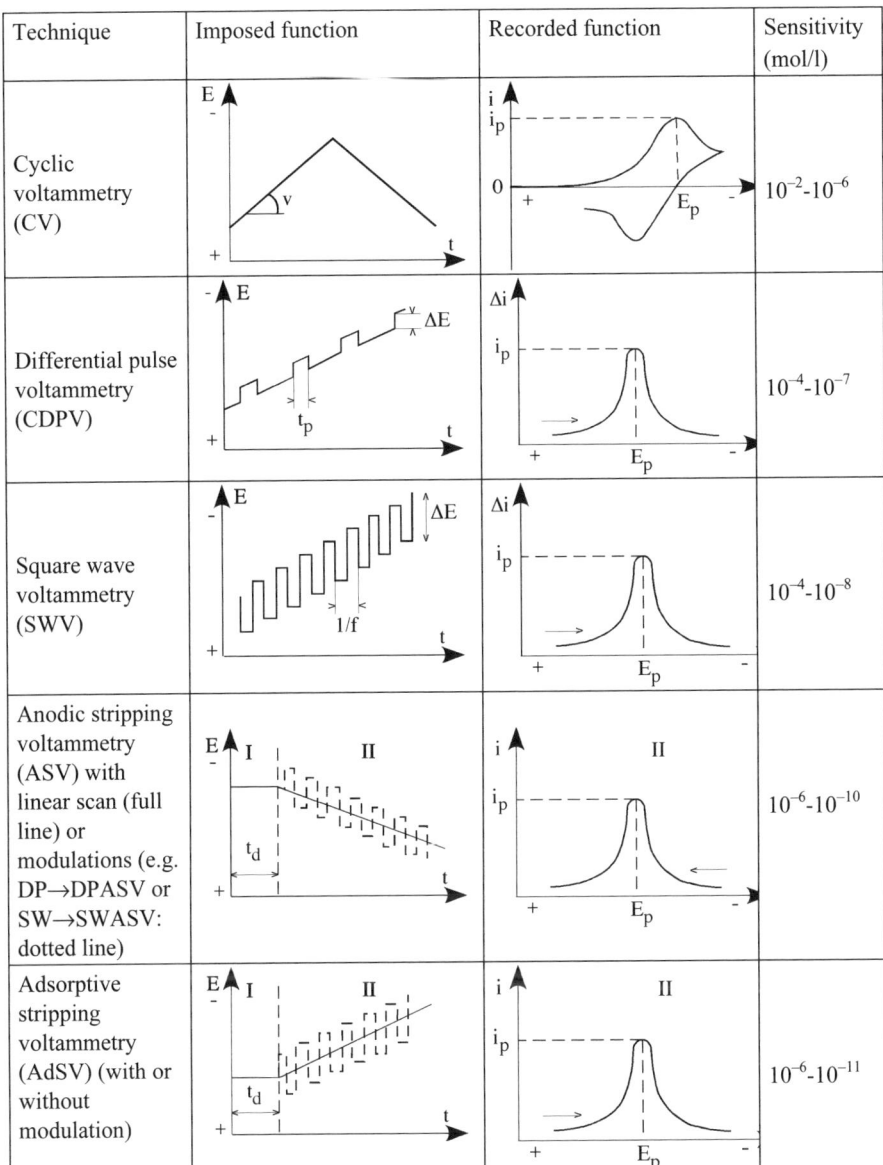

Table 8.1. *Main voltammetric techniques used for the analysis of trace elements in aquatic media, along with their sensitivities. ΔE = pulse amplitude (mV), f = frequency (Hz), t_d = deposition (or pre-concentration) time, v = scan rate (mV/s), i_p = peak current (A), E_p = peak potential (mV). (Adapted from [BUF 00])*

8.2.1.2. Polarization range: potential limit

In the case of mercury working electrodes (full curve in Figure 8.2a), the potential window within which the oxidation or the reduction of an electroactive species is measurable is limited i) on the negative potentials side, by the reduction current of the electrolyte (sodium ions, Na^+) or of the water proton, and ii) on the side of the positive potentials, by the oxidation current of mercury. Thus, the potential domain within which the potential imposed to a mercury electrode can vary is typically limited from +0.2 to –1.7 V with respect to the saturated Ag/AgCl/KCl reference electrode (section 8.4.2). The polarization ranges of electrodes other than mercury are presented in Figure 8.2b. The negative potential limit is similar for all the types of working electrodes (reduction of H^+ or Na^+). The positive limit is due to oxidation either of the substrate (platinum, silver) or of the solvent water. Let us note that, as the reduction and oxidation potentials of water both depend on the pH of the solution, the corresponding polarization limits also depend on this parameter. The silver electrode is seldom used because its polarization range is narrow. The platinum, gold, iridium, and carbon electrodes are complementary to the mercury electrode: the electroactive species reduced at negative potentials can generally be analyzed with mercury electrodes, whereas the electroactive species having a redox potential superior to 0.2 V are analyzed at solid electrodes, in spite of the lower reproducibility of the latter (section 8.4.1).

8.2.1.3. Electroactive species

The main characteristics of an electroactive species are: i) its half wave potential (Figure 8.2a) or its peak potential (Table 8.1), depending on the technique used, which are directly linked (and sometimes very close) to the standard potential E_0 of the corresponding couple [BUF 00]; and ii) the limiting current of the wave plateau (Figure 8.2a) or the peak current (Table 8.1), which are proportional to the electroactive species concentration. The *in situ* analysis of surface waters is done in the presence of oxygen, with a concentration (3×10^{-4} mol/l for saturated water), which is much larger than that of the trace metals to analyze (typically $<10^{-7}$ mol/l). Specific precautions (section 8.4.4) must, therefore, be taken so that the oxygen reduction current does not mask the currents due to the analyzed components, which have a negative redox, potential (Figure 8.2a). Similarly, in highly anoxic media, hydrogen sulfide (H_2S) and HS^- may be present in high concentrations ($\geq 10^{-6}$ mol/l) and screen the signals linked to the target elements at potentials <-0.6 V in the case of mercury electrodes (Figure 8.2a).

8.2.2. Influence of the transport properties of the electroactive species on the voltammetric signal

The transport mode of the analyzed species, M, from the bulk of the solution towards the electrode surface, has a strong influence upon the voltammetric signal and must consequently be well controlled. In a quiescent medium, and in the presence of a complex ML (Figure 8.3a), the important parameters are the charge transfer rate constant, k_0, the diffusion coefficients of M and ML, D_M and D_{ML}, the rate constants for the formation/dissociation reactions of the metallic complexes, k_f and k_d. If the chemical reaction is fast with respect to the diffusion processes, the overall transport of M + ML is characterized by a mean diffusion coefficient, $\overline{D}$, which depends on D_M and D_{ML} [BUF 00]. Moreover, if $E-E_0 \ll 0$, the faradaic current (see section 8.3.1), noted i, at constant potential E for a stationary spherical electrode is given by [HEY 66]:

$$i = nFA.\overline{D}.c.\left(\frac{1}{\delta} + \frac{1}{r}\right) \quad [8.1]$$

where n = number of electrons exchanged per mole of M, A = electrode area, c = total metal concentration in the bulk of the solution, δ = width of the diffusion layer and r = electrode radius. Equation [8.1] strictly holds for a spherical electrode in a motionless solution [HEY 66], but it can be shown [AOK 86] that, for times $t > 10$ s, it may also be applied to disc electrodes, by replacing the term $1/r$ with $1.27/r$. It thus permits to discuss the role of the size and the geometry of the electrodes, as well as the influence of the hydrodynamic conditions for *in situ* applications. The important points are the following (fore more details, see reference [BUF 00]).

The macro-electrodes correspond to the condition: $1/r \ll 1/\delta$ (it is the case for classic commercial electrodes with a typical size r >100 µm). Therefore, I depends essentially on δ ($1/r$ is negligible). It can be shown [BUF 00, VAN 00, LEV 62] that:

– in quiescent solutions, $\delta = \sqrt{(\pi.\overline{D}.t)}$, so that i decreases with t if E is kept constant; an additional influence of t may occur when a modulation $E = f(t)$ is imposed;

– in a stirred solution, (for instance, rotating disc electrode or stirring with a magnetic rod): $\delta \approx (\overline{D}/\omega)^p$ where ω = rotation speed and $p \approx 0.3\text{-}0.5$; i is then a function of the stirring conditions and is time-dependent only if a modulation $E = f(t)$ is used.

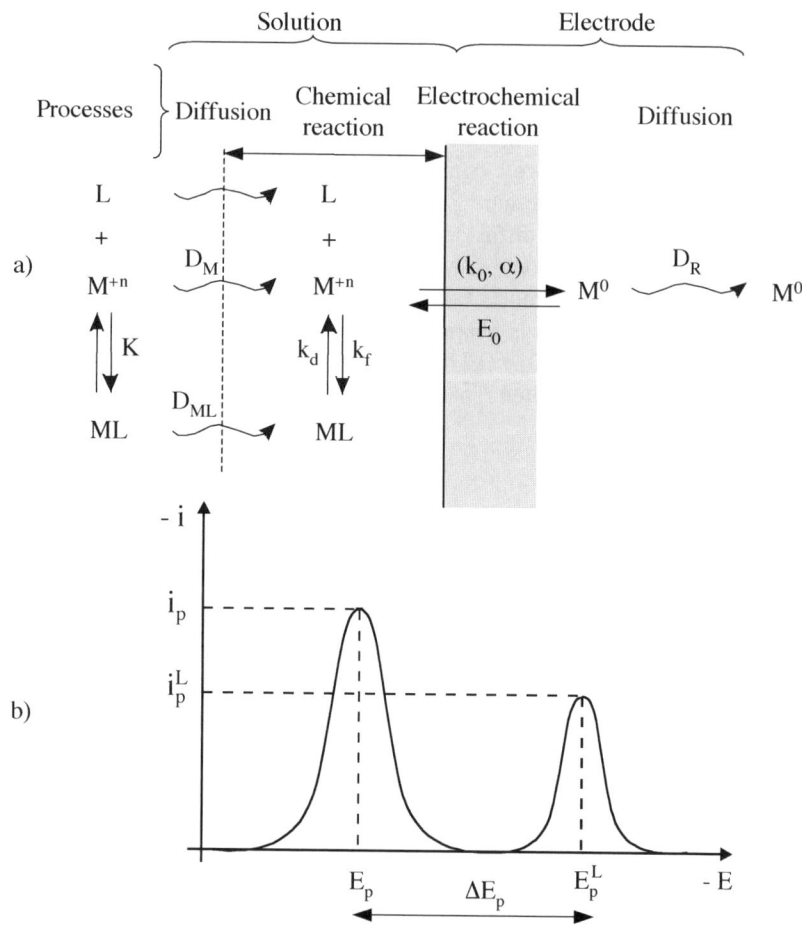

Figure 8.3. *a) The major processes and the parameters influencing the voltammetric signal resulting from the reduction of a metal ion, M^{+n}, complexed by a ligand, L. b) Example of a possible variation of the i_p and E_p parameters of the voltammetric curves for the reduction of M^{+n} in the presence (exponent L) and in the absence (no exponent) of a ligand L*

The microelectrodes correspond to $1/r \gg 1/\delta$ (typically $r \leq 10$ µm). Therefore, i does not depend on time and on the stirring conditions ($1/\delta$ is negligible). Due to this property, linked to other characteristics [PLE 91, BON 94], they are first-rate electrodes for *in situ* applications in aquatic media (sections 8.4.1.2 and 8.4.6).

8.2.3. Influence of the speciation of the electroactive compounds on the voltammograms

The nature and influence of the transport processes and of the chemical reactions, linked to the reduction of the electroactive species (Figure 8.3a), upon the $i = f(E, t)$ corresponding curve, may be diverse and complicated. This has been discussed in many books, in particular [GAL 94, BAR 82, HEY 66, BUF 88] for general approaches; [DRY 77] for applications to biomolecules; [MAI 68] for applications in organic chemistry. These reactions must be taken into account for interpreting voltammograms undertaken *in situ* in natural water. These aspects can only be briefly discussed here; more details will be found in [BUF 88, BUF 00].

For a metal ion, M^{n+}, forming a complex, ML, in solution and reducible to M^0 at the electrode (Figure 8.3a), the interpretation of the peak current and the peak potential (Figure 8.3b; or of the limiting current and the half wave potential, depending on the voltammetric technique used), depends at least on three types of processes in series (Figure 8.3a): i) diffusion in solution of M^{n+} and ML (diffusion coefficients = D_M, D_{ML} and $\overline{D}$; see below); ii) the chemical reactions involving M^{n+} (equilibrium constants K and rate constants, k_f and k_d); and iii) the redox reaction (standard potential, E_0, rate constant k_0), as well as the diffusion of M^0 in mercury (diffusion coefficient D_R) for metals forming an amalgam with mercury. It is important to understand some limiting cases for the interpretation of the voltammetric results in water [BUF 00]:

– the redox reaction is called rapid or reversible if the rate of the redox process is much larger than that of the diffusion process ($k_0 \gg \overline{D}/\delta$) for macro-electrodes and $k_0 \gg \overline{D}/r$ for microelectrodes). In such a case, $E_p \sim E_o$ and the potential shift, $\Delta E_p = E_p - E_p^L$ (Figure 8.3b), simply attain the equilibrium constant, K, or the complexation degree of M, $\alpha = [M]_{tot}/[M]$, in solution, if the ML complexes are labile or inert. In natural waters, the complexes of Cu(II), Pb(II), Cd(II), Zn(II), Tl(I), with the inorganic simple ligands (carbonate, sulfate, chloride) are often labile and can be analyzed in this way;

– as far as the chemical process is concerned, the ML complex is called labile, non-labile, or inert, respectively, if the formation/dissociation of ML is very fast (labile complexes) or slow (non-labile complexes) in comparison to diffusion or, when the dissociation does not occur at all (inert complexes). An unambiguous interpretation of the speciation is generally possible only in the case of labile or inert complexes, for which the voltammetric curves do not depend on k_d or k_f;

– as far as the physical transport is concerned (diffusion) the complex ML is considered as mobile if $\varepsilon = D_{ML}/D_M$ is close to 1. It is immobile if $\varepsilon = 0$. Natural waters often contain colloidal complexing agents (clays, biopolymers), in this case, $\varepsilon < 1$.

Typically, when voltammograms of a reversible M^{+n}/M° redox couple in a non-complexing solution ($[L] = 0$) and in a complexing solution ($[L] > 0$) are compared, for a same value of the total metal concentration, $[M]_{tot}$, the following observations can be made:

– $i_p^L/i_p < 1$, and $\Delta E_p = 0$ suggest the existence of inert complexes and $[M]_{tot}/[M]$ can be deduced from $i_p/i_p^L = [M]_{tot}/[M]$,

– $i_p^L/i_p < 1$, and $\Delta E_p > 0$ (Figure 8.3b) may result from the existence of complexes i) either labile and hardly mobile, or ii) mobile but not totally labile,

– $i_p^L/i_p = 1$, and $\Delta E_p > 0$ suggest labile complexes with $D_M = D_M$.

In all cases, a non-ambiguous interpretation of the voltammograms necessitates preliminary tests concerning the reversibility of the redox couples, and the labile and mobile characters of the complexes of the metal studied. Such tests have been described in detail in the literature [BUF 88, BUF 00].

A brief calculation of the degree of complexation of M is given hereafter for labile ML complexes (for the inert complexes: $\alpha = i_p/i_p^L$) of a metal, M, in a reversible redox couple, and when is in excess with respect to M. The complexation degree $\alpha = [M]_{tot}/[M]$, can then be calculated from the following equation [GAL 94, BAR 82, HEY 66, BUF 88, BUF 00, CRO 69]:

$$\Delta E_p = \frac{RT}{nF} \ln \frac{i_p^L}{i_p} + \frac{RT}{nF} \ln \alpha \qquad [8.2]$$

with:

$$\frac{i_p^L}{i_p} = \sqrt{\frac{\overline{D}}{D_M}} \qquad [8.3]$$

for macro-electrodes (linear diffusion), and:

$$\frac{i_p^L}{i_p} = \frac{\overline{D}}{D_M} \qquad [8.4]$$

for microelectrodes (spherical diffusion). $\overline{D}$ is the mean diffusion coefficient of M and ML, applicable to labile complexes:

$$\overline{D} = D_M \frac{1}{\alpha} + D_{ML}\left(\frac{\alpha-1}{\alpha}\right) \quad [8.5]$$

Equations [8.2] to [8.5] yield the calculation of α, but also of D_{ML} and the equilibrium constant, K, if the concentration of L is in excess and is known.

In the complicated aquatic media, the overall concentration of M, $[M]_{tot}$, is the sum of free M and of a great number of M-containing species with different labilities and different mobilities. $\overline{D}$ includes then a high number of corresponding terms. By analogy with equation [8.5], if all the complexes are labile:

$$\overline{D} = D_M f_M + \sum_m D_m f_m \quad [8.6]$$

where f_m is the fraction of the metal complex of rank m ($f_m = [M]_m/[M]_{tot}$), sufficiently mobile to contribute to the total current. Information in the literature on the diffusion and the lability of various natural complexes is currently too scarce for detailed values of f for each specific M species to be obtained. Nevertheless, an estimation of the proportions i) of free M, $f_M = 1/\alpha$, ii) of the labile and mobile species containing M, f_m; and iii) of the M-containing immobile species, $f_i = \Sigma[M]_i/[M]_{tot}$, can be obtained by assuming that most of the non-mobile M species are also inert (for more details, see reference [BUF 00]). Under these conditions, $[M]_{tot}$ is given by:

$$[M]_{tot} = [M] + \sum_m [M]_m + \sum_i [M]_i \quad [8.7]$$

Moreover, combining equations [8.4] and [8.6] leads to equation [8.8], which holds for the microelectrodes:

$$\frac{i_p^L}{i_p} = \frac{[M]}{[M]_{tot}} + \frac{\sum_m D_m [M]_m}{D_M [M]_{tot}} \quad [8.8]$$

In aquatic systems, most of the mobile and labile species are complexes with small inorganic ligands (carbonate, sulphate, chloride) for which $D_m \sim D_M$. This helps us to approximate a value of f_m, even in these media [BUF 00], provided that the corresponding total concentration of M, $[M]_{tot}$, be measured by another analytical technique (for example, atomic absorption spectroscopy or inductively coupled plasma mass spectrometry (ICP-MS)) or by voltammetry after a ML-decomplexation pre-treatment of the sample. It should be noted that the speciation of M is only possible from precise and reliable measured values of i_p^L/i_p and $E_p - E_p^L$

(errors ≤ 5% and 1 mV, respectively), which necessitates the development of reliable working and reference electrodes (sections 8.4.1 and 8.4.2), and also of optimized cell geometry and electrode positioning (section 8.4.3). It is also important to mention that the calculation of the speciation of M is only possible if no secondary phenomenon liable to alter the values of i_p and E_p occurs. A description of such secondary phenomena, tests and possible "remedies" are proposed in the references [BUF 88, BUF 00]. The phenomena encountered most frequently in *in situ* voltammetric measurements are presented in sections 8.4.4 and 8.4.5.

8.3. Voltammetric techniques used for the analysis of trace elements in waters

8.3.1. *Sensitivity limit of the voltammetric techniques*

Even if no electroactive species is present, the current is not zero in the polarization range of the working electrodes (full curve in Figure 8.2a). This is due to the existence of a capacitor formed at the electrode–solution interface by the accumulation of opposite charges at the surface of the electrode and in the adjacent solution layer [HEY 66, BUF 00]. As in the case of any capacitor, a change in the potential difference at the interface gives rise to a current, called capacitive current, i_c, due to the reorganization of the charges on both planes of the capacitor. In opposition to the faradaic current, (noted i_f in this section), which is due to the redox processes of the electroactive species, i_c implies no charge transfer through the interface. For any value of the potential at which an electroactive species is reduced or oxidized, the total measured current is thus:

$$i_t = i_f + i_c \qquad [8.9]$$

As i_c depends on the whole composition of the medium, but not on the concentration of the specifically analyzed compound, it is a major limitation to the sensitivity of the voltammetric techniques. i_c can be minimized i) by using a modulation $E = f(t)$ (see hereafter) and an appropriate timescale during voltammetry; and ii) by controlling the nature of the electrode–solution interface, and in particular, avoiding any adsorption of compounds at the surface of the sensor. Numerous modulations of $E = f(t)$ have been developed in view of maximizing the ratio i_f / i_c, and thus decreasing the detection limit of voltammetry down to ~10^{-11} mol/l (Table 8.1). Regarding interface control, that of the mercury electrode is much more reproducible than that of solid electrodes (section 8.4.1), which is why mercury still remains the best electrode substrate in voltammetry. Adsorption can be minimized by protecting the electrode by using membranes or gels (section 8.4.6).

8.3.2. Various voltammetric techniques

The main voltammetric techniques used for trace elements analysis in natural waters are summarized in Table 8.1 (for more details see [GAL 94, BAR 82, BUF 88, BUF 00, KIS 84, BAR 66, OST 85, CHA 97]). Cyclic voltammetry (CV) does not have enough sensitivity for *in situ* analysis of water, but it permits the study of electrode processes and optimizes the analytic conditions. The direct reduction methods, such as DPP (differential pulse polarography), DPV (differential pulse voltammetry), or SWV (square wave voltammetry), are used when the concentration of the analyzed element is not too low. For the detection of elements with concentrations $\leq 10^{-8}$ mol/l, techniques such as anodic stripping (ASV) or adsorptive (AdSV) voltammetry with various modulations are necessary. These techniques are based on a pre-concentration step (also called deposition step) during a time, t_d, followed by a re-dissolution step. The pre-concentration step, which may last up to 15 min, causes a significant increase of the measurement time, but, in return, it enables very high sensitivities (down to 10^{-11} mol/l). For the detection of trace metals, the pre-concentration step can be carried out in two ways: i) electroreduction of the analyzed ion to metal atoms deposited on (or in in the case of mercury) the electrode for ASV; ii) adsorption of a complex ML, previously formed by the addition of a ligand, L, complexing with the metal ion under analysis, at the electrode surface, for AdSV. In both cases, the re-dissolution step is operated by a potential scan, linear or modulated as a function of time. For the determination of trace metals, differential pulse modulations (differential pulse: DP) or mainly SWV are the most frequently used methods because they offer the highest detection sensitivities (section 8.4.1).

8.3.3. Voltammetric determinations of natural samples in the laboratory or on the field

An exhaustive list of the voltammetric technical applications for analyzing metals in natural water samples cited since 1984 is given in reference [BUF 00]. Similar tables for applications before 1984 can be found in [KIS 84 (chap. 19), KAL 87, DAV 77, WHI 75]. The direct reduction methods, such as DPP, DPV, and SWV, are mainly used for the determination of Mn(II), Fe(II) and S(-II). However, in most cases, their sensitivity is too low for measuring trace metals in natural water. Techniques comprising a pre-concentration step, such as ASV, AdSV (with various modulations, Table 8.1), allow sensitivities that are 10^3 to 10^4 times higher to be reached. The applications of AdSV for water analysis are rather new but very promising and they should, in particular, considerably extend the range of measurable elements by voltammetry [KAL 87, WAN 88, VAN 89, KAL 88, ABU 98]. Nevertheless, studies are still required to evaluate the reliability and the usefulness of these techniques for routine analysis. The ASV technique has been used in laboratories for over 30 years for trace metals analysis in natural water samples [WAN 89, NÜR 84, VAL 88, NÜR 85, WHI 81, NÜR 83] and instruments have been developed for routine measurements. Figure 8.4 gives a general view of the measurable elements in waters, by ASV or AdSV, with or without modulation.

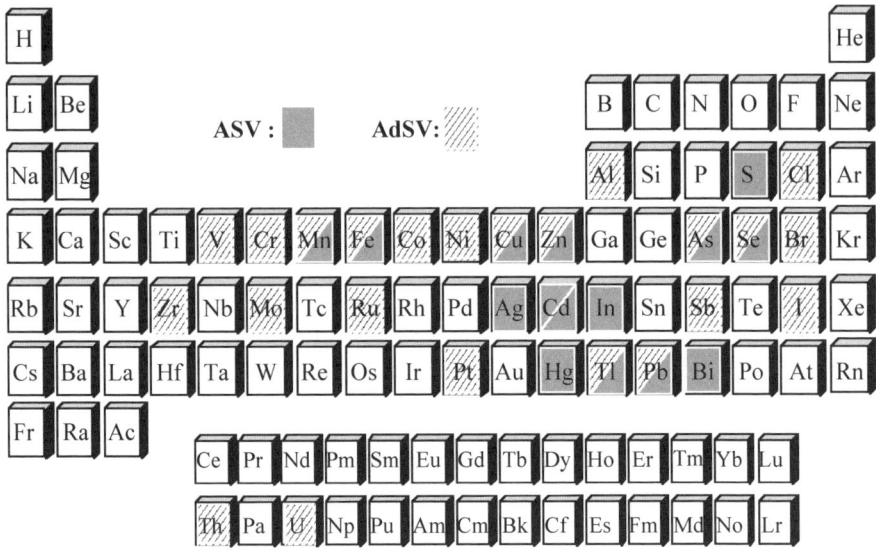

Figure 8.4. *General view of the elements measurable in waters by ASV and AdSV*

In comparison with the numerous analytic procedures developed for the laboratory determinations of trace metals and for their speciation, scarce efforts have been invested in the development of voltammetric equipment specifically designed for direct measurement in the field or *in situ* (section 8.5). The few systems described for measurements in the field can be classified as portable field instruments [MAN 84, MAN 87, BEH 90] and stationary instruments installed on the shore or on board boats for automated continuous measurements, *ex situ*, of trace metals by ASV or AdSV [COL 97, CLA 85, ACH 99]. These systems present a major drawback: they are based upon standard commercial equipments developed for laboratory use (electrodes, voltammetric analyzers, pumps, and valves) and therefore, are not directly adaptable, and even less optimized for *in situ* measurements. In fact, the major improvements undertaken have been i) for the portable instruments: power supply via a battery and an optimization of the size of the instrument for making the device transportable with the reactants inside a box [VAL 82, BON 93]; and ii) for stationary instruments: the development of continuous flow cells in a few cases [COL 97, CLA 85, ACH 99].

8.4. Development of reliable submersible voltammetric probes

The main components of a complete submersible voltammetric analyzer, called a voltammetric probe, are the electronic components and the software (which will not be discussed in this section) and the voltammetric cell including the three electrodes, and, in particular, the working electrode or voltammetric sensor. In this section, the development of sensors and of cells for continuous *in situ* voltammetric determinations is discussed. The conditions that must be fulfilled for developing a reliable submersible voltammetric probe are:

– for the voltammetric cell and the probe as a whole:

- as little transformation as possible of the sample during analysis, to minimize analytical artifacts, as well as the number and the volume of reactants,

- optimization and simplification of the analytic steps for minimizing the complexity of the system and, in particular, the number of valves and pumps, the energy required, as well as the electronic control system,

- flexibility and easy calibration,

- simple setting and maintenance in field conditions;

– for the sensor:

- selection of the most reliable working and reference electrodes,

- choice of optimal cell geometry and electrode positioning,

- optimization of the electrochemical conditions to avoid any alteration of the surface of the sensor and to ensure re-conditioning of the surface of the sensor if necessary,

- short analysis time,

- coping with the problem of pollution of the sensor surface,

- elimination of the interference due to the oxygen.

These aspects are discussed in more detail below.

8.4.1. *Working electrodes*

The working electrode is the key part of any voltammetric system, as the quality and the reliability of measurements depend mainly on this sensor. An ideal working electrode should have a reproducible surface area, and a low residual current. These requirements are not so easy to comply with. Reviews of the working electrodes used in voltammetry [BON 90] and more specifically for the speciation of trace metals in natural waters [GOY 93, FLO 86, SMA 87] can be found in the literature. As already mentioned, they can be divided into two main groups: the mercury electrodes and the solid-state electrodes (including chemically modified electrodes), with two types of geometry: the macro-electrodes (size >100 µm) and the microelectrodes (size ≤10 µm). In this section, the discussion is limited to the important aspects concerning *in situ* voltammetric determinations.

8.4.1.1. *Macro-electrodes: characteristics and limitations.*

The most commonly used materials, for voltammetric macro-electrodes, are: mercury, gold, platinum, carbon [HEY 66, KIS 84, BAR 66, YOS 81], their choice depending on the desired polarization range (Figure 8.2b). Mercury electrodes remain the most frequently used due to their very high reliability and their very low capacitive current. The classic mercury electrodes are: the dropping mercury electrode (DME), the static mercury drop electrode (SMDE), the hanging mercury drop electrode (HMDE), and the mercury film electrode (MFE). All of them are used in analysis of waters [BUF 00].

The advantage of the mercury drop electrodes comes from the fact that the electrode is renewed prior to each measurement, which minimizes memory effects. However, they are not easy to use for *in situ* measurements: the mercury drop-renewal device is not very adaptable, and the necessary volume of mercury is rather large.

The MFE is obtained by depositing a mercury layer on a substrate, by electrolysis of a Hg(II) solution. MFE offers lower detection limits and higher selectivity than HMDE, and also a better selectivity in ASV, because the peaks are sharper. But they are more sensitive to the problems of adsorption of compounds at the surface of the sensors, and to the problems of formation of intermetallic compounds and of mercury over-saturation by the analyzed metals, due to the low volume of mercury in use. Among the typical metals analyzed, indium, cadmium, thallium, zinc, tin, lead, and bismuth are highly soluble in mercury (> 1% (w/w)); copper, manganese, and nickel have an intermediate solubility (2 to 8×10^{-3} % (w/w)); and other metals, such as cobalt and iron have too low solubility ($<10^{-3}$% (w/w)) which preclude their analysis by ASV [KIS 84, GAL 84]. The formation of many intermetallic compounds in mercury has been reported. The most pertinent ones, concerning water analysis, are: Cu-Cd, Cu-Mn, Cu-Zn, Fe-Mn, Ni-Zn, Pt-Zn, and Pt-Cd. The existence of the last two compounds causes a severe limitation of the use of platinum as a substrate for the MFE (see below). Due to the very low concentration of metals generally observed in natural waters, the other intermetallic compounds are not formed in significant quantities. However, Cu-Zn may be an important exception, depending on the conditions [ZIR 81, PEI 00]. The formation of insoluble CuZn compounds and of $CuZn_2$ complexes soluble in mercury depends on the concentration of these metals in the medium analyzed, and on the duration of the pre-concentration steps in ASV techniques [PEI 00, PIC 87]. The most frequently used substrates for the MFE are: platinum, carbon (glassy carbon or graphite), iridium, gold, and silver. All these substrates have drawbacks, which cause the MFE to have too low a reliability for long-term *in situ* applications. Platinum, gold, and silver are soluble in mercury. In the case of gold and silver, the mercury film is then converted into a film of concentrated amalgam, the nature of which is time-dependent [YOS 81, STO 77]. They may also form intermetallic compounds with the analyzed metals (for example, cadmium and zinc with platinum [COX 67]). A MFE with a gold substrate has been used for measurements in sediments, with a reported typical stability of one day [BRE 95b]. Inert substrates, such as glassy carbon and graphite, also have limitations. On these substrates in particular, mercury is deposited as small droplets instead of a true film. In spite of their frequent use for daily laboratory analyzes, such films cannot be used for continuous *in situ* measurements over long periods. Moreover, graphite may become oxidized when the redox potential of the medium is too high, which limits its lifetime [BRA 90a]. Iridium is currently the only substrate having low solubility in mercury, being hardly oxidizable, and on which true mercury films can be deposited [KOU 86, KOU 87]. But the mercury film on large iridium surface areas is not very stable and turns to mercury droplets once the electrode is disconnected, which also hinders applications in *in situ* measurements. However, it must be noted that a very stable mercury film is obtained on microsubstrates of iridium (see section 8.4.1.2).

Solid state electrodes, such as platinum, gold, silver, and various types of carbon have also been used, mainly for the analysis of elements having redox potentials higher than the potential of mercury oxidation (Figure 8.2 b). Such electrodes are more and more often suggested for replacing mercury. However, voltammetric analyzes of metals, using solid-state electrodes, are less reproducible and sensitive than when mercury electrodes are used. This is due to a higher capacitive current, and to the fact that the electrochemical reactivity of the test compounds depends on the crystalline plane of the solid electrode in contact with the solution. Moreover, for techniques with a pre-concentration step, the crystalline form of the metal (ASV), or of the deposited compound (AdSV), may also significantly influence the shape of the re-oxidation peak. Intermetallic compounds are also formed much more readily than on mercury. A major problem linked to the long-term use of solid-state electrodes is how to regenerate their surface. Various methods have been described: mechanical polishing [BER 69], laser activation [POO 87], treatments in a plasma [EVA 79], chemical [RIC 83], electrochemical techniques [ENG 84], or heating under vacuum [FAG 85]. Except for the electrochemical (and possibly chemical) treatments, these methods are not applicable *in situ*. Despite these difficulties, solid-state electrodes remain necessary for certain *in situ* applications and their development continues. The most promising solid electrode is the gold electrode

[BUF 00]. Iridium is also interesting because it is hardly oxidizable, but it has not yet been systematically tested for trace analyzes.

Chemically modified electrodes (CMEs) constitute a vast category of mercury-free electrodes. They have given rise to much interest over the last 20 years, and numerous review articles have been published [GUN 89, WAN 89, TAY 96, ABR 88, DON 89, KAL 90, BAL 91, WIG 89]. The CMEs are generally quite specific in the analysis of a given compound, only their general principle will be mentioned here. The electrode surface (a carbon paste-based substrate in most cases) is modified by: direct chemical bonding of a substance to the surface, adsorption, reaction in a plasma, or other processes. The modification may result in a molecular change at the surface or in the formation of a rather thick layer. Surface-chemical modification may have various functions: preferential pre-concentration of a compound at the surface, immobilization in the layer of a catalyst, which can enhance the electrochemical response or alter the physical properties of the electrode surface, for instance, to shift and separate overlapping signals. Although these CMEs have potential interest, their applications remain limited to the research field.

Let us finally note that an important limitation of any type of macroelectrodes is that *in situ* measurements can only be performed on natural waters with high enough ionic strength, e.g. sea waters, unless the sample is modified with the addition of electrolytes.

8.4.1.2. *Microelectrodes: properties and advantages*

Mercury, carbon, platinum, gold, silver and iridium microelectrodes (size ≤10 µm), with different geometries, have been developed [BUF 00, BON 94]. The simple microelectrodes are produced using conventional mechanical techniques [WIG 89, FLE 87, TER 95]. Electrode networks are generally fabricated using microtechnical methods such as: thin-layer deposition, photolithography, and LIGA (*Lithographie Galvanoformung Abformung*) [FIA 00]. These techniques offer important advantages, in particular: automated production at reduced costs at the industrial scale and a flexible choice of microelectrode geometry Individually addressable microelectrode arrays (see example, Figure 8.5c, section 8.4.6), with a center-to-center spacing as short as 10 µm, can be produced using such techniques. Voltammetric, amperometric, or potentiometric measurements can be done successively using each electrode [MEY 95, FRI 92], or by groups of microelectrodes, by means of a multiplexer [HER 94, FIA 94, TER 00b]. This type of sensor is of great interest for measurements at interfaces [TER 00b].

The considerable interest for microelectrodes since the end of the 1970s is due to the remarkable characteristics obtainable by decreasing the size of the sensor [WIG 89, FLE 87, TER 95, MAG 89, STR 90]. In particular: i) a small ohmic drop and a reduced double layer capacitance allowing determinations in low conductivity media, and for very high sweep rates ($\leq 10^6$ V/s); ii) due to a higher mass transfer rate (which increases when the size of the electrode decreases) and to a reduced capacitive current, the microelectrodes have a higher signal:noise ratio than the macro-electrodes, and therefore, a higher sensitivity; iii) the current intensities at microelectrodes are controlled by spherical diffusion and reach, for a constant potential and long times, values different from zero; iv) the signals measured at microelectrodes during the pre-concentration step are independent of the hydrodynamic conditions and of time (section 8.2.3) and also negligible for species having sizes over a few nanometers.

Due to these unique properties, microelectrodes are the electrodes of choice for *in situ* measurements of trace metals, and their speciation in natural waters. In particular, the characteristics i) to iv) permit, respectively:

– direct measurements, even in soft waters with low ionic strength (typically HCO_3^- concentration $\leq 5 \times 10^{-3}$ M) with no addition of electrolyte;

– detection of elements present in very low concentration (10^{-11}-10^{-9} mol/l) [TER 95, DAN 89, WAN 87, SOT 87, BEL 96] with shorter pre-concentration times than with a macro-electrode;

– the avoidance of stirring for the pre-concentration steps of the ASV and AdSV techniques; the measured currents are thus more reproducible, which helps interpreting the signals in terms of speciation, and the instrumentation and the cell geometry are much simplified;

– an improved discrimination between the truly dissolved and colloidal species [BUF 00], which is essential in quality control and (eco-)toxicity assessment.

However, their use for routine applications and continuous automated measurements is limited, mainly because of their poor stability and of their lack of reproducibility. This is mainly due to the fact that they are produced using the same type of materials (platinum, carbon, iridium, gold, and silver) as for the macro-electrodes and, therefore, present the same limitations (see above). The only exceptions today are mercury film-electrodes on an iridium substrate: the surface tension of mercury and the cohesion forces between mercury and iridium favor an optimum stability of the mercury deposits on iridium microdiscs. These electrodes can be transferred from one solution to another, stirred, and rinsed with no significant modification of their performances [TER 95]. A reliability and a reproducibility close to 100% have been obtained with this type of sensor for the electrodeposition of mercury films and continuous determinations of trace metals using SWASV (square wave anodic stripping voltammetry) in low ionic strength media (0.1 to 10 mmol/l) during a period up to 14 days without renewing the mercury film [TER 95, TER 00a]. Let us note that such a reliability of the sensor necessitates not only the optimization of all the analytic procedures, but also the systematic control of the fabrication steps [TER 95, BEL 96]. The key fabrication steps that must be perfectly controlled are: perfect welding between the iridium wire and the glass at the tip of the simple microelectrodes, the morphology of the surface of the iridium substrate, good adhesion of the passivation layer, and the absence of re-deposition of materials used for the fabrication on the surface after the photolitographic operation in the case of microelectrodes produced by microtechniques, minimized resistance of the electric contact Ir-Cu.

8.4.2. *Reference electrodes*

The reference electrode is the second key-component of any voltammetric cell. A detailed description of the principle of reference electrodes for various laboratory physico-chemical studies is given in several books (for instance [CAT 73, CAT 74, COV 69]). It is essential, for certain applications, that the reference electrode potential be well-defined and stable within ± 0.1 mV. For *in situ* analysis of trace elements, a potential stability of ± 1 mV is generally sufficient. Other important criteria must be considered for selecting the reference electrode of a submersible voltammetric probe [BUF 00]:

1) the electric resistance of the electrode must be stable and low (typically ≤50 kΩ) when the techniques using fast modulations, such as SWASV are used at microelectrodes;

2) the influence of pressure or temperature variations upon the stability of the potential must be negligible;

3) contamination of the studied medium due to the reference electrode is a major potential problem for any determination of compounds with very low concentrations ($< 10^{-8}$ mol/l). This is linked to the fact that any reference electrode includes highly concentrated chemical reagents (for example, AgCl, KCl, for the silver chloride reference electrodes). These reagents may contain certain impurities in low proportions, although sufficient for contaminating the samples by diffusion through the liquid junction. Moreover, test compounds may also diffuse in the internal electrolyte of the reference electrode and later be released in samples (memory effect) in subsequent analyzes;

4) the size of the electrode must be small for integrating the small voltammetric cells which are desirable for *in situ* measurements (see section 8.4.3). However, a size-limit exists, which depends on the nature and the fabrication of the electrode, under which the performance of the electrode is not controlled any more;

5) the reproducibility of the fabrication procedure, the mechanical strength, and the long-term stability are also important characteristics.

The various types of available reference electrodes, their advantages and limitations for *in situ* measurements, are summarized in Table 8.2 and briefly discussed hereafter:

– classic commercial reference electrodes with conventional liquid junctions of the a) type are not suitable because they are sensitive to pressure and are difficult to handle because of the internal solution;

– pseudo-reference electrodes of the (d) type are not suitable because their potential depends on the medium and on the pH (significant variations of these parameters can be observed in natural waters as a function of distance and/or depth);

– metal/metal oxide electrodes of type (e) covered by a buffered gel are potentially promising pseudo-reference electrodes, but more systematic studies are necessary to confirm this;

– solid-state reference electrodes of type c) have a stable potential, non-depending on pressure, but their lifetime is limited to 2 to 3 weeks;

– electrodes of type b) are stable, reliable, and long-lasting. They can withstand pressures as high as 60,000 kPa (= 600 bars) and can, therefore, be used at depths of 6,000 m. Their drawback is their large size, except for type b(5), and possible problems linked to contamination and memory effects. The use of a ZrO_2 low porosity ceramic junction (type b(5)) helps to solve the problem of contamination.

Electrode type	Advantages	Disadvantages	Ref.
a) Reference electrodes with internal liquid 1) Ag/AgCl/x M KCl 2) Hg/Hg$_2$Cl$_2$/x M KCl (Calomel)	- well controlled and stable E - commercially available	- large size - contamination of the media by Cl$^-$ - liquid junction problem due to the pressure frequently encountered for measuring in low ionic strength (even for P compensated electr.)	[CAT 73, CAT 74]
b) Reference electrodes with polymers 3) Ag/AgCl/KCl sat. in solid gel (no ceramic junction) 4) Ag/AgCl/KCl in Xerolyt polymer (no ceramic junction)	- well controlled and stable E - withstand P up to: 3) 7×10^7 Pa (700 bars) and 4) 1.6×10^6 Pa (16 bars) without P compensation - commercially available - well controlled and stable E - small size - low diffusion through the ceramic junction - withstand P up to 3×10^6 Pa	- large size - risk of contaminations due to diffusion through the gel - influence of T not tested	*† [TER 98a, TER 98b]

5) Ag/AgCl/KCl sat. in gel (Metrohm or 1% LGL agarose); zirconium oxide ceramic junction	(30 bars; max. P tested) without P compensation - readily prepared - lifetime > 1 year		
c) Solid-state reference electrodes 6) Ag/AgCl/vinyl ester resin doped with KCl (Reflex low resistance material) 7) Hg/Hg$_2$Cl$_2$/KCl/graphite paste in a PVC tube closed with a composite Teflon glass membrane 8) Ag/AgCl wire inserted in a pellet of KCl-alumina-Teflon pressed around it	- stable E - small size - readily prepared -stable E - small size - systematic tests of the influence of T (5-25°C) and P (10^5 to 3×10^6 Pa, i.e. 1 to 30 bars) on E	- limited lifetime: 6) ~ 3 months, 7) ~ 2 weeks - influence of T and P not tested - limited lifetime: ~ 2 weeks	[DIA 94, REH 95, RUZ 72] [BOU 77, JER 92]
d) Ag/AgCl covered with a thin film 9) Ag/AgCl/Nafion film (~ 20 μm) 10) Ag/AgCl/polyacrylamide hydrogel (30 μm) 11) Ag/AgCl/melted glass layer (few 100 μm)	- small size - readily prepared - no internal electrolyte	- E = f(media, pH) - limited lifetime: 9) – 2 weeks, 10), 11) ~ 1 month 11) fragile	[ARQ 94, BRE 95, MOU 94a, MOU 94b]
e) Metal/metallic oxide pseudo references 12) Sb/Sb$_2$O$_3$; 13) Ru/RuO$_2$; 14) Ti/TiO$_2$; 15) Ir/IrO$_2$; 16) Pd/PdO; 17) Si/MO with M=Pt, Pd, Ru, Ir	- small size - no internal electrolyte	- test media must be pH buffered or the elect. covered with a pH buffered gel - E = f (pH) - oxide layer preparation tricky	[ARD 81, GAL 74, GLA 89, KRE 95]

Table 8.2. *Various types of reference electrodes, and in particular those of type b)-e) which have been suggested for miniaturization and suppression of the internal solution (adapted from reference [BUF 00]). * (3) Idronaut Srl (Via Monte Amiata 10, I-20047 Brugherio, MI) †(4) Ingold Electrodes Inc. (261 Ballardvale St, Wilmington, MA, USA)*

The type b(5) and c(8) electrodes have been successfully used for *in situ* voltammetric monitoring of trace metals for periods of several days and seem to be the most reliable electrodes.

8.4.3. *Voltammetric cells*

Two types of arrangements for three-electrode systems (working, reference, and auxiliary electrodes) can be considered for *in situ* measurements (see section 8.5): direct immersion of the three electrodes in the natural medium, with the electrodes being isolated from each other or combined to form an integrated multisensor (system a); or integration of the three electrodes in a continuous flow cell (system b). System a is simple and potentially more compact that continuous flow cells (system b), but it is much more limited in terms of the number of compounds that can be analyzed and long-term operation (see below). This is because, despite its complicated technical aspect, a continuous flow cell has many advantages for *in situ* measurements in the water column, in particular, for automated measurements, continuously, over long periods of time (≥ 1 week). These electrodes are protected and clogging by large particles, algae, plankton cells, etc, is easy to avoid using a filter with large pores (~100 µm) at the cell inlet.

The problems of contamination and of losses by adsorption on the walls of the cell and the electrodes can be minimized by careful rinsing of the cell with the sample. Another major advantage is the possibility of connecting a flow injection analysis (FIA) system [RUZ 88] for performing *in situ*: i) chemical transformations of the medium and/or of the test compounds before the voltammetric determination, ii) the ASV technique with medium exchange, or iii) the AdSV. Such possibilities may considerably extend the number of compounds measurable *in situ* by voltammetry, including anions and organic compounds [BUF 00].

Detailed discussions on the advantages and drawbacks of the numerous configurations of continuous flow cells, for given applications, can be found in references [KIS 84 (chap.22), STU 84, STU 92, STU 87], and for *in situ* measurements more specifically in reference [BUF 00]. The important characteristics to consider for the development of a submersible voltammetric cell for *in situ* determination of trace elements are summarized hereafter:

– selection of suitable working (section 8.4.1) and reference electrodes (section 8.4.2);

– well-controlled cell geometry (and hydrodynamic conditions for macroelectrodes) allowing a theoretical prediction of the faradaic current. This implies optimal positioning of the three electrodes in order to minimize: i) the ohmic drop and to ensure a perfect distribution of the potential on the surface of the working electrode, which is particularly important when the measurements are made in low conductivity media, such as fresh waters; and ii) interference of the compounds produced by the auxiliary electrode;

– conditions ensuring a high mass transfer rate and a high sensitivity; for macroelectrodes, it depends only on the hydrodynamic conditions; for microelectrodes, it depends only on the electrode geometry;

– volume of the cell as small as possible (ideally: 10-100 µL); this is specifically important for *in situ* analysis with chemical pre-treatment to minimize the volume of the consumed reagent and the volume of the collected solution after each measurement;

– minimum noise, necessitating minimized electric resistance and optimized position of the electrodes. It must be noted that an important reduction in the size of the cell channel may entail an increased noise [STU 81] with an intensity depending strongly on the quality of the polishing of the cell material and on the type of hydrodynamic flow (laminar or turbulent). More detailed information concerning the various noise sources and their effects is given in references [STU 81, JOH 86];

– robustness, easy maintenance, and/or replacement of the working and reference electrodes on the field;

– pressure compensation (for use at depths >10 m);

– automated valves and pumps working in depth, with low energy consumption and simple maintenance.

The development of a cell complying with all these characteristics is not an easy task. The submersible cell of the VIP (voltammetric *in situ* profiling) system presented in section 8.5.4 (Figure 8.8b) [TER 98a] was developed with consideration of all of them. It is described hereafter for as an illustration. A three-electrode configuration, consisting of an integrated gel microelectrode (see section 8.4.6) and a mini-reference electrode Ag/AgCl/KCl gel saturated (Table 8.2, type b5) and a platinum ring as the auxiliary electrode, were used. A microelectrode was selected because of the advantages mentioned in section 8.4.1.2, in particular for the fact that it permits making ASV determinations, even in fresh waters with low ionic strength, and without stirring. This last characteristic significantly increases the reproducibility of the measurements and reduces the energy consumption, which is important for automated measurements over long periods of time.

The positioning of the electrodes is presented in Figure 8.8b of section 8.5.4. This configuration i) prevents the products generated at the auxiliary electrode from interfering upon the measurement, ii) minimizes accumulation, at the electrode surfaces, of gas microbubbles due to temperature changes, iii) minimizes the ohmic drop thanks to the very low distance between the three electrodes. The compartment between the internal cell A and the external cell B (Figure 8.8b, section 8.5.4) is filled with an agarose gel in 1 M $NaNO_3$.

This gel is in contact with the solution and the reference electrode, fixed at the bottom of the external cell B, by means of porous ceramic ZrO_2 junctions. It plays several important roles: i) double bridge for the reference electrode minimizing Cl^- diffusion towards the sample; ii) screening of the working and of the auxiliary electrodes and iii) pressure equalizer by means of an rubber pressure compensator (Figure 8.8b). The cell is screwed, via "O-ring" joints, on the lid of a waterproof sub-box containing the preamplifier of the voltammetric sensor and the motor of the submersible peristaltic pump. The advantages of this configuration are the following: i) the voltammetric cell is protected against shocks; ii) the maintenance and the replacement of the working and the reference electrodes are easy as these electrodes are simply screwed via "O-ring" joints; iii) the water tightness of the electric connections is possible for normal conductors and iv) the preamplifier is very close to the sensor and isolated from the main electronic part, which minimizes important sources of noise.

8.4.4. *Interference due to the dissolved oxygen*

Oxygen is a major electroactive compound in natural oxygenated waters and can thus interfere in the measurement of other elements. In particular, its concentration in oxygen-saturated waters is $\sim 3 \times 10^{-4}$ mol/l, which is much larger than the trace metal concentrations. Oxygen electroreduction yields two successive reduction waves at mercury electrodes, corresponding to the de oxygen reduction to H_2O_2 (~ -0.1 V versus Ag/AgCl/KCl sat.; equation [8.10]) and the H_2O_2 reduction to water (~ -0.9 V versus Ag/AgCl/KCl sat.; equation [8.11]) successively:

$$2\ O_2 + 4\ H^+ + 4\ e^- \rightarrow 2\ H_2O_2 \qquad [8.10]$$

$$H_2O_2 + 2\ H^+ + 2e^- \rightarrow 2\ H_2O \qquad [8.11]$$

Similar reduction processes occur at non-mercury electrodes. For laboratory voltammetric measurements, dissolved oxygen is generally eliminated by bubbling nitrogen (N_2), helium or argon, or, in the case of natural water samples, N_2/CO_2 in well-controlled proportions [LEC 81] for preventing eliminating CO_2 (which would entail a pH change, and thus a possible metal speciation change).

Oxygen reduction may induce two types of interferences during *in situ* analysis of trace metals in oxygenated waters: a) generation of a signal much larger than those due to the analyzed trace metals, which consequently masks them, b) pH increase at the surface of the electrode resulting from H^+ consumption by the oxygen reduction reactions (equations [8.10]–[8.11]).

It has been shown that, for measurements in well-buffered aquatic media, such as sea water, where the interference b) is not observed, interference a) can be minimized using techniques of subtractive voltammetry or SWASV with a rather high frequency [TER 90, WOJ 85], and eliminated by combining both techniques [TER 90, TER 98a]. But, during measurements in fresh waters, which have a low buffering capacity (natural HCO_3^- concentration typically $\leq 5 \times 10^{-3}$ M), oxygen reduction, during the pre-concentration step, may entail a rise in pH up to ~11 (oxygen-saturated samples) at the electrode surface and induce the formation and the precipitation, of hydroxides and/or carbonates of the analyzed metals during the anodic re-dissolution step [BUF 81a]. The voltammetric signals are then strongly distorted, or even suppressed. To avoid this difficulty, either i) a pH buffer is added to the sample, which is not recommended for trace analyzes because of contamination risks, or ii) oxygen must be eliminated prior to the measurement. Several laboratory systems have been proposed for online removal of oxygen from the solutions. They are based either upon chemical [PER 81, MAC 84, OLS 83], photochemical [BAR 92], or electrochemical [HAN 80] reduction techniques, or on removal techniques by a) diffusion through hydrophobic membranes [COL 97, TER 00a, BES 96, MOS 95, PED 94, ROL 83, CHA 96] or b) transport, via a ligand, through supported liquid membranes [JOH 87, NIE 84].

Only the latter techniques are promising for on field and *in situ* applications. In type a) techniques, the sample is pumped along a semi-permeable membrane [BES 96, MOS 95] or a tube surrounded by a reducing agent or an inert gas under low pressure. Such systems have been used for measurements onboard boats [COL 97, ACH 99] for analyzing the total concentration of metals after acidification and ultraviolet (UV) irradiation.

For *in situ* analyzes and speciation of non-treated samples, a system based on the selective trapping of oxygen, in view of increasing the rate of oxygen removal, while minimizing CO_2 diffusion, and thus the pH variations in the test samples, is necessary. Such a development has been recently reported in the literature [TER 00a]. This system is based on oxygen diffusion through a silicon tube surrounded by a reticulated enzymatic gel trapping oxygen selectively. Such a system, based on a gel, is also more suitable in *in situ* measurements because of the easier transport and pressure compensation of gels. The type b) technique should also permit selective removal of oxygen, but it necessitates more development before being used for routine applications.

8.4.5. Interferences due to the adsorption of organic and inorganic compounds

The major problem encountered in direct voltammetric measurements in complicated matrixes such as natural waters, is the covering of the working electrode surface by biopolymers and organic or inorganic colloids (fouling effect). Adsorption of these compounds may: i) increase significantly the capacitive current [BUF 81b]; ii) stop the electron transfer of redox reactions and thus decrease, or even suppress the voltammetric signals [TER 96]; and iii) induce a complexation of metal ions at the surface of the electrode thus modifying the current and peak potentials [BUF 88, BUF 87]. Methods have been proposed for minimizing these effects. They are briefly presented hereafter.

The removal of organic compounds by oxidation by UV irradiation is commonly used in laboratories. However, this procedure is too intricate for *in situ* applications. Brainina and collaborators have proposed an electrochemical method for oxidizing organic compounds, which can be applied *in situ* to sea waters [BRA 90a, BRA 84]. Oxidation of organic compounds is produced by the oxidized species generated at a graphite anode by the following reactions:

$$2\ Cl^- \rightarrow Cl_2 + 2\ e^- \qquad [8.12]$$

$$Cl_2 + H_2O \rightarrow HCl + HClO \qquad [8.13]$$

$$2\ H_2O \rightarrow O_2 + 4H^+ + 4\ e^- \qquad [8.14]$$

The trace metals can then be measured by voltammetry in the organic compounds-free solution. This process is very useful for the determination of the total concentration of metals in seawater. However, it is not applicable to measurements and speciation in fresh waters. Moreover, it does not protect the surface against adsorption of inorganic compounds.

Thin, protecting, semi-permeable membranes (with a typical thickness from 5 to 20 μm) covering the mercury film electrodes have been proposed for preventing diffusion of interfering compounds at the surface of the electrode by size-exclusion effects and/or electric charge-repulsion effects. The most commonly used are thin membranes of cellulose acetate or of Nafion [BAL 91, ALD 93, VID 92] (for more details see reference [BUF 00]). Although interesting results have been obtained with these two types of membranes for direct measurements in synthetic solutions containing various surfactants [SMA 85, DAL 94] and biologic media [HOY 87], these membranes are not efficient enough in the case of direct measurements in natural samples containing humic and fulvic low molecular weight compounds

260   Chemical and Biological Microsensors

[DAL 94, ZEN 95]. Moreover, most of the proposed membranes are not perfectly inert towards the analyzed compounds and can thus entail memory effect problems [ALD 93, DAL 94].

It has been demonstrated recently that the use of a "thick" membrane present specific advantages for protecting the surface of voltammetric microsensors [TER 96, BEL 98]. A mercury film microelectrode has been covered with a layer of agarose gel, rather thick (200–500 µm) in comparison with the electrode size and the width of the diffusion layer existing during voltammetric measurements (see section 8.4.6). This very pure agarose gel has been proved to be totally inert towards monovalent and divalent cations and anions. Consequently, these compounds diffuse freely through the gel. Conversely, this gel efficiently protects the electrode surface against the polysaccharides produced by organisms [BEL 98], clay particles [TER 96, BEL 98], autochthonous iron oxyhydroxides [TER 98b], and certain humic compounds [TER 96]. The properties of this gel have been used for the development of new gel-integrated sensors described below.

8.4.6. Gel-integrated microsensors

In fact, the abovementioned agarose gel membrane plays a much more important role than that of a mere protection against adsorption of compounds at the sensor surface. The combination of this gel with a microelectrode (Figure 8.5d) should instead be considered as an integrated analytical microsystem, which performs the two distinct following operations:

– the gel ensures i) the separation of the electroactive species from the majority of the other species by dialysis during the establishment of the equilibrium between the gel and the test solution, and ii) well-controlled physical and chemical conditions at the electrode surface (see below). Both actions are important in facilitating the interpretation of voltammetric data,

– the voltammetric sensor detects species inside the gel.

It is important to mention that macro-electrodes cannot be used in such a system (for more details, see reference [TER 96]).

Three types of voltammetric gel-integrated microelectrodes (GIME) have been developed (Figure 8.5a–c; [TER 00b, TER 96, BEL 98]): a gel-integrated mercury film microelectrode on a single iridium substrate covered with an agarose membrane called µ-AMMIE (agarose gel membrane mercury-plated iridium-based microelectrode; Figure 8.5a) and two mercury film microelectrodes on iridium substrates in a network covered by an agarose membrane, one interconnected called µ-AMMIA (agarose gel membrane mercury-plated iridium-based microelectrode aray;

Figure 8.5b) and the second individually addressable called IA' μ-AMMIA (Figure 8.5c). Details on their fabrication steps are given in the references [TER 95, TER 00b, BEL 96, TER 96, BEL 98].

Basically, the single iridium microelectrode is produced by sealing an iridium wire, with a tip having be electro-etched to a diameter of a few μm, in a glass capillary covered with a palladium layer ensuring electric screening, followed by a mechanical polishing.

The iridium microelectrodes arrays are produced using thin-layer microtechniques [FIA 00]. Their geometry can be easily changed. The geometry of the microelectrode array used in reference [BEL 98] consists of an array of 5 × 20 interconnected iridium microdiscs having a diameter of 5 μm and a center-to-center distance of 150 μm, surrounded by a 300 μm-thick epoxy SU 8-ring acting as a reservoir for the agarose gel (Figure 8.5b).

The geometry of the individually addressable microelectrode array used in reference [TER 00b] consists of 64 lines of three iridium microdiscs having a diameter of 5 μm and a center-to-center distance of 150 μm, surrounded by nine 300 μm-thick epoxy SU 8-rings. The spacing between the sensor line numbers 1 to 23 decreases from 2000 μm to 220 μm and keeps a constant value of 200 μm between the sensor lines numbered 23 to 32. The geometry of lines numbered 33 to 64 is the mirror-image of that of the lines 1 to 32. This configuration has been selected to enable real-time measurements of concentration profiles at interfaces over a total distance of 4 cm (2 cm in each phase) with a maximum resolution of 200 μm at the interface (see [PEI 01] for more details).

Before use, the surfaces of the three types of microelectrodes are covered with a 1.5% LGL type of agarose gel membrane (high-purity gel; sulfide and impurities < 0.03%)). Mercury hemispheres are then deposited on the iridium substrate(s) and re-oxidized electrochemically through the agarose gel using solutions of Hg^{2+} and KSCN, respectively. For optimal morphology of the iridium discs, the electrochemical yields for the formation of the mercury hemispheres are close to 100%, which ensures a very high reproducibility of the size of the mercury deposit (variability ≤5% [TER 95, BEL 96, TER 96]). A given agarose membrane can be used over a period of more than one month, and it has been demonstrated that the diffusion through the membrane does not depend on the pressure even for pressures as high as 6×10^4 kPa (600 bars) [BEL 98]. Mercury deposits are stabilized by the gel and can be used during periods up to 14 days (maximum tested time) without being renewed [TER 00a].

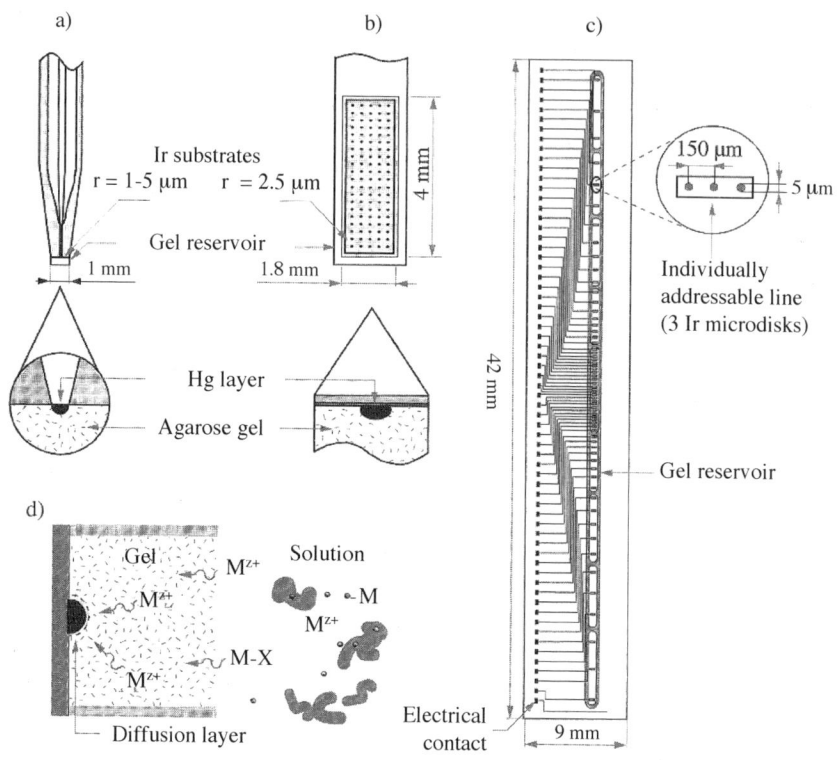

Figure 8.5. *Schematic representation of gel-integrated microsensors based on: a) a simple microelectrode, b) an interconnected microelectrode array, c) an individually addressable microelectrode array, and d) gel-integrated microsensor principle (adapted from reference [TER 01])*

Measurements using gel-integrated microsensors are performed in two successive steps: 1) establishment of equilibrium between the gel and the test solution (typically 5 min for a 300 μm-thick membrane) and 2) voltammetric measurement inside the gel. Examples of sensitivity and reproducibility obtained with these sensors are given in Figures 8.6, 8.9 and 8.10. The major advantages of these sensors, in addition to those reported in section 8.4.1.2, for *in situ* measurements in natural media, are the following:

a) bio-polymers and natural inorganic colloids are efficiently excluded from the gel and thus cannot interfere in the voltammetric measurements ([TER 98b, TER 96, BEL 98], section 8.5.4);

b) well-controlled molecular diffusion of the mobile species settles inside the gel and thus the uncontrolled hydrodynamic conditions in the bulk of the aquatic medium have no influence on the voltammetric signals [TER 96];

c) the voltammetric signals measured inside the gel depend on the diffusion coefficients of the elements in the gel and not in the medium, which is particularly important for interpreting correctly the measurements *in situ* in sediments [PEI 01, TER 01];

d) well-controlled chemical conditions are conceivable, such as pH control inside the gel, and thus at the surface of the sensor;

e) the natural medium is not modified by the voltammetric measurement as the voltammetric detection is made once the equilibrium between the gel and the test medium is reached, and as the voltammetric diffusion layer is small in comparison with the gel thickness;

f) because of the advantages a)–e), the interpretation of voltammetric data in terms of speciation is significantly simplified with respect to voltammetric measurements in the absence of gel [BUF 00, PEI 00, TER 01].

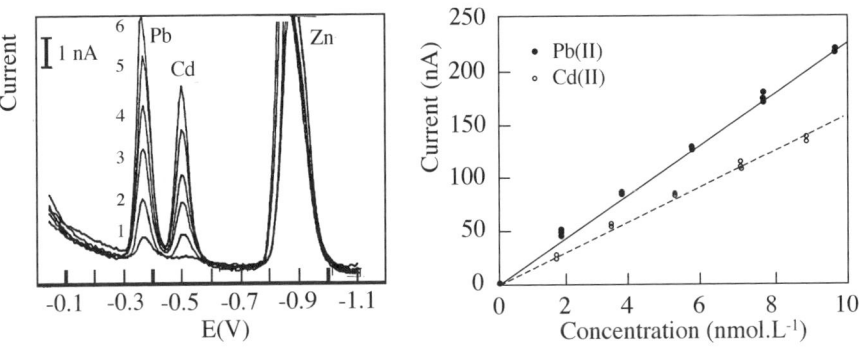

Figure 8.6. *Example of data for Pb(II) and Cd(II) determinations at nanomolar concentration levels. a) SWASV curves for (curve 1) measurements in samples of the river Arve, Switzerland, filtered on a 0.2 µm thick membrane ([Pb] = 0.8 nmol/l; [Cd] = 0.35 nmol/l); and (curves 2–6) successive standard additions of 1 nmol/l of each metal; µ-AMMIE. b) Triplicate calibration curves measured in the same type of medium; µ-AMMIA* [TER 95, BEL 96]

8.5. Submersible voltammetric probes reported in the literature

To date, very few efforts have been devoted to developing submersible voltammetric probes. Most of the systems reported were prototypes based on the immersion of a continuous flow cell or of a sensor connected via a long cable to a standard commercial instrument located on the surface. Their use was generally restricted to daily measurements on the surface (depth ≤ 20 m). The very first prototypes were reported in 1990 by Tercier *et al.* [TER 90] and Brainina *et al.* [BRA 90b]. All the reported systems are briefly presented below.

8.5.1. *Continuous-flow probe based on a microelectrode and a pre-treatment of the sample*

Brainina *et al.* [BRA 90b] developed a probe for concentration measurements of total trace metals in seawater. Although a few descriptions of this system have been given, no detailed article has ever been published. It mainly consists of two units performing *in situ*: i) oxidation of organic compounds by electrochemical generation of chlorine and simultaneous acidification of the sample (section 8.4.5; [BRA 84, BRA 90a]) and ii) voltammetric measurements of trace metals (copper, lead, cadmium and zinc) in the treated sample. A thin layer cell integrating an electrode consisting of a mercury film pre-deposited on a graphite substrate is used for the voltammetric detection. Both units are integrated in a submersible continuous flow device. The system also includes the possible addition of a standard solution.

8.5.2. *Continuous-flow probe based on a macro- or a microelectrode with no sample treatment*

This probe was reported by Tercier *et al.* [TER 90]. It was based on three sub-units: a Plexiglas continuous-flow voltammetric cell, a submersible unit also made of Plexiglas, and a control box with its communication cord. Various types of electrodes have been integrated into this probe: a SMDE, a mercury film on a glassy carbon substrate macroelectrode (MFE) and an MFE on an iridium substrate. For ensuring, good functioning of the electromagnetic valves to a depth of 20 m, the pump and the step motor (used for forming the mercury drops when using a SMDE) were located in the submersible unit, and a nitrogen counter-pressure of 0.5 to 0.7 bar in excess of the pressure at the measurement depth was imposed inside the unit via the communication cord. This probe was successfully used for *in situ* measurements of copper, cadmium, lead, and zinc in seawater without pre-treating the sample [TER 90].

This prototype has shown the advantages and the perspectives of real-time *in situ* measurements. It has also indicated the developments and criteria to consider (section 8.4) for the achievement of a probe usable in routine applications. In particular, trace metal concentrations as low as 3×10^{-11} mol/l could be determined in oxygen-saturated seawater by SWASV using a mercury film on glassy carbon macro-electrode.

The importance of *in situ* measurements for minimizing contamination problems and measuring daily variations of metal concentrations has been established. This probe has shown that *in situ* measurements without any sample pre-treatment are more difficult in fresh waters than in sea waters because of the low ionic strength, the poor buffering capacity, and the presence of higher concentrations of colloidal and particle compounds liable to adsorb at the sensor surface in fresh waters. It evidenced the necessity of developing a microsensor with a protecting membrane (section 8.4.6) and of an oxygen-removal system coupled to the voltammetric cell (section 8.4.4; [TER 00a]) for measurements in fresh waters.

8.5.3. *Probes based on direct immersion of the electrodes*

In situ analyzes by potentiometric stripping analysis (PSA) at constant current by direct immersion of a combined sensor including the working, reference, and auxiliary electrodes have been reported by Wang *et al.* [WAN 95]. The combined sensor was fixed to a polyvinyl chloride (PVC) tube connected to a commercial potentiostat on the surface via a 15 m long, screened communication cable (Figure 8.7a). A working macroelectrode consisting of a gold cylinder (100 μm in diameter, 5 mm in length), characterized in the laboratory for the measurement of mercury, copper, and lead by PSA [GIL 94], has been used. The system had been calibrated previously in the laboratory by standard additions in synthetic seawater. Linear relationships were observed for the three metals in the concentration range of 50 to 150 nmol/l. Detection limits of 5, 10, and 25 nmol/l (corresponding to a signal:noise ratio of 3) have been estimated, for copper, lead, and mercury, respectively, after a pre-concentration time of 5 min, with a standard deviation of 2.8% for Cu^{2+} (based on 20 repeated measurements in a solution containing 200 nmol/l Cu^{2+}).

This system has been used for the *in situ* determination of Cu(II) in the surface waters of various stations situated in the bay of San Diego. The sensor was immersed at 0.6 m depth on the side of a small boat. Cu(II) concentrations in the range of 20–60 nmol/l have been measured. It is interesting to note that electrode surface poisoning (i.e. fouling problems) have been reported in the case of similar gold electrodes during PSA measurements in river waters, but not in sea waters [GIL 94].

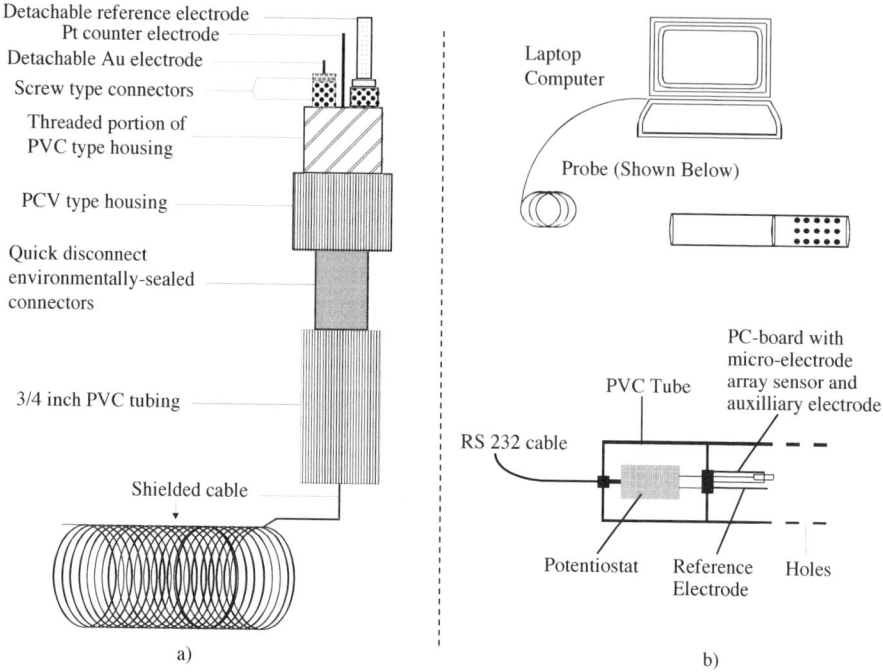

Figure 8.7. *Schematic representations a) of the combined sensor reported by Wang et al. [WAN 95] and b) of the probe reported by Herdan et al. [HER 98]*

A similar system has been reported by Luther et al. [LUT 99] for *in situ* measurements of Fe(II), Mn(II) and S(-II) by SWV in the sediments. The combined sensor consisted of a 100 µm in diameter Hg-Au amalgam working electrode, a Ag/AgCl pseudo-reference electrode and a platinum auxiliary electrode. This combined sensor was fixed to a micromanipulator allowing the insertion of the working electrode in sediments within a millimeter scale resolution; the whole set being mounted on a small, autonomous vehicle remote-controlled from a boat. The combined sensor and the micromanipulator were respectively connected via a 30 m cable to a potentiostat and a control motor located on board. Detection limits of 5, 10, and less than 0.2 µmol/l, with a standard deviation of 5%, were estimated for Mn(II), Fe(II) and S(-II), respectively, from calibrations performed in the laboratory by standard additions in synthetic sea water. The system was applied to *in situ* measurements of the elements cited above in interstitial waters of sediments, over a total depth of 5 cm, in the Raritan bay of New York (depth 5–6 m). Concentration profiles have been reported for these three elements. Note that if currents can easily be measured for these three elements, the interpretation of the concentration values is very complicated and should only be made on the basis of a detailed understanding of the electrode processes. The important aspects to consider are discussed in detail in reference [BUF 00] (see also section 8.4.6).

Another type of probe based on direct immersion of the electrodes and the SWASV has been proposed by Herdan et al. [HER 98], for a quick and *in situ* screening of heavy metal concentrations in underground waters. The working electrode, similar to that of Figure 8.5b, consisted of a network of 20 interconnected iridium microdiscs electrochemically coated with mercury hemispheres. An iridium microband, integrated on the same chip as that of the working electrode, and a Ag/AgCl commercially available electrode were used as auxiliary and pseudo-reference electrodes, respectively (Figure 8.7b). The three electrodes were protected by a PVC tube with holes drilled into it to allow water circulation (Figure 8.7b). The complete submersible probe was made by this tube connected via O-ring joints and screws to a PVC case and a microcontroller. The power supply of the potentiostat used four 9 V batteries. The performance of this probe for a quick screening for trace metals has been evaluated by testing in three wells of a contaminated site (Bedford, MA-USA). More specifically, the probe was used for i) undertaking surface and *in situ* measurements of the mobile fraction of Cu(II) and Pb(II); and ii) undertaking surface and/or in the laboratory measurements of the total concentrations of these two metals after acidification of the samples. Cu(II) and Pb(II) concentrations in the 0.06–0.2 μmol/l and 0.15–1.5 μmol/l ranges, respectively, were obtained for *in situ* measurements performed in the three different wells at depths between 6 and 9 m. The comparison of the results obtained *in situ* and on the surface with the electrochemical and ICP analyzes made in the laboratory showed that the developed probe allows the estimation of the order of magnitude of the concentrations and evolution trends of the metals investigated. The innovative entity about this probe was the miniaturized potentiostat, which permitted a considerable reduction in the size of the system (probe diameter: 5 cm), which is particularly interesting for *in situ* measurements in underground waters. However, the electric noise of such a potentiostat is much larger than that of classic potentiostats, which limits considerably the sensitivity of this system. Moreover, the reproducibility and the precision of the measurements were also lower than those reported for classic systems.

8.5.4. *Continuous-flow probe based on a gel-integrated microsensor with no sample pretreatment: the VIP system*

The systems described in sections 8.5.1 to 8.5.3 have demonstrated the feasibility and the utility of submersible voltammetric probes for *in situ* monitoring of trace metals in aquatic media. In most cases, however, these probes are not sensitive enough for the determination of trace metals in non-polluted waters, and there are no probes available for continuous measurements over a period of time superior to one day or for the determination of concentration profiles at important depths. The VIP system was developed to solve these problems (Figure 8.8a). It is based on a sophisticated sensor, an advanced microprocessor, and a telemetric data transfer technology. The submersible part, comprising a voltammetric and a multiparameter probe, was fabricated to enable measurements down to a depth of 500 m [TER 98a, TER 98b]. The core of the voltammetric probe is a µ-AMMIE or a µ-AMMIA, which are described in section 8.4.6 (Figures 8.5a and 8.5b). These gel-integrated microsensors were developed specifically to permit direct voltammetric measurements in complex media, and more specifically in aquatic media. The voltammetric probe is comprised of several parts: an electronic housing, a Plexiglas flow-through voltammetric cell with pressure compensation (internal volume of 1.5 ml) based on a three-electrode system (Figure 8.8 b; for more detail, see section 8.4.3), a watertight subcasing incorporating a pre-amplifier for the microsensor and a submersible peristaltic pump. The electronic housing contains all the hardware and the software necessary for managing: the voltammetric measurements; the data acquisition, storage and transfer, on operator request, by RS-232 (max. depth 100 m) or telemetry (depth >100 m); the interfacing of a multi-parameter probe, a calibration deck unit, a surface telemetry deck unit, and the submersible peristaltic pump (Figure 8.8a).

User-friendly Windows *management software* allows the user to configure the operating conditions for the voltammetric techniques, to calibrate the voltammetric and multiparameter probes, to control the data transfer from the internal probe memory to the PC, to perform the automatic data treatment and maintenance operations, and to perform *in situ* measurements on the manual or automatic mode according to a pre-programmed sequence. The measurement files are stored in an internal memory powered by its own battery, which guarantees the preservation and protection of the data. The multiparameter probe ensures the accurate control of the depth of the voltammetric probe and the simultaneous measurement of basic major physico-chemical parameters necessary for interpreting the trace metal measurements, such as temperature, pH, oxygen, conductivity, salinity, and redox potential. The calibration deck unit performs the following procedures in the laboratory or on the field: i) renewing the sensor mercury layers, ii) probe calibration with standard solutions, and iii) analysis of the samples taken and chemically treated. The surface deck unit powers and interfaces the voltammetric probe with the PC by telemetry (depth >100 m). The telemetric signal is superimposed to the power supply all along the cable sustaining the probe. This unit has autonomy of about 35 hours and can be recharged, either after use, or in a continuous mode via solar cells. A detailed technical description of the VIP system is given in [TER 98a].

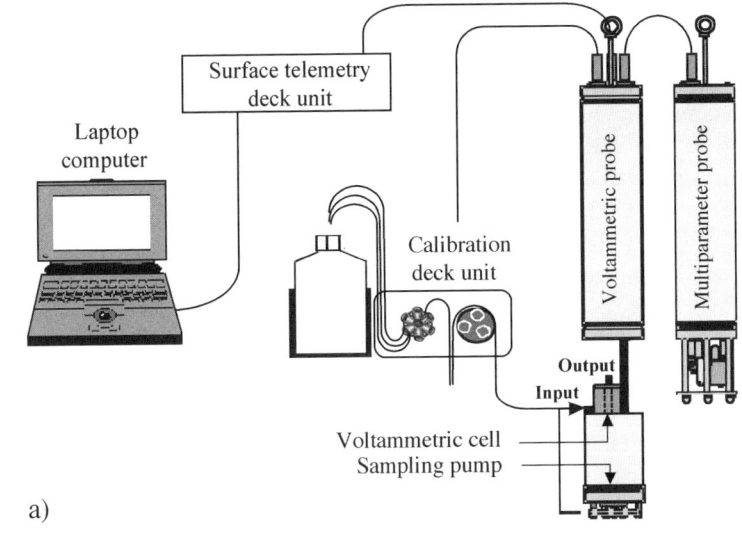
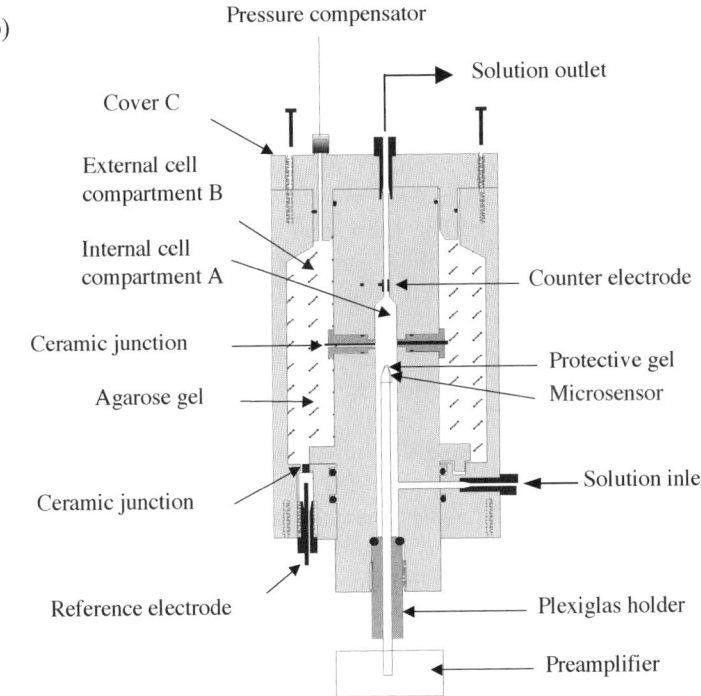

Figure 8.8. *a) Schematic representation of the VIP system and b) of its flow-through cell [TER 98]*

At the present time, systematic tests of *in situ* voltammetric measurements (sea waters [TER 98a], fresh waters [TER 98b] and underground waters [PAU 01]), and of the influence of parameters such as: temperature, pressure and ionic strength upon the voltammetric signals (section 8.6; [TER 00a, TER 98b]), have only been reported for the VIP system. A few examples are briefly presented below and used for discussing evaluation and performance of the voltammetric technique *in situ*.

8.5.4.1. Mn(II) in anoxic lakeside medium

Mn(II) concentration profiles have been measured in the (anoxic) hypolimnion of Lake Bret, Switzerland (depths: 11–18 m) and of Lake Lugano, Switzerland (depths: 80–95 m) with the aim of testing: i) the stability, the precision, and reliability of the measurements performed *in situ* with the VIP system, and ii) of studying the speciation of Mn(II) in both lakes [TER 98b]. For doing this, the *in situ* voltammetric measurements from the VIP probe have been compared with: i) voltammetric measurements done at the surface on a raft, using sensor arrays with and without protecting gel; and ii) laboratory measurements done using AAS, ICP-AES and/or colorimetry in acidified samples (pH = 2) of raw water, filtered water (on membranes with pore sizes of 0.2 μm or 0.45 μm) and ultra-centrifuged water (30,000 revolutions/min for 15 h, corresponding to a limit size of the order of 5 nm [PER 94]). The *in situ* measurements were done with no renewing of the mercury layers of the sensor during three to five days of operation, including the calibrations of the probe the day before and after each field deployment. The calibration procedure is an important aspect, which is discussed in more details in section 8.6. An example of the manganese concentration profiles obtained in Lake Bret with the various techniques is given in Figure 8.9. The major results gained during these studies were the following. The calibration curves were reproducible before and after a field measurement operation, and from one deployment to another (standard deviations of the calibration curve slopes of 5.2–6.9% (95% probability) for seven calibration cycles [TER 98b]). This demonstrates the reproducibility of the formation of the mercury layers on the gel-integrated microsensors, their stability, and the absence of memory effect during the *in situ* field operation. The good correspondence (Figure 8.9) between the concentration profiles obtained from *in situ* measurements, after correcting for the effect of temperature, and the surface voltammetric measurements, done at 20°C, as well as the ICP determinations done in the ultracentrifugated samples, demonstrates the validity of the information gained from *in situ* voltammetric measurements. In particular, theses tests confirmed the validity of the correction factors, determined in the laboratory, for temperature effect on the voltammetric signal [TER 99], and that pressure has no effect on the results [BEL 98]. Comparing the voltammetric measurements done with sensors that were covered with an agarose gel membrane or not covered (Figure 8.9) shows clearly the importance and the efficiency of the protecting effect of the agarose gel. In particular, the concentrations obtained with a non-protected sensor were systematically too low, due to the adsorption, on the mercury surface, of colloidal autochthonous Fe(III) hydroxides [TER 98b]. So, the measurements done using sensors without protection membranes would have led to: i) significant underestimations of Mn(II) concentrations and erroneous values and shape profiles of the Mn(II) concentrations as a function of the depth; and ii) an incorrect interpretation of the speciation of Mn(II) in Lake Bret in that season (see below). The inter-comparison between the voltammetric *in situ* measurements and the spectrometric determinations in the laboratory, of raw and filtered or centrifuged samples, obtained information on the nature of the manganese species present in both lakes. In Lake Bret, Mn(II) was mainly present as Mn^{2+}, while in Lake Lugano a significant proportion of Mn(II) was adsorbed on colloidal species [TER 98b].

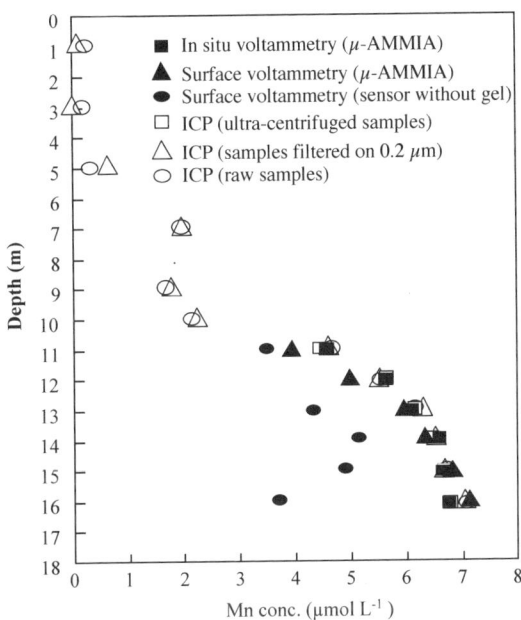

Figure 8.9. *Concentration profiles of Mn(II) measured in Lake Bret, Switzerland (August 20, 1997; adapted from reference [TER 98])*

8.5.4.2. *In situ measurements of trace metals in oxygenated sea and fresh waters*

As mentioned in section 8.4.4, oxygen is an electroactive compound, which may cause two types of interferences upon the measurement of metals in oxygenated media: a) a signal with an order of magnitude 4 to 6 times higher than that due to the analyzed trace metals, and b) a pH increase at the electrode surface, due to oxygen reduction, in particular during the preconcentration step of the techniques involving anodic re-dissolution. The applications of the VIP system to *in situ* measurements in oxygenated sea waters (Lagoon of Venice, Italy and Gullmar Fjord, Sweden) showed that no pH increase (interference b) was observed in sea water because of the large buffering capacity of this medium, and that interference a) can be eliminated by using SWASV, with a rather high frequency, coupled with the subtraction of the background noise current [TER 98]. Based on a signal:noise ratio of 2, the detection limits obtained for Cu(II), Pb(II), Cd(II) and Zn(II) in sea waters saturated with oxygen are 0.1, 0.05, 0.05 and 0.3 nmol/l, respectively, using a preconcentration time of 15 min.

Station	Metal	C_m [nM]	C_{tot} [nM]	C_m/C_{tot} [%]
1	Cu	1.50	12.96	11.6
	Pb	0.65	3.98	16.3
	Cd	0.02	0.06	33.3
2	Cu	1.47	12.65	11.6
	Pb	0.40	2.25	17.8
	Cd	0.02	0.05	40.0
3	Cu	0.94	19.79	4.7
	Pb	0.42	1.55	27.1
	Cd	<LD	0.04	----
4	Cu	0.84	18.57	4.5
	Pb	0.30	0.89	33.7
	Cd	<LD	0.03	----
5	Cu	1.54	19.58	7.9
	Pb	0.24	0.87	27.6
	Cd	<LD	0.04	----
6	Cu	2.00	12.41	16.1
	Pb	0.11	0.61	18.0
	Cd	0.02	0.04	38.0
7	Cu	1.17	10.34	11.3
	Pb	0.09	0.34	26.5
	Cd	0.01	0.04	25.0
8	Cu	1.87	12.95	14.4
	Pb	0.11	0.60	18.3
	Cd	0.02	0.06	33.3
9	Cu	1.13	10.22	11.0
	Pb	0.10	0.35	28.5
	Cd	0.01	0.03	33.3
10	Cu	0.96	10.92	8.8
	Pb	0.11	0.46	23.9
	Cd	0.01	0.04	25.0

Figure 8.10. *Coastal measurements of trace metals in the lakes of Lucerne and Alpnach, Switzerland (June 2000). C_m = concentrations of the mobile fraction of the metals measured in situ with the VIP system (Fig. 8.8a); C_{tot} = total concentrations of the metals measured in the laboratory by ICP-MS in samples acidified to pH = 2. <LD = below the detection limit [TER 01]*

In oxygenated fresh waters, interference b) may entail drastic deformations, or even suppression, of the voltammetric signal of trace metals, because of the formation of hydroxide and/or carbonate metal precipitates during the anodic re-dissolution step. To solve this problem, a submersible oxygen removal system has been developed [TER 00a]. The efficiency of this module, coupled to the VIP system for *in situ* voltammetric measurements in oxygenated fresh waters, has been tested and demonstrated in various Swiss lakes: Leman [TER 00a], Lucerne, and Alpnach. The *in situ* measurements using the VIP probe were done on surface waters (depth: 2–5 m) in various stations. Samples were also collected (exactly at the same depths), then acidified to pH = 2 (HNO_3 suprapure) for the determination by ICP-MS of the total extractable metal concentrations. As an example, the concentrations in the waters of the lakes of Lucern and Alpnach determined using the VIP system *in situ* and using ICP-MS in the laboratory are given in Figure 8.10. The detection limits for the various metals were similar to those obtained in sea waters (see above). The ratios of the mobile species concentrations to those of the total extractible metals were typically ranging from 5 to 40%. Similar proportions have been observed in the River Arve [PEI 00, TER 95] and in the Lagoon of Venice [TER 98a]. Standard deviations ≤10%, determined from three successive measurements in the stations 1, 3, 6 and 8, were obtained for the three metals, demonstrating that concentrations and concentration variations at levels as low as 10^{-11}–10^{-10} mol/l can be measured *in situ* with good reproducibility. This allows: i) the detection of rapid changes in the concentration of mobile metal species as a function of the variation of the physicochemical conditions of the medium; and ii) the interaction of these species with the aquatic "biota" to be studied, as observed during automated *in situ*, continuous measurements in Gullmar Fjord in Sweden [TER 98a], and during the determinations of the trace metal concentration profiles at Lake Leman [TER 00a], respectively. These characteristics are essential for any analytical tool used for quality control of waters and for metal speciation studies.

8.6. Conclusion

The voltammetric techniques allow the determination of a large number of trace elements (Figure 8.4, section 8.3.2) with a sensitivity matching those of the most sensitive spectrometric techniques, but with shorter overall analysis duration [TER 98b]. Moreover, they are potentially able to measure a large number of other inorganic and organic compounds [BUF 00]. Due to their basic principle, they can yield information on the speciation [PEI 00, TER 95, TER 00a, TER 01] with no necessary coupling with additional separation techniques, in contrast to most of the other classic detection techniques. Miniaturized instruments, with low energetic consumption, can be developed. Due to these unique characteristics, the voltammetric techniques are the method of choice for the automated monitoring of chemical compounds *in situ* and in real-time in aquatic media. In addition, to the understanding of their theoretical principles, two important factors must be considered to optimize their development and for their application in routine analysis [BUF 00].

8.6.1. Calibration of voltammetric procedures

Because numerous factors may affect the voltammetric signals, they must be carefully calibrated, and the physico-chemical processes involved must be fully considered. In particular, several factors must be studied in detail:

– *pH, temperature and ionic strength.* The pH acts strongly upon numerous redox reactions, either directly or indirectly. For example, potential variations from 0 to – 60 mV per pH unit (sometimes more) may occur according to the redox reaction considered. The voltammetric peak potential of an element may thus vary as a function of the depth during the recording of the concentration profiles, which must be taken into account in the interpretation of the data. The influence of temperature on the voltammetric peak currents is also important. Typically, current intensity variations of 1 to 10%/°C, depending on the element, may be observed. It is, therefore, essential to take these influences into account for a rigorous conversion of the peak currents to concentrations, in particular to monitor concentration profiles (i.e. the temperature in lakes and oceans may vary typically from 25 to 4°C as a function of the depth). Ionic strength at values $\leq 10^{-2}$ M, and the nature of the electrolyte, may also influence the current and potential peak values. Therefore, the influence of this parameter must also be studied for each environmental system;

– *hydrodynamic convection* is a major problem when sensors ≥ 100 µm are used; this parameter can also play an important role even in the case of microelectrodes with a diameter in the order of 10 µm [TER 95]. Nowadays, the gel-integrated microsensors of a few micrometers are the only microsensors able to ensure that the currents measured *in situ* are controlled by pure diffusion;

– *the voltammetric parameters,* in particular when they are linked to the measuring time of the technique, (sweep rate, pulse duration, frequency, pre-concentration time), may have a strong influence upon the peak current, potentials and the peak width. These factors must, therefore, be studied in detail and optimized. Such a study also obtains useful information on electrode processes [GAL 94].

Finally, it is important to note that it is generally necessary to obtain linear relations with reproducible slopes for classic calibration curves (peak current as a function of the concentration of the test compound) in solutions of a pure electrolyte, but that it is certainly not a sufficient condition for using these curves to compute the concentrations of the analyzed compounds from their peak currents measured in environmental media. Understanding the electrode mechanism and the effects of the major compounds of the medium studied on the voltammetric signals is a necessity (see discussion in details in [BUF 00]).

8.6.2. Development of robust and reliable sensors and probes

The development of sensors that are not influenced by adsorption phenomena is necessary for *in situ* measurements. Currently, four types of electrodes have been used for measurements *in situ*: the Hg-Au amalgam electrode, the major limitation of which is that it remains stable for a maximum period of 12 hours [BRE 95b]; the gold electrode, the reliability of which has not been clearly defined [WAN 95]; the mercury film on iridium substrate microelectrode, which is influenced by adsorption of inorganic or organic compounds [TER 96, BEL 98, HER 98], and the mercury film on iridium substrate microelectrode integrated in a gel [TER 00a, TER 98a, TER 98b, BEL 98, PEI 01, TER 99], which has been proved to be reliable for continuous *in situ* measurements over periods of several days in various types of natural waters (sea waters, fresh waters, underground waters). In all cases, additional studies are necessary for a better understanding of the advantages and limitations of these electrodes. The chemically modified electrodes based on nanotechnologies are promising, but much work is required for this type of sensors to become usable in routine applications. Finally, the main reported limitation of most of the submersible probes described is that they are mainly based upon an adaptation of laboratory tools. The only exception, up to now, is the VIP system (section 8.5.4), which was developed specifically for automated long-term measurements at important depths. It is the only system commercially available (Idronaut Srl, Italy; www.Idronaut.it) with capacity for *in situ* trace metal monitoring and profiling, down to 500 m, with sub-nanomolar sensitivity. Such developments necessitate close collaboration between academic scientists and industrial partners.

8.7. Bibliography

[ABR 88] ABRUÑA H.D., "Coordination chemistry in two dimensions: chemically modified electrodes", *Coord. Chem. Rev.*, vol. 86, p. 135-189, 1988.

[ABU 98] ABUZUHRI A.Z., VOELTER W., "Applications of adsorptive stripping voltammetry for trace analysis of metals, pharmaceuticals and biomolecules. A review", *Fres. J. Anal. Chem.*, vol. 360, p. 1-9, 1998.

[ACH 99] ACHTERBERG E.P., BRAUNGARDT C.B., "Stripping voltammetry for the determination of trace metal speciation and distributions in marine waters", *Anal. Chim. Acta*, vol. 400, p. 381-397, 1999.

[ALD 93] ALDSTADT J.H., DEWALD H.D., "Effect of model organic compounds on potentiometric stripping analysis using a cellulose acetate membrane-covered electrode", *Anal. Chem.*, vol. 65, p. 922-926, 1993.

[ARD 81] ARDIZZONE S., CARUGATI A., TRASATTI S., "Properties of thermally prepared iridium dioxide electrodes", *J. Electroanal. Chem.*, vol. 126, p. 287-292, 1981.

[ARQ 94] ARQUINT P., KOUDELKA-HEP M., DE ROOIJ N.-F., BÜHLER H., MORF W.E., "Organic membranes for miniaturized electrochemical sensors: fabrication of a combined PO_2, PCO_2 and pH sensor", *J. Electroanal. Chem.*, vol. 378, p. 177-183, 1994.

[BAL 91] BALDWIN R.P., THOMSEN K.N., "Chemically modified electrodes in liquid chromatography detection. A review", *Talanta*, vol. 38, p. 1-16, 1991.

[BAR 66] BARD A.J. (ed)., *Electroanalytical Chemistry*, Marcel Dekker, 1966-1998.

[BAR 82] BARD A.J., FAULKNER L.R., *Electrochemical Methods: Fundamentals and Applications*, Masson, 1982.

[BAR 92] BARISCI J.N., WALLACE G.G., "Removal of oxygen in flowing solutions using a photochemical process", *Electroanalysis*, vol. 4, p. 323-326, 1992.

[BEH 90] BEHNERT J., RAEZKE K.P., "Schwermetallanalytik mittels Voltammetrie", *LaborPraxis*, vol. 14, p. 508-511, 1990.

[BEL 96] BELMONT-HEBERT C., TERCIER M.-L., BUFFLE J., FIACCABRINO G.C., KOUDELKAHEP M., "Mercury-plated iridium-based micro-electrode arrays for trace metals detection by voltammetry: optimum conditions and reliability", *Anal. Chim. Acta.*, vol. 329, p. 203-214, 1996.

[BEL 98] BELMONT-HEBERT C., TERCIER M.-L., BUFFLE J., FIACCABRINO G.C., KOUDELKA-HEP M., "Gel-integrated micro-electrode arrays for direct voltammetric measurements of heavy metals in natural waters and other complex media", *Anal. Chem.*, vol. 70, p. 2949-2956, 1998.

[BER 69] BERGE H., STRÜBING B., "Verwendung von Kohleelektroden bei der voltammetrischen Anordnung mit kontinuierlich aktivierter Oberfläche", *Fresenius Z. Anal. Chem.*, vol. 247, p. 12-15, 1969.

[BES 96] BESSARABOV D.G., JACOBS E.P., SANDERSON R.D., BECKMAN I.N., "Use of nonporous polymeric flat-sheet gas-separation membranes in a membrane-liquid contactor: experimental studies", *J. Memb. Sci.*, vol. 113, p. 275-284, 1996.

[BON 90] BOND A.M., SCHOLZ F., "A survey of electrodes used for voltammetric analysis", *Z. Chem.*, vol. 30, p. 117-129, 1990.

[BON 93] BOND A.M., SVESTKA M., "Developments, trends and commercial availability of instrumentation (hardware and software) in microcomputer based voltammetry. Review", *Collect. Czech. Chem. Commun.*, vol. 58, p. 2769-2812, 1993.

[BON 94] BOND A.M., "Past, present and future contributions of micro-electrodes to analytical studies employing voltammetric detection. A review", *Analyst*, vol. 119, p. R1-R21, 1994.

[BOU 77] BOUND G.P., FLEET B., "The development of a solid state reference electrode for use in soil measurements", *J. Sci. Food. Agric.*, vol. 28, p. 431-437, 1977.

[BRA 84] BRAININA K.Z., STOZHKO R.M., YU N., CHERNYSHEVA A.V., "Electrochemical wet ashing and separation of interfering elements in stripping voltammetry of natural waters", *Zh. Anal. Khim.*, vol. 39, p. 1660-1663, 1984.

[BRA 90a] BRAININA K.Z., VILCHINSKAYA E.A., KHANINA R.M., "Influence of the redox potential of the medium on stripping voltammetric measurement results", *Analyst*, vol. 115, p. 1301-1304, 1990.

[BRA 90b] BRAININA K.Z., KHANINA R.M., FORHSTADT V.M., VILCHINSKAYA E.A., GAPONENKO G.L., in: *Proc. of J. Heyrovsky Centenial Congress on Polarography*. 41th Meeting of International Society of Electrochemistry, Prague, p.175, 1990.

[BRE 95a] BREHIER D.C., BELFORD R.E., "Thick-film reference electrodes for solid-state pH measurement", *Anal. Proc. Ind. Anal. Comm.*, vol. 32, p. 323-326, 1995.

[BRE 95b] BRENDEL P.J., LUTHER III G.W., "Development of a gold amalgam voltammetric micro-electrode for the determination of dissolved Fe, Mn, O_2 and S(-II) in porewater sediments", *Environ. Sci. Technol.*, vol. 29, p. 751-761, 1995.

[BUF 81a] BUFFLE J., "Calculation of the surface concentration of the oxidized metal during the stripping step in the anodic stripping techniques and its influence on speciation measurements in natural waters", *J. Electroanal. Chem.*, vol. 125, p. 273-294, 1981.

[BUF 81b] BUFFLE J., COMINOLI A., "Voltammetric study of humic and fulvic substances. Part IV: behaviour of fulvic substances at the mercury–water interface", *J. Electroanal. Chem.*, vol. 121, p. 273-299, 1981.

[BUF 87] BUFFLE J., VUILLEUMIER J.J., TERCIER M.-L., PARTHASARATHY N., "Voltammetric study of humic and fulvic substances. Part V: Interpretation of metal ion complexation measured by anodic stripping voltammetric methods", *Sci. Tot. Environ.*, vol. 60, p. 75-96, 1987.

[BUF 88] BUFFLE J., "Studies of complexation properties by voltammetric methods", Chapter 9 in *Complexation Reactions in Aquatic Systems. An Analytical Approach*, Horwook, 1988.

[BUF 00] BUFFLE J., TERCIER-WAEBER M.-L., "*In situ* voltammetry: concepts and practice for trace analysis and speciation", Chapter 9 in BUFFLE J., HORVAI G. (eds), *In situ Monitoring of Aquatic System; Chemical Analysis and Speciation*, IUPAC Series in Analytical and Physical Chemistry of Environmental Systems, 2000.

[CAT 73] CATON Jr. R.D., "Topics in chemical instrumentation. LXXIII. Reference electrodes", *J. Chem. Educ.*, vol. 50, p. A571-A578, 1973.

[CAT 74] CATON Jr. R.D., "Topics in chemical instrumentation. LXXIII. Reference electrodes" (concluded), *J. Chem. Educ.*, vol. 51, p. A7-A8, A14-A19, 1974.

[CHA 96] CHAI X.S., DANIELSSON L.G., "Approaches to in-line removal of dissolved oxygen in flow systems for process analysis", *Anal. Chim. Acta*, vol. 332, p. 31-38, 1996.

[CHA 97] CHAMPAGNE G.Y., CHEVALET J., "High sensivity multiple waveform voltammetric method and instrument.", PCT International Applied Patent No. Au 97-40057, US 97-798016, p. 101, 1997.

[CLA 85] CLAVELL C., ZIRINO A., "On-line shipboard determination of trace metals in seawater with a computer-controlled voltammetric instrument", Chapter 8 in ZIRINO, A., (ed.), *Mapping Strategies in Chemical Oceanography*, ACS, 1985.

[COL 97] COLOMBO C., VAN DEN BERG C.M.G., DANIEL, A., "A flow cell for on-line monitoring of metals in natural waters by voltammetry with a mercury drop electrode", *Anal. Chim. Acta*, vol. 346, p. 101-111, 1997.

[COV 69] COVINGTON A.K., "Reference electrodes", Chapter 4 in DURST R.A., (ed), *Ionselective Electrodes*, NBS Special publication 314, US Governmental Printing Office, 1969.

[COX 67] COX J.A., Analytical utility of mercury film electrode, PhD Thesis, University of Illinois, 1967.

[CRO 69] CROW D.R., *Polarography of Metal Complexes*, Academic Press, 1969.

[DAL 94] DALANGIN R.R., GUNASINGHAM H., "Mercury(II) acetate-Nafion modified electrode for anodic stripping voltammetry of lead and copper with flow-injection analysis", *Anal. Chim. Acta*, vol. 291, p. 81-87, 1994.

[DAN 89] DANIELE S., BALDO M.A., UGO P., MAZZOCCHIN G.A., "Determination of heavy metals in real samples by anodic stripping voltammetry with mercury micro-electrodes. Part 2: Application to rain and sea waters", *Anal. Chim. Acta*, vol. 219, p. 19-26, 1989.

[DAV 77] DAVISON W., WHITFIELD M., "Modulated polarographic and voltammetric techniques in the study of natural water chemistry", *J. Electroanal. Chem.*, vol. 75, p. 763-789, 1977.

[DIA 94] DIAMOND D., MCENROE E., MCCORRICK M., LEWENSTAM A., "Evaluation of a new solid-state reference electrode junction material for ion-selective electrodes", *Electroanalysis*, vol. 6, p. 962-971, 1994.

[DON 89] DONG S., WANG Y., "The application of chemically modified electrodes in analytical chemistry. A review", *Electroanalysis*, vol. 1, p. 99-106, 1989.

[DRY 77] DRYHURST G., *Electrochemistry of Biological Molecules*, Academic Press, 1977.

[ENG 84] ENGSTROM R.C., STRASSER V. A., "Characterization of electrochemically pretreated glassy carbon electrodes", *Anal. Chem.*, vol. 56, p. 136-141, 1984.

[EVA 79] EVANS J., KUWANA T., "Introduction of functional groups onto carbon electrodes via treatment with radio-frequency plasmas", *Anal. Chem.*, vol. 51, p. 358-365, 1979.

[FAG 85] FAGAN D.T., HU I.F., KUWANA T., "Vacuum heat treatment for activation of glassy carbon electrodes", *Anal. Chem.*, vol. 57, p. 2759-2763, 1985.

[FIA 94] FIACCABRINO G.C., KOUDELKA-HEP M., JEANNERET S., VAN DEN BERG A., DE ROOIJ N.F., "Array of individually addressable micro-electrodes", *Sens. Actuators B*, vol. 19, p. 675-677, 1994.

[FIA 00] FIACCABRINO G.C., DE ROOIJ N.F., KOUDELKA-HEP M., HENDRIKSE J., VAN DEN BERG A., "Microtechnology for the development of *in situ* microanalytical systems", Chapter 12 in BUFFLE J., HORVAI G. (eds), *In situ Monitoring of Aquatic Systems; Chemical Analysis and Speciation*, IUPAC Series in Analytical and Physical Chemistry of Environmental Systems, Wiley, 2000.

[FLE 87] FLEISCHMANN M., PONS S., ROLISON D.R., SCHMIDT P.P., *Ultramicroelectrodes*, Datatech Systems Inc., Morganton, 1987.

[FLO 86] FLORENCE T.M., "Electrochemical approaches to trace element speciation in waters. A review", *Analyst*, vol. 111, p. 489-505, 1986.

[FRI 92] FRISBIE C.D., FRITSCH-FAULES I., WOLLMAN E.W., WRIGHTON M.S., "Preparation and characterization of redox active molecular assemblies on microelectrode arrays", *Thin Solid Films*, vol. 210-211, p. 341-347, 1992.

[GAL 74] GALIZZOLI D., TANTARDINI F., TRASATTI S., "Ruthenium dioxide: a new electrode material. Part I: Behaviour in acid solutions of inert electrolytes", *J. Appl. Electrochem.*, vol. 4, p. 57-67, 1974.

[GAL 84] GALUS Z., "Diffusion coefficients of metals in mercury", *Pure Appl. Chem.*, vol. 56, p. 635-644, 1984.

[GAL 94] GALUS Z., *Fundamentals of Electrochemical Analysis*, 2nd edn, Ellis-Horwood, 1994.

[GIL 94] GIL E.P., OSTAPCZUK P. "Potentiometric stripping determination of mercury (II), selenium (IV), copper (II) and lead (II) at a gold film electrode in water samples", *Anal. Chim. Acta*, vol. 293, p. 55-65, 1994.

[GLA 89] GLAB S., HULANICKI A., EDWALL G., INGMAN F., "Metal/metal oxide and metal oxide electrodes as pH sensors", *Crit. Rev. Anal. Chem.*, vol. 21, p. 29-47, 1989.

[GOY 93] GOYAL R.N., MITTAL A., "Application of electroanalytical techniques in monitoring metal ions pollution in water. A review", *J. Sci. Ind. Res.*, vol. 52, p. 607-623, 1993.

[GUN 89] GUNASINGHAM H., FLEET B., "Hydrodynamic voltammetry in continuous-flow analysis", in BARD A.J., (ed), *Electroanalytical Chemistry*, vol. 16, Marcel Dekker, p. 89, 1989.

[HAN 80] HANEKAMP H.B., WOOGT W.H., BOS P., FREI R.W., "An electrochemical scrubber for the elimination of eluent background effects in polarographic flow-through detection", *Anal. Chim. Acta*, vol. 118, p. 81-86, 1980.

[HER 94] HERMES T., BUEHNER M., BUECHER S., SUNDERMEIER C., DUMSCHAT C., BORCHARDT M., CAMMANN K., KNOLL, M., "An amperometric micro-sensor array with 1024 individually addressable elements for two-dimentional concentration mapping", *Sens. Actuators B*, vol. 21, p. 33-37, 1994.

[HER 98] HERDAN J., FEENEY R., KOUNAVES S.P., FLANNERY A.F., STORMENT C.W., KOVACS G.T.A., "Field evaluation of an electrochemical probe for *in situ* screening of heavy metals in groundwater", *Environ. Sci. Technol.*, vol. 32, p. 131-136, 1998.

[HEY 66] HEYROWSKY J., KUTA, J., *Principles of Polarography*, Academic Press, 1966.

[HOY 87] HOYER B., FLORENCE T.M., "Application of polymer-coated glassy carbon electrodes to the direct determination of trace metals in body fluids by anodic stripping voltammetry", *Anal. Chem.*, vol. 59, p. 2839-2842, 1987.

[JER 92] JERMANN R., TERCIER M.-L., BUFFLE J., "Pressure insensitive solid state reference electrode for *in situ* voltammetric measurements in lake water", *Anal. Chim. Acta*, vol. 269, p. 49-58, 1992.

[JOH 86] JOHNSON D.C., WEBER S.G., BOND A.M., WIGHTMAN R.M., SHOUP R.E., KRULL I.S., "Electroanalytical voltammetry in flowing solutions", *Anal. Chim. Acta*, vol. 180, p. 187-250, 1986.

[JOH 87] JOHNSON B.M., BAKER R.W., MATSON S.L., SMITH K.L., ROMAN I.C., TUTTLE M.E., LONSDALE H.K., "Liquid membrane for the production of oxygen enriched air. II: Facilitated-transport membranes", *J. Memb. Sci.*, vol. 31, p. 31-67, 1987.

[KAL 87] KALVODA R., "Polarographic determination of adsorbable molecules", *Pure Appl. Chem.*, vol. 59, p. 715-722, 1987.

[KAL 88] KALVODA R., "Ecoelectrochemistry: Prospects for selective electrochemical sensing and removal/destruction of pollutants", *Selective Electrode Rev.*, vol. 10, p. 127-183, 1988.

[KAL 90] KALCHER K., "Chemically modified carbon paste electrodes in voltammetric analysis. A review", *Electroanalysis*, vol. 2, p. 419-433, 1990.

[KIS 84] KISSINGER P.T., HEINEMAN W. R. (eds), *Laboratory Techniques in Electroanalytical Chemistry*, Marcel Dekker, 1984.

[KOU 86] KOUNAVES S.P., BUFFLE J., "Deposition and stripping properties of mercury on iridium electrodes", *J. Electrochem. Soc.*, vol. 133, p. 2495-2498, 1986.

[KOU 87] KOUNAVES S.P., BUFFLE J. "An iridium-based mercury film electrode. Part I: Selection of substrate and preparation", *J. Electroanal. Chem.*, vol. 216, p. 53-69, 1987.

[KRE 95] KREIDER K.G., TARLOV M.J., CLINE J.P., "Sputtered thin-film pH electrodes of platinum, palladium, ruthenium and iridium oxides", *Sens. Actuators B*, vol. 28, p. 167-172, 1995.

[LEC 81] LECOMTE J., MERICAM P., ASTRUC A., ASTRUC M., "Oxygen elimination in the direct polarographic determination of trace metals in natural waters", *Anal. Chem.*, vol. 53, p. 2371-2374, 1981.

[LEV 62] LEVICH V.G., *Physicochemical Hydrodynamics*, Prentice Hall, 1962.

[LUT 99] LUTHER III G.W., REIMERS C.E., NUZZIO D.B., LOVALVO D. "*In situ* deployment of voltammetric, potentiometric and amperometric micro-electrodes from a ROV to determine dissolved = 2, Mn, Fe, S(-II), and pH in porewaters", *Env. Sci. Tech.*, vol. 33, p. 4352-4356, 1999.

[MAC 84] MACCREHAN W.A., MAY W.E., "Oxygen removal in liquid chromatography with a zinc oxygen-scrubber column", *Anal. Chem.*, vol. 56, p. 625-628, 1984.

[MAG 89] MAGEE JR. L.J., OSTERYOUNG J., "Fabrication and characterization of glassy carbon linear array electrodes", *Anal. Chem.*, vol. 61, p. 2124-2126, 1989.

[MAI 87] MAIRANOVSKII S.G., *Catalytic and Kinetic Waves in Polarography*, Plenum Press, 1968.

[MAN 84] MANN A.W., LINTERN M.J., "Field analysis of heavy metals by portable digital voltammeter", *J. Geochem. Expl.*, vol. 22, p. 333-348, 1984.

[MAN 87] MANN A.W., "Portable digital voltammetry and its application to ultratrace analysis in exploration", *Symp. Ser.- Australas Inst. Min. Metall.*, vol. 54, p. 119-122, 1987.

[MEY 95] MEYER H., DREWER H., GRUENDIG B., CAMMANN K., KAKEROW R., MANOLI Y., MOKWA W., ROSPERT M., "Two-dimensional imaging of O_2, H_2O_2 and glucose distributions by an array of 400 individually addressable micro-electrodes", *Anal. Chem.*, vol. 67, p. 1164-1170, 1995.

[MOS 95] MOSKVIN L.N., RODINKOV O.V., KATRUZOV A.N., GRIGOREV G.L., KROMOVBORISOV S.N., "Dissolved oxygen removal from aqueous media by the chromatomembrane method", *Talanta*, vol. 42, p. 1707-1710, 1995.

[MOU 94a] MOUSSY F., HARRISON D.J., "Prevention of the rapid degradation of subcutaneously implanted Ag/AgCl reference electrode using polymer coatings", *Anal. Chem.*, vol. 66, p. 674-679, 1994.

[MOU 94b] MOUSSY F., JAKEWAY S., HARRISON D.J., RAJOTTE R.W., "*In vitro* and *in vivo* performance and lifetime of perfluorinated ionomer-coated glucose sensors after hightemperature curing", *Anal. Chem.*, vol. 66, p. 3882-3888, 1994.

[NIE 84] NIEDERHOFFER E.C., TIMMONS J.H., MARTELL A.E., "Thermodynamics of oxygen binding in natural and synthetic dioxygen complexes", *Chem. Rev.*, vol. 84, p. 137-203, 1984.

[NÜR 83] NÜRNBERG H.W., "Investigations on heavy metal speciation in natural waters by voltammetric procedures", *Fresenius' Z. Anal. Chem.*, vol. 316, p. 557-565, 1983.

[NÜR 84] NÜRNBERG H.W., "The voltammetric approach in trace metal chemistry of natural waters and atmospheric precipitation. Review", *Anal. Chim. Acta*, vol. 164, p. 1-21, 1984.

[NÜR 85] NÜRNBERG H.W., "Applications and potentialities of voltammetry in environmental chemistry of ecotoxic metals", in KALVODA R., PARSONS R. (eds), *Electrochemistry in Research and Development*, Plenum Press, p. 121, 1985.

[OLS 83] OLSSON B., ÖGREN L., JOHANSSON G., "An enzymatic flow injection method for the determination of oxygen", *Anal. Chim. Acta*, vol. 145, p. 101-108, 1983.

[OST 85] OSTERYOUNG J.G., OSTERYOUNG R.A., "Square-wave voltammetry", *Anal. Chem.*, vol. 57, p. 101A-110A, 1985.

[PAU 01] PAUWELS H., TERCIER-WAEBER M.-L., ARENAS M., CASTROVIEJO R., DESCHAMPS Y. LASSIN A., GRAZIOTTIN F., ELORZA F.-J., "Chemical characteristics of groundwaters at two massive sulphide deposits in an area of previous mining contamination, South Iberian Pyrite Belt, Spain", *J. Geochem. Expl.*, vol. 75, p. 17-41, 2002.

[PED 94] PEDROTTI J.J., ANGNES L., GATZ G.R., "A fast, highly efficient, continuous degassing device and its application to oxygen removal in flow-injection analysis with amperometric detection", *Anal. Chim. Acta*, vol. 298, p. 393-399, 1994.

[PEI 00] PEI J., TERCIER-WAEBER M.-L., BUFFLE J., "Simultaneous determination and speciation of zinc, cadmium, lead and copper in natural waters with minimum handling and artefacts by voltammetry on gel-integrated micro-electrode array", *Anal. Chem.*, vol. 72, p. 161-171, 2000.

[PEI 01] PEI J., TERCIER-WAEBER M.-L., BUFFLE J., FIACCABRINO G.C., KOUDELKA-HEP M., "Individually addressable gel-integrated voltammetric microelectrode array for high resolution measurements of concentration profiles at interfaces", *Anal. Chem.*, vol. 73, p. 2273-2281, 2001.

[PER 81] PERSSON B., ROSEN L., "Flow injection determination of isosorbide nitrate with polarographic detection", *Anal. Chim. Acta*, vol. 123, p. 115-123, 1981.

[PER 94] PERRET D., NEWMAN M., NEGRE J.C., CHEN Y., BUFFLE, J., "Submicron particles in the Rhine river. I: Physico-chemical characterisation", *Water Res.*, vol. 28, p. 91-106, 1994.

[PIC 87] PICCARDI G., UDISTI R., "Intermetallic compounds and the determination of copper and zinc by anodic stripping voltammetry", *Anal. Chim Acta*, vol. 202, p. 151-157, 1987.

[PLE 91] PLETCHER D., "Why micro-electrode", Chapter 1 in MONTENEGRO M.I., QUEIROS M.A., DASCHBACH J.L. (eds), *Micro-electrodes. Theory and Applications*, NATO AST Series, 1991.

[PON 87] PONS S., FLEISCHMANN M., "The behavior of micro-electrodes", *Anal. Chem.*, vol. 59, p. 1391A-1399A, 1987.

[POO 87] POON M., MC CREERY R.L., "Repetitive *in situ* renewal and activation of carbon and platinum electrodes: applications to pulse voltammetry", *Anal. Chem.*, vol. 59, p. 1615-1620, 1987.

[REH 95] REHM D., MCENROE E., DIAMOND D., "An all solid-state reference electrode based on a potassium chloride doped vinyl ester resin", *Anal. Proc. Ind. Anal. Comm.*, vol. 32, p. 319-322, 1995.

[RIC 83] RICE M.E., GALUS Z., ADAMS R.N., "Graphite paste electrodes. Effects of paste composition and surface states on electron-transfer rates", *J. Electroanal. Chem.*, vol. 143, p. 89-102, 1983.

[ROL 83] ROLLIC M.E., HO C.-N., WARNER I.M., "Sample deoxygenation for fluorescence spectrometry by chemical scavenging", *Anal. Chem.*, vol. 55, p. 2445-2448, 1983.

[RUZ 72] RUZICKA J., LAMM C.G., TJELL J.C., "SelectrodeTM – the universal ion-selective electrode. Part 3: Concept, constructions and materials", *Anal. Chim. Acta*, vol. 62, p. 15-28, 1972.

[RUZ 88] RUZICKA J., HANSEN E.H., (eds), *Flow Injection Analysis*, 2nd edn, John Wiley, 1988.

[SMA 85] SMART R.B., STEWART E.E., "Differential pulse anodic stripping voltammetry of cadmium (II) at a membrane-covered electrode: measurements in the presence of model organic compounds", *Environ. Sci. Technol.*, vol. 19, p. 137-140, 1985.

[SMA 87] SMART R.B., "Electrode systems for measurement of environmental pollutants", *Hazard Assess. Chem.*, vol. 5, p. 1-27, 1987.

[SOT 87] SOTTERY J.P., ANDERSON C.W., "Short-pulse rapid-scan stripping voltammetry at a thin mercury film carbon fiber electrode", *Anal. Chem.*, vol. 59, p. 140-144, 1987.

[STO 77] STOJEK Z., BUBLIK Z., "Silver based mercury film electrode. Part III: Comparison of theoretical and experimental anodic stripping results obtained for lead and copper", *J. Electroanal. Chem.*, vol. 77, p. 205-224, 1977.

[STR 90] STROBHEN W.E., SMITH D.K., EVANS D.H., "Characterization of arrays of microelectrodes for fast voltammetry", *Anal. Chem.*, vol. 62, p. 1709-1712, 1990.

[STU 81] STULIK K., PACÀKOVÀ V., "Comparison of several voltammetric detectors for high-performance liquid chromatography", *J. Chromatogr.*, vol. 208, p. 269-278, 1981.

[STU 84] STULIK K., PACÀKOVÀ V., "Electrochemical detection in high-performance liquid chromatography", in *CRC Critical Reviews in Analytical Chemistry*, CRC Press, vol. 14, p. 297-351, 1984.

[STU 87] STULIK K., PACÀKOVÀ V. (eds), *Electrochemical Measurements in Flowing Liquids*, Ellis Horwood, 1987.

[STU 92] STULIK K., PACÀKOVÀ V., "Selectivity in flow electroanalysis", *Selective Electrode Rev.*, vol. 14, p. 87-142, 1992.

[TAY 96] TAYLOR R.F., SCHULTZ J.S. (eds), *Handbook of Chemical and Biological Sensors*, IOP Publishing Ltd, 1996.

[TER 90] TERCIER M.-L., BUFFLE J., ZIRINO A., DE VITRE R.R., "*In situ* voltammetric measurement of trace elements in lakes and oceans", *Anal. Chim. Acta*, vol. 237, p. 429-437, 1990.

[TER 95] TERCIER M.-L., PARTHASARATHY N., BUFFLE J., "Reproducible, reliable and rugged Hg-plated Ir-based micro-electrode for *in situ* measurements in natural waters", *Electroanalysis*, vol. 7, p. 55-63, 1995.

[TER 96] TERCIER M.-L., BUFFLE J., "Antifouling membrane-covered voltammetric microsensor for *in situ* measurements in natural waters", *Anal. Chem.*, vol. 68, p. 3670-3678, 1996.

[TER 98a] TERCIER M.-L., BUFFLE J., GRAZIOTTIN F., "A novel voltammetric *in situ* profiling system for continuous, real-time monitoring of trace elements in natural waters", *Electroanalysis*, vol. 10, p. 355-363, 1998.

[TER 98b] TERCIER-WAEBER M.-L., BELMONT-HEBERT C., BUFFLE J., "Real-time continuous Mn(II) monitoring in lakes using a novel voltammetric *in situ* profiling system", *Envi. Sci. Techn.*, vol. 32, p. 1515-1521, 1998.

[TER 99] TERCIER-WAEBER M.-L., BUFFLE J., CONFALONIERI F., RICCARDI G., SINA A., GRAZIOTTIN F., FIACCABRINO G.C., KOUDELKA-HEP M., "Submersible voltammetric probes for *in situ* trace element measurements in surface water, groundwater and sediment-water interface", *Meas. Sci. Technol.*, vol. 10, p.1202-1213, 1999.

[TER 00a] TERCIER-WAEBER M.-L., BUFFLE J., "Submersible on-line oxygen removal system coupled to an *in situ* voltammetric probe or trace element monitoring in freshwater", *Environ. Sci. Technol.*, vol. 34, p. 4018-4024, 2000.

[TER 00b] TERCIER-WAEBER M.-L., PEI J., BUFFLE J., FIACCABRINO G.C., KOUDELKA-HEP M., RICCARDI G., CONFALONIERI F., SINA A., GRAZIOTTIN, F., "A novel voltammetric probe with individually addressable gel-integrated microsensor arrays for real-time high spatial resolution concentration profile measurements", *Electroanalysis*, 2000.

[TER 01] TERCIER-WAEBER M.-L., BUFFLE J., KOUDELKA-HEP M., GRAZIOTTIN F., "Submersible voltammetric probes for *in situ* real-time trace element monitoring in natural aquatic systems", Chapter 4 in TAILLEFERT M., ROZAN T. (eds), *Environmental Electrochemistry: Analysis of Trace Element Biogeochemistry*, ACS Series Symposium Volume, American Chemical Society, 2001.

[VAL 82] VALENTA P., SIPOS L., KRAMER I., KRUMPEN P., RÜTZEL H., "An automatic voltammetric analyzer for the simultaneous determination of toxic trace metals in water", *Fresenius Z. Anal. Chem.*, vol. 312, p. 101-108, 1982.

[VAL 88] VALENTA P., "Potentialities and applications of voltammetry in the determination of ecotoxic trace metals in natural waters body fluids and foods", *GIT Fachz. Lab.*, vol. 32, p. 312-320, 1988.

[VAN 89] VAN DEN BERG C.M.G., "Electroanalytical chemistry of sea water", Chapter 51 in RILEY J.P. (ed), *Chemical Oceanography*, vol. 9, Academic Press, 1989.

[VAN 00] VAN LEEUWEN H.P., "Dynamic aspects of *in situ* speciation processes and techniques", Chapter 8 in BUFFLE J., HORVAI G. (eds), *In situ Monitoring of Aquatic Systems; Chemical Analysis and Speciation*, IUPAC Series in Analytical and Physical Chemistry of Environmental Systems, Wiley, 2000.

[VID 92] VIDAL J.C., VIÑAO R.B., CASTILLO J.R., "Binding capacity of casein to lead and voltammetric speciation of lead in milk with a Nafion coated electrode", *Electroanalysis*, vol. 4, p. 653-659, 1992.

[WAN 81] WANG J., ARIEL M., "Subtractive anodic stripping voltammetry with twin identical mercury film electrodes differing in their convection transport during deposition", *Anal. Chim. Acta*, vol. 128, p. 147-153, 1981.

[WAN 87] WANG J., TUZHI P., ZADEII J.M., "Evaluation of differential-pulse anodic stripping voltammetry at mercury-coated carbon fiber electrodes. Comparison to analogous measurements at rotating disk electrodes", *Anal. Chem.*, vol. 59, p. 2119-2122, 1987.

[WAN 88] WANG J., "Adsorptive stripping voltammetry. A new electroanalytical avenue for trace analysis", *J. Res. Nat. Bar-Stand.*, U.S., vol. 93, p. 489-490, 1988.

[WAN 89] WANG J., "Voltammetry following nonelectrolytic preconcentration", in BARD, A.J. (ed.), *Electroanalytical Chemistry*, vol. 16, Marcel Dekker Inc., p. 1-88, 1989.

[WAN 95] WANG J., FOSTER N., ARMALIS S., LARSON D., ZIRINO A., OLSEN K., "Remote stripping electrode for *in situ* monitoring of labile copper in the marine environment", *Anal. Chim. Acta*, vol. 310, p. 223-231, 1995.

[WHI 75] WHITFIELD M., "The electroanalytical chemistry of sea water", Chapter 20 in RILEY J.P., SKIRROW G. (eds), *Chemical Oceanography*, vol. 4, Academic Press, 1975.

[WHI 81] WHITFIELD M., JAGNER D., *Marine Electrochemistry. A Practical Introduction*, Wiley-Interscience, 1981.

[WIG 89] WIGHTMAN R.M., WIPF D.O., "Voltammetry at ultramicro-electrodes", in BARD, A. J. (ed.), *Electroanalytical Chemistry*, vol. 15, Marcel Dekker, p. 267-353, 1989.

[WOI 85] WOJCIEWCHOSKI M., GO W., OSTERYOUNG J., "Square-wave anodic stripping analysis in the presence of dissolved oxygen", *Anal. Chem.*, vol. 57, p. 155-158, 1985.

[YOS 81] YOSHIDA Z., "Structure of mercury layer deposited on platinum and hydrogenevolution reaction at the mercury-coated platinum electrode", *Bull. Chem. Soc. Jpn*, vol. 54, p. 556-561, 1981.

[ZEN 95] ZEN J.M., HSU F.S., CHI N.Y., HUANG S.Y., CHUNG M.J., "Effect of model organic compounds on square-wave voltammetric stripping analysis at the Nafion/chelating agent mercury film electrodes", *Anal. Chim. Acta*, vol. 310, p. 407-417, 1995.

[ZIR 81] ZIRINO A., "Voltammetry of natural sea water", Chapter 10 in WHITFIELD, M., JAGNER, D. (eds), *Marine Electrochemistry*, Wiley, 1981.

Chapter 9

Chemometrics

9.1. Introduction

Chemometrics is a tool for extracting pertinent information from crude physicochemical data, which may be either measured or already known. It is based upon the construction, followed by the exploitation of a behavior model with statistic tools. It can deal with complex, therefore, generally multivariate, systems.

Chemometrics is a new discipline initially associating data analysis and analytical chemistry. Currently, it covers the whole field of applications of chemistry, physics, life sciences, economy, sociology, statistical methods, and computing. This is why the less restrictive term (with respect to chemistry) of multivariate analysis or pattern recognition is often preferred.

We will focus on the original applications of chemometrics in the field of analytical chemistry, and more specifically in the domain of multisensors or multivariate sensors.

Chemometrics (or multivariate analysis) in instrumentation consists of modeling the variations of a certain number of variables, which will be called Y variables, which are difficult to obtain (for instance, necessitating a chemical analysis), as a function of other variables, called X variables that are "easily" measurable (from physical sensors for instance), whereby the latter is obtainable without measuring the Y variables.

Chapter written in French by Philippe BREUIL.

Two operations are distinguished:

– *calibration* done in the laboratory (or *modeling*), where all the variables must be measured and the model (or "predictor") is calculated;

– *prediction*, or in typical use "on the ground", where only the "easy" X variables are measured, with the Y variables being calculated by the model.

9.1.1. *The problem of multivariate analysis*

Let us consider m samples, called calibration samples, each of them being characterized by c characteristic Y variables, which are by definition "non-measurable easily" (such as chemical concentrations for instance), but for which an evaluation y_j is supposed to be known during the calibration phase:

$$Y = y_j^i, \quad 1 \leq i \leq m \quad 1 \leq j \leq c$$

Conversely, n X variables are "easily" measurable (for instance, potentials), and their measure is called x_k:

$$X = x_k^i, \quad 1 \leq i \leq m \quad 1 \leq k \leq n$$

From these calibration samples, the problem is finding a suitable model, i.e. a multivariable function, F, allowing, *for any other sample*, to evaluate Y knowing only the X variables X:

$$Y = F(X) \quad \text{where } F \text{ is called predictor.}$$

Two types of models are conceivable:

– "knowledge model", constructed using the available physicochemical data concerning the problem. This approach is already complex and approximate (from an instrumentation point of view) so when the X variables have to be expressed as a function of the Y variables, it is usually impracticable in the inverse case (which is the pertinent case for us: $Y = F(X)$), because of the complex character of the functions, of noise arising from the measurement, and of the non-measured effects of impurities;

– "behavior model", which is typically preferred, is only concerned with the mathematical model reproducing at best the relations between the X variables and the calibrating Y variables. The physicochemical knowledge of the problem is now no longer necessary; the model is of the "black box" type.

However, in the case of the behavior model, it is generally preferable that a minimum amount of theoretical knowledge be available (for instance: is the problem linear?); however at the end of this chapter, we will see neuronal networks that allow this without any knowledge model.

9.1.2. *Example: Beer-Lambert law of light absorption*

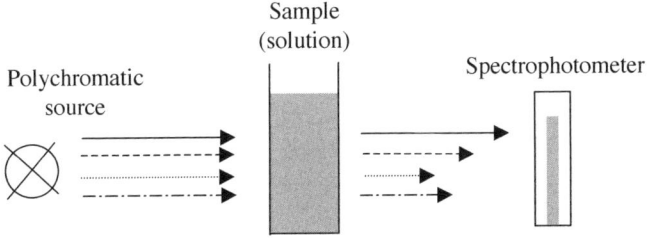

Figure 9.1. *Spectrophotometry*

Spectrophotometry (Figure 9.1) consists of measuring the attenuation of the light intensity of a luminous source through a solution containing a dissolved compound that absorbs light: the absorbance, at the wavelength, λ_k, is given by:

$$A_k = log_{10}\left(\frac{I_{k\,ref}}{I_k}\right) \quad \text{where } I_{k\,ref} \text{ is the reference intensity.} \tag{9.1}$$

The Beer-Lambert law indicates that this quantity is proportional to the concentration of the compound:

$$A_k = m_k C. \tag{9.2}$$

For a solution containing several compounds, the absorbances are additive for each wavelength λ_k:

$$a_k = \sum_j m_k^j c_j. \tag{9.3}$$

After calibration with samples having known concentrations, multivariate analysis enables the determination of the concentrations of other samples by measuring the absorbances only.

9.1.3. *General method*

We want to find $Y = F(X)$ with the utmost possible precision. We can write, for the calibration samples:

$$Y = F(X) + E \qquad [9.4]$$

where E is the error matrix on the variables Y, which has to be minimized:

$$E = Y - F(X) = \left(y_j^i - f_j(x_1^i, ..., x_j^i)\right) \text{ minimal} \qquad [9.5]$$

The concept of matrix minimizing necessitates the introduction of a distance. The Euclidian distance is generally used:

$$\sum_{i,j} \left(y_j^i - f_j(x_1^i, ..., x_k^i)\right)^2 \text{ minimal} \qquad [9.6]$$

It is thus a least squares method; in fact, other types of distances could be used, but the linear case would not then have a simple analytic solution.

The literal resolution of equation [9.6] is only possible in certain particular cases, and particularly the linear case.

9.2. A particular case: the linear case

9.2.1. *Notations and preliminary considerations*

In the linear case, equation [9.4] can be written in a matrix form:

$$Y = XA + E \qquad [9.7]$$

or:

$$y_j^i = \sum_k x_k^i a_j^k + e_j^i \qquad [9.8]$$

with:

$1 \leq i \leq m,$ sample number,

$1 \leq k \leq n$,   X variable number,

$1 \leq j \leq c$,   Y variable number.

NOTE: The chemometrics convention for the writing of matrices will be used: each line represents a sample (or an experiment) and each column represents a variable.

9.2.2. *Simple least square methods*

9.2.2.1. *First method: ILS (inverse least square) or P matrix*

The Beer-Lambert law is linear, it can be written in its inversed form (*hence the name of the method, the so-called "classic method", which is more complex, is explained later*) by expressing the concentrations as a function of the absorbances:

$$C = AP + E \qquad [9.9]$$

For the sake of illustration, it is interesting to make a representation in the so-called *sample space*. Let us suppose that we have three samples characterized by two absorbance X variables, A_1 and A_2, and one concentration Y variable, C. Each of these variables can be represented by a vector in the *sample space* where each dimension represents a sample. We, therefore, have a three-dimensional space represented in perspective. The three components of each vector ($\vec{A}$ or $\vec{C}$) represent the values of the variable for the three samples. So, two independent variables will be represented by two orthogonal vectors, as shown in Figure 9.2.

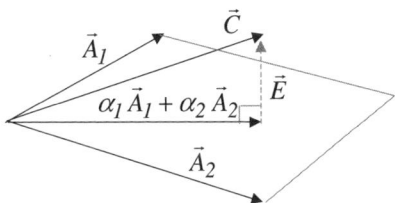

Figure 9.2. *ILS method*

When doing a linear regression (regression of C on A), we try to express C as a linear combination of A_1 and A_2 ($c = \alpha_1 A_1 + \alpha_2 A_2$), which is impossible *a priori*, as $\vec{C}$ does not belong to the ($\vec{A}_1, \vec{A}_2$) plane, but the solution will be approached by minimizing $\vec{E}$, which is the difference between $\vec{C}$ and the approached linear combination (in the plane ($\vec{A}_1, \vec{A}_2$)). $\vec{E}$ will be minimum when it is perpendicular to the plane ($\vec{A}_1, \vec{A}_2$) and the modeling of $\vec{C}$ will be its projection on this plane.

Making a regression of C on A consists of projecting $\vec{C}$ on the plane of the $\vec{A}$ vectors. P is, therefore, the projection matrix on the plane of the $\vec{A}$ vectors.

The standardization will be all the better as $\vec{C}$ will be closer to the plane of the $\vec{A}$ vectors. $\vec{C}$ represents the C variations non-correlated with those of A, which are, consequently, not explained by the model. It is called the "absorbance residue".

Here the error E on the concentrations is minimized; therefore, the concentrations are regressed on the absorbances. It can be shown that the matrix, P, projection operator, can be written:

$$P = (A'A)^{-1}A'C \quad \text{which is the prediction matrix} \quad [9.10]$$

Advantage of this method

The major advantage of this method is that it is easy to observe that the calculation of the P terms concerning the component j (= columns of P) does not depend on the concentrations of other compounds during the calibration.

Therefore, knowing the concentrations of all the compounds is not compulsory. The compounds present, but with an unknown concentration during standardization, may be considered as impurities.

The method is said to solve the problem of impurities, but these must be present in significant proportions in the calibration samples.

In a more general way, it solves the problem of variations due to external causes, which do not have to be quantified during the calibration phase, but must as already mentioned, be significantly present.

Drawbacks

The disadvantage of this method is that the $A'A$ matrix, which is the covariance matrix of absorbances, must be inversed; we must, therefore, have:

Number of wavelengths $\leq$ number of samples

The number of samples prepared must be at least equal to the number of wavelengths. But, with the modern spectrophotometers with CCD (charge coupled device) or photodiode arrays, the number of wavelengths is very high (100 to 1,000) and it is not conceivable to decrease the number of X variables by screening or selection because this would result in a loss of precious information.

Moreover, if absorbances that have very similar wavelengths are "almost" collinear, AA', even though it is invertible mathematically, it will have a small determinant. The matrix, P, will the have high (absolute) value coefficients and instability problems will occur during the prediction phase.

These drawbacks are partially eliminated by the classic least squares (CLS) method.

9.2.2.2. *Second method: CLS or K matrix, or MLR (multiple linear regression)*

If the Beer-Lambert relation is written $A = C\,K$, it expresses the X variables as a function of the Y variables; therefore, it cannot yield direct evaluation of the Y variables (concentrations). This problem will be circumvented in two steps, as follows.

The first step consists of evaluating the matrix K of the Beer-Lambert coefficients:

$$A = C K + E \qquad [9.11]$$

To minimize E, i.e. the errors on the absorbances, we will therefore regress A on C.

The matrix K is then obtained:

$$K = (C'C)^{-1} C'A \qquad [9.12]$$

In the prediction phase, we have: $A = C\,K$, but it is the matrix, C, that we want to determine. Unfortunately, K is generally not square, therefore, it cannot be inverted. We can, however, write:

$$C = AK'(KK')^{-1} = AM \qquad [9.13]$$

M is then the prediction matrix.

We have to make two inversions of low-dimension square matrixes (dimension = number of compounds):

– $C'C$, matrix of covariance of the concentrations; we must, therefore have:

number of compounds $\leq$ number of samples

$-KK'$; we must therefore have:

number of compounds $\leq$ number of wavelengths.

These two conditions, which are *a priori* obvious, are easily fulfilled.

Advantages of the MLR method

Theoretically, as many wavelengths as required can be used. A high number of wavelengths results in an "averaging" effect, which is beneficial to the ratio signal:noise.

It is interesting for the good understanding of the phenomena: the matrix, K, gives the Beer-Lambert coefficients directly.

Drawbacks of the method

The calculation of the prediction parameters for a compound uses the concentrations of all the compounds: this is because we have to make an orthogonal projection of each absorbance on the hyperspace of concentrations. The result (projector operator) will, therefore, depend on the consideration of the compounds acting upon the absorbances. All the compounds or interferents, or even all the events liable to be present in the prediction phase, will have to be introduced during the calibration phase, and their concentration or numerical value must be known in the form of a Y variable.

Therefore, this method does not solve the problem of impurities. For similar reasons, the method cannot take into account non-quantified or non-quantifiable variations (for instance variations of the base line) or interactions between the constituents.

9.2.2.3. *Common problems of both methods: overfitting*

Overfitting and modeling error

Overfitting occurs when the model is *too precise*; therefore there is a tendency to model "peculiarities" of the calibration samples, whereas only the common and useful information is interesting. A *modeling error* is introduced (difference between the model and the real law) *a priori* unknown.

As such a model is very complex, it can be applicable correctly only to the samples used for its creation.

However, this problem of "noise modeling" tends to be attenuated when the number of calibration samples increases.

Overfitting and error linked to the measurement

When the prediction calculation is done from the model and the X variables the measurement error on the X variables (characterized by the uncertainty) spreads through the model and contributes (with the modeling error) to the error on the Y variables. This component of the error is called "error linked to the measurement".

Measure of X: $\dot{X} + \Delta X$ ($\dot{X}$ = real value, ΔX = error)

Contribution of ΔX to the error on Y: ΔY_{meas}

Error on Y: $\Delta Y_{meas} + \Delta Y_{mod,}$

Therefore, reasoning on standard deviations:

$$\sigma_Y = \sqrt{\sigma_{Y_{meas}}^2 + \sigma_{Y_{mod}}^2}$$

During the prediction phase, it can be noticed that, even if the model is physically correct (negligible modeling error), the more complex a model is, the more amplified the (relative) errors linked to the measurement will be during Y variable calculation (Figure 9.3):

$$\frac{\Delta Y_{meas}}{\|Y\|} = \alpha \frac{\Delta X}{\|X\|},$$

α increases when the complexity of the model increases.

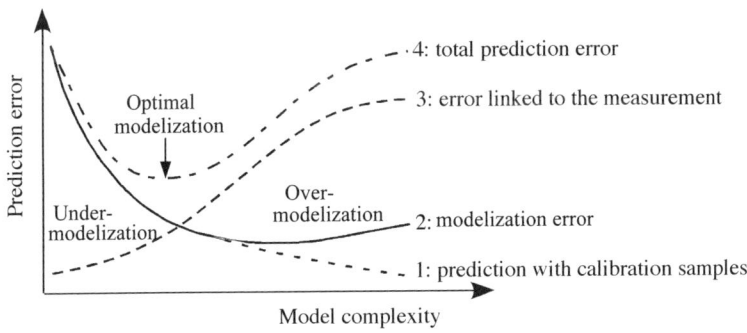

Figure 9.3. *Complexity and prediction error*

However, the effects of overfitting, which are preponderant if the number of calibration samples is small, tend to decrease as the calibration samples increase, which is costly in time and money.

Conversely, if the model is too simple, there will be little propagation of the measurement error, but the prediction will be less successful because the model will not reflect sufficiently the physical reality, i.e. underfitting. So, the model is a compromise between a simple and robust model (withstanding measurement errors), but lacking precision, and a precise, but fragile complex model.

It appears thus interesting to have the possibility of choosing the complexity of the model in relation with the phenomenon studied and with the metrologic qualities of the measurements.

NOTE: these considerations on the over- or under-modeling, although they are given in the section dealing with "linear" phenomena, have a more general character, as it will be observed with the neuronal networks.

9.2.3. *Factor analysis*

Factor analysis consists of making a variable change at the X variables level. Fewer linear combinations of the old variables are taken as new X variables to concentrate useful information.

Initially, there are n X variables, generally more or less correlated to each other. The aim is to find new independent X variables by a base change, with a reduction of dimension. These variables are called factors or latent variables and their value for a given sample is called score (Figure 9.4).

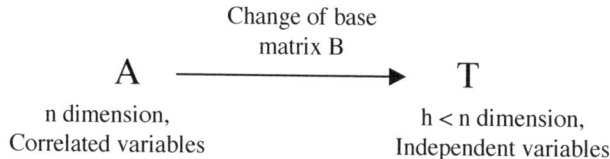

Figure 9.4. *Factor analysis*

If $h = n$, there is no reduction of data, the problem is equivalent to the previous one and the method presents little interest.

Several methods exist for determining the base change matrix B ("loadings vectors") and will be explained later.

During the prediction phase, the base change, T, that allows the scores to be known is performed as follows:

$$T = B*A \quad (B : h \times n \text{ matrix}) \quad [9.14]$$

It is clear that, as $h < n$, there will be a loss of information during the passage from A to T. We will, therefore, control this so that the information retained will be useful information. We can then calculate the concentrations by regression of T starting from a prediction matrix, V, determined, during the calibration, by ILS (see section 9.2.2.1) from the scores:

$$C = TV + E \quad [9.15]$$

A regression of C on V is then performed.

The interest of this data compression is that measurement noise is not removed. Only useful information is retained (which could initially be distributed on the whole spectrum) in the first factors, and therefore, used, but the noise remains uniformly distributed for all the factors. The "abandoned" information is, therefore, practically only noise, and the latter is, therefore, globally reduced.

The drawbacks are thus eliminated and the advantages of the two first methods are added:

– *ILS*: the number of wavelengths used is only limited by the calculation capacity. There is no co-linearity problem left as only the combinations of absorbances that are orthogonal to each other are taken. The number of factors must, however, be inferior or equal to the number of samples;

– *MLR*: there is no impurity problem as the concentration predictor is determined by ILS from scores, therefore, independently for each compound.

In both cases, the risk of "*overfitting*" (noise modeling) is reduced as the number of variables is reduced.

Two principal factor analysis methods exist; they are described in the following sections.

9.2.3.1. PCR (principal component regression)

This method is based upon the decomposition into principal components [MAR 89, MAN 86].

It is very efficient, but it does not take into account the concentration information in the first calibration phase (choice of the new base), which is only used in the phase of regression of concentrations on the scores.

However, important absorbance variations that do not correlate with the concentrations may occur. Therefore, the following method is often preferred, which is more recent and more complex.

9.2.3.2. PLS (partial least squares)

This more recent method (1980) consists of jointly constructing the matrixes of base change, W, ("*loadings vectors*") and of predicting V by using the absorbances and the concentrations together.

We will schematically explain the simplest algorithm (called "non-orthogonal") for only one component (PLS1). For a more complete explanation or for the algorithm with several components (PLS2), the interested reader can refer to the literature [MAR 89, TEN 98].

PLS calibration

A_0 and C_0 are the initial, normed, and centered values of absorbance and concentration:

– A: matrix m lines, k columns;

– C: vector (only 1 component) m values;

– m: number of calibration samples;

– k: number of X variables (or of wavelengths).

1. Try to find the A component that is most correlated to the concentration variations by regressing A_0 on C_0:

$$A_0 = C_0 W_1 + E. \qquad [9.16]$$

W_1 will then be the first component of the base change matrix, W. It is the "average projector" operator of A_0 on C_0 (and will probably not project each component of A_0 on C_0 perfectly):

$$W_1 = (C'_0 C_0)^{-1} C'_0 A_0. \qquad [9.17]$$

$W_1 C_0$ is thus the projection of A_0 on C_0 in the sample space, and W_1 is now normalized: we have $W_1 W'_1 = I_d$.

2. Search for the "*score*" (projected) corresponding to this "average projector" by regressing A_0 on W:

$$A_0 = T_1 W_1 + E \qquad [9.18]$$

therefore:

$$T_1 = A_0 W'_1 (W_1 W'_1)^{-1} = A_0 W_1. \qquad [9.19]$$

3. Search by regressing C_0 on T_1 for the component of the predictor V_1 corresponding to this factor; an ILS is then made:

$$C = T_1 v_1 + E \qquad [9.20]$$

thus:

$$v_1 = (T'_1 T_1)^{-1} T'_1 C. \qquad [9.21]$$

4. Calculate the absorbance and concentration residuals (non-used information, orthogonal to the previous one) by subtracting the already modeled information:

$$A_1 = A_0 - T_1 W_1 \qquad [9.22]$$

$$C_1 = C_0 - T_1 v_1. \qquad [9.23]$$

5. Resume step (1) with these new values and carry on constructing W and V until the desired number of factors is reached.

It is thus an iterative method (one iteration per factor). For each iteration, the information (A or C) used for constructing the model is subtracted before the subsequent iteration.

Once the n iterations done, the remaining (A or C) information is called residual (of absorbance or of concentration).

9.2.3.3. *PLS prediction*

It is almost the inverse operation. One extracts from A the information corresponding to each factor successively, with which the concentration C (initially zero) is constructed, thus, for each factor with index a:

– Initialization: A_0 normalized, centered, $C_0 = 0$.

– Calculation of contribution of this score to the concentration:

$$c_a = c_{a-1} + T_a V_a. \qquad [9.24]$$

– Calculation of the new absorbance residual:

$$A_{a+1} = A_a - T_a W_a. \qquad [9.25]$$

Resume equation [9.25] with the new absorbance residual until the desired number of factors is reached.

The final concentration, once "no longer centered and no longer normalized", will thus be the sum of the contribution of the various factors.

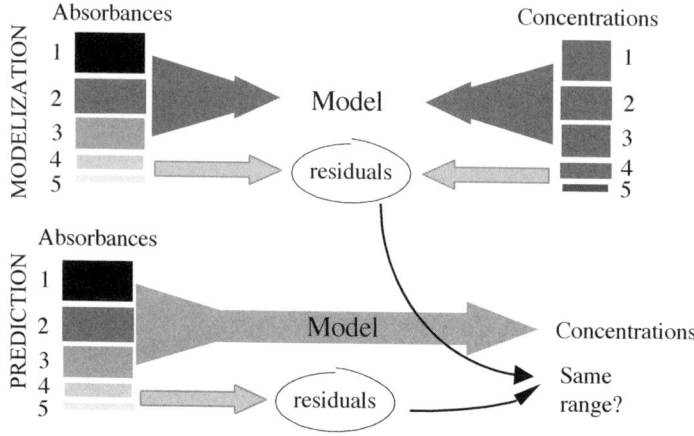

Figure 9.5. *Use of the residuals*

In addition to the concentration, this prediction method has the advantage of yielding the absorbance residuals, which must theoretically be of the same order of magnitude as those obtained during the calibration (Figure 9.5). Examining the ratio (prediction residual : calibration residual) for each wavelength can then allow the detection of anomalies, in particular the presence of impurities, which were not present during the calibration. The prediction calculation may then be supposed to be possibly erroneous.

At this stage, it is worth citing the example of the detection of nitrates in the presence of chromium only in the prediction phase (Figure 9.6).

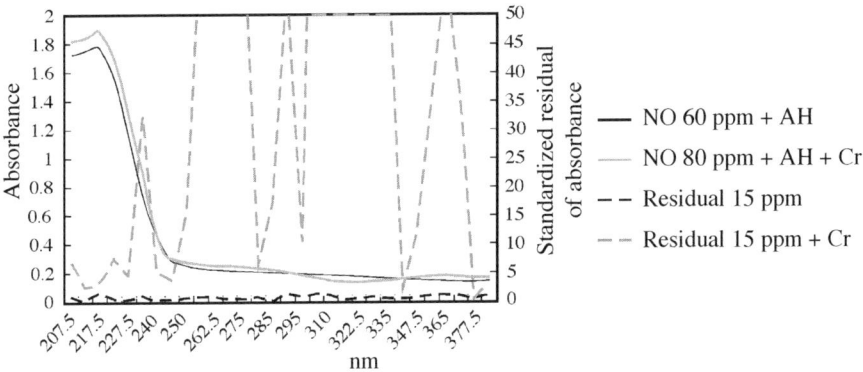

Figure 9.6. *Example of use of residuals*

It can be noticed that with less than 5 ppm of chromium (III), the residues are multiplied by more than 10 between 245 and 370 nm, which leaves no doubt about the presence of impurities. But it is then reasonable to doubt the validity of the calculation of concentrations.

In reality, for each prediction calculation, it is often preferable to calculate a term of "probable error" called deviation [MAR 89]:

$$Ydev = \sqrt{\frac{Var(Y_{calib.})}{2}\left(\frac{Var(Xresiduals_{predict.})}{Var(Xresiduals_{calib.})} + \frac{1}{nb\ calib.samples.}\right)} \quad [9.26]$$

This term, which must be multiplied by the norm of the calibration concentrations, if the latter have been normalized, has the same order of magnitude as the standard deviation of the prediction error, in most cases.

9.2.3.4. *Common problem in both PCR and PLS methods: choice of the number of factors*

We have noted that it was necessary to stop the modeling for a number of factors giving optimal prediction results, in other words when, in the case of PLS, the absorbance residues reach the same order of magnitude as the measurement noise, once the useful information has been extracted.

For each new factor, it is thus necessary to perform tests to minimize the variance of the prediction errors. These tests must not be made using the calibration samples; otherwise an optimal number of factors equal to the maximum number of factors would be found. Again the calibration concentrations would be established, but the measurement noise would be restored and the performances would not be good for other samples.

It is thus necessary:

– either to have a set of samples reserved for the prediction tests, but it may then be great pity to waste samples in this manner, because calibration is improved as the number of samples increases;

– or to use the cross–validations method. If m samples are available, $m-1$ is used for calibration, the last one is reserved for tests. The ultimate calibration is done using m samples with the optimal number of factors thus determined.

9.3. Least squares methods: non-linear case

9.3.1. *Case when transformations can reduce the problem to linear functions*

If the variables $x_1, ..., x_n$ are separable, it is possible to use the linear regression methods after having transformed these variables:

$$y = f(x_1, ..., x_k)$$
$$y = a_0 + a_1^1 f_1^1(x_1) + a_1^2 f_1^2(x_1) + ... + a_2^1 f_2^1(x_2) + ... a_k^{b_k} f_k^{b_k}(x_k) \quad [9.27]$$

The variables have to be transformed before doing the calibration, as in the case of prediction:

$$y = a_0 + a_1^1 X_1^1 + a_1^2 X_1^2 + ... + a_2^1 X_2^1 + ... a_k^{b_k} X_n^{b_k}. \quad [9.28]$$

Classic particular case: the polynomial regression. A single variable x is replaced by n variables $x^1 ... x^n$:

$$y = a_0 + a_1 x + a_2 x^2 + ... + a_n x^n \quad\quad [9.29]$$
$$y = a_0 + a_1 X + a_2 X_2 + ... + a_n X_n$$

This method can also be applied to the Y variables.

9.3.2. PLS can model non-linear phenomena

Although it may seem paradoxical for a method based upon linear algebra, PLS (like PCR does) can model non-linear phenomena.

In practice, if the various X variables have *"different"* *non-linearities* in their relations with the Y variables, then the PLS modeling can implicitly combine these different non-linearities in order to express a linear relation between X and Y, an this without any additional "non-linear" term.

However, if all the relations between X and Y variables have the same type of non-linearity, then any rigorous modeling is impossible. But it is often possible to approach it by adding terms of non-linear functions of the X variables as additional X variables, consequently linearly independent of the ancient ones (this is what is done in polynomial regression) (see section 9.3.1).

However, this is always done at the expense increasing the complexity of the model (increased number of factors) and it must always be preferred, when possible, to resort to a linearizing treatment of the X and/or Y variables (this is what is done implicitly in spectrophotometry when transforming the luminous intensities into absorbances).

9.4. Neural networks

Neural networks may be considered as an algorithm of multivariate analysis. In fact it is a universal behavior model in that it is no longer necessary to make the initial mathematical hypothesis. Only the structure of the network may change (number and size of the intermediate layers, transfer functions).

9.4.1. *General structure of the network*

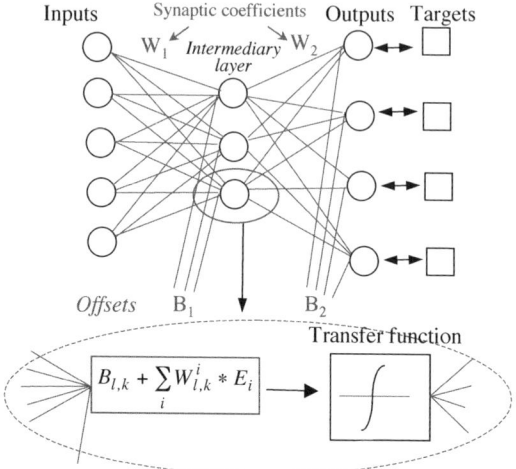

Figure 9.7. *Neural network*

The neural network is composed of an "input layer" corresponding to the X variable which is to be transformed, of an "output layer" yielding the modeled Y variable, and possibly of one or several intermediate layers, with sizes selected by the user (Figure 9.7). The output and intermediate layers are linear combinations of the preceding layer, the coefficients, $W_{i,j}$, being called "synaptic coefficients". A constant term B_i called "offset" is sometimes added. What makes the richness of neural networks is the transformation of this linear combination by a transfer function, which is generally non-linear (often a sigmoid function, or truncated linear).

It can be noticed that, if the transfer function is linear, then the network is equivalent to the linear methods described earlier, with the intermediate layers corresponding to the factors; only the mode of calculation of the coefficients, iterative and, here, called learning, is different.

9.4.2. *Learning (i.e. calibration)*

Two sets of variables, as in the case of any multivariate analysis, are used:
– *input variables* (X variables),
– *target variables* (Y variables).

The aim is then to find, *by successive approximation*, the synaptic coefficients (*W*) and the offsets (*B*) for which the targets and the outputs are *as close as possible in the sense of the least squares*. The sum of the squares of the errors (*target-output difference*) will, therefore, be optimized. It is a problem of optimization, which is solved by iterations. If the transfer functions are linear, then the problem can be reduced to a classic multivariate analysis problem (simple or with factor analysis if an intermediate layer of neurons exists).

Otherwise, the most frequently used optimization process is *back-propagation*. It consists of propagating the errors (differences between target and corresponding output) from the output towards the input and modifying coefficients W and B "imperceptibly" on the way with the aim of decreasing the error. The method is applied many times using the whole set of samples until the desired convergence degree is obtained.

It is not necessary to make the network converge excessive, because of the risk of overfitting, which may be harmful for the prediction, as explained previously. It is, therefore, advisable to test the network, along with the convergence process, with a set of samples that are not used for the learning phase.

9.4.3. Prediction

The network (W, B) can then be stimulated with new input X variables in order to find the corresponding output Y variables (which are *a priori* unknown).

Essentially pattern recognition (in a broad sense), chemometrics may then be defined as spectra or curve recognition. But neural networks are mainly famous for their use in image, characters, and voice recognition, or in identification of dynamic systems.

9.5. Conclusion

Chemometrics, or multivariate analysis, allows a behavior model of physicochemical phenomena to be constructed to exploit these phenomena in instrumentation. Once the methods have been understood, the main difficulty is adapting the complexity of the model to the metrological qualities of the measurements, and, therefore, in finding a compromise between the precision and reliability of this model.

Generally speaking, although factor analysis methods seem more successful for modeling linear or slightly non-linear phenomena, neural networks are successful for strongly non-linear phenomena, which they are able to model in a less complex manner.

9.6. Bibliography

[MAR 89] MARTENS H., NAES T., *Multivariate Calibration*, John Wiley & Sons, 1989.

[HAS 92] HASWELL S.J., *Practical Guide to Chemometrics*, Marcel Dekker, Inc, 1992.

[MAN 86] MANLY B.F.J., *Multivariate Statistitcal Methods: A Primer*, Chapman & Hall, 1986.

[TEN 98] TENNENHAUS M., *La régression PLS, théorie et pratique* , Technip, 1998.

[SHA 86] SHARAF M.A., ILLMAN D.L., KOWALSKI B.R., "Chemometrics", *Chemical Analysis* series, vol. 82, John Wiley & Sons, Chichester, 1986.

Chapter 10

Impedancemetric Sensors

10.1. Introduction

Impedancemetric sensors are *passive* devices, i.e. they require an external power supply to generate the signal, and they work as electrical receptors. *Modulation* principles are employed because these devices require that an electric current is driven through the tested sample, and the electrical impedance of the electrolyte modulates the resulting voltage. Basically, the cell impedance depends both on the size geometry of the cell and on the electrical properties of the electrolyte. The conductivity of the solution depends on various factors: concentration, mobility and valencies of ions, temperature and, to a lesser extent, pressure. Although the signal is not selective, with the exception of pH measurements, these sensors are the most widely used chemical sensors, mainly for water quality control and process monitoring.

As this chapter cannot be an exhaustive review of all the impedancemetric sensors that have been developed, only some examples will be given. The main aims of this chapter are to address the working principles and briefly describe the main devices with emphasis on their utilization and sources of error.

10.2. Fields of application

Direct conductivity measurements of the analyzed medium are used routinely in many industrial and environmental applications as a fast, inexpensive, and reliable way of measuring the ionic content of a solution. Many fields are concerned: quality

---

Chapter written by Jacques FOULETIER and Pierre FABRY.

of water (drinking, river and sea waters), monitoring of water treatments, chemical industries, processing (boiler blow-down, leak detection, corrosion monitoring), foods, pulp and paper, textiles, biomedical applications, etc. It should be noted that whilst individual ions cannot be differentiated, except for chromatography devices, the conductivity gives a measure of the total charged impurities that are present in the analyzed medium.

More recently, new devices based on changes of conductivity of a material as a function of its surrounding medium have been proposed. Such a transduction mode is also referred to as *second kind transduction*, or *indirect transduction*. They will be described briefly in this chapter.

The most classical type of conductivity design is a two-electrode cell, with flat or cylindrical electrodes. The electrical conductivity σ is defined as:

$$\sigma = \frac{1}{R_{cell}} \frac{d}{A} \qquad [10.1]$$

where R_{cell} is the resistance of the cell with the tested liquid, d and A are the electrode separation and the area of the inter-electrodes section, respectively. The term $K = d/A$ is called the *cell constant*.

The SI unit of conductivity is Siemens per meter (S/m) (K in per meter) and, unless otherwise specified, it refers to 25°C (reference temperature). In practice, depending on the analyzed solutions, more traditional units are µS/cm or mS/cm (K in per centimeter). Table 10.1 lists typical conductivity ranges of waters and chemicals.

Water	Conductivity (µS/cm)	Chemicals (solutions in water)	Conductivity (mS/cm)
Ultra pure water	0.055	KCl 0.01 mol/l	1.41
Distilled water	0.5–5	KCl 0.1 mol/l	5.81
Rain water	20–100	KCl 1 mol/l	112
Mineral water	50–200	NaOH 50%	150
River water	250–800	NaOH 5%	223
Tap water	100–1,500	NaCl saturated	251
Surface water	30–7,000	HCl 10%	700
Waste water	700–7,000	HNO$_3$ 31% (highest known)	865
Ocean water	40,000–55,000		

Table 10.1. *Conductivity of waters and chemicals at 25°C*

A wide variety of equipment is available to measure conductivity in water ranging from ultrapure water to concentrated solutions. Figure 10.1 compares the conductivity ranges of waters with the conductivity of KCl solutions (from 1 mol/l to 10^{-4} mol/l). The adequate cell constant is given as a guideline: for very dilute solutions the distance between the electrodes can be very small, for more conductive solutions, the distance between the electrodes has to be increased.

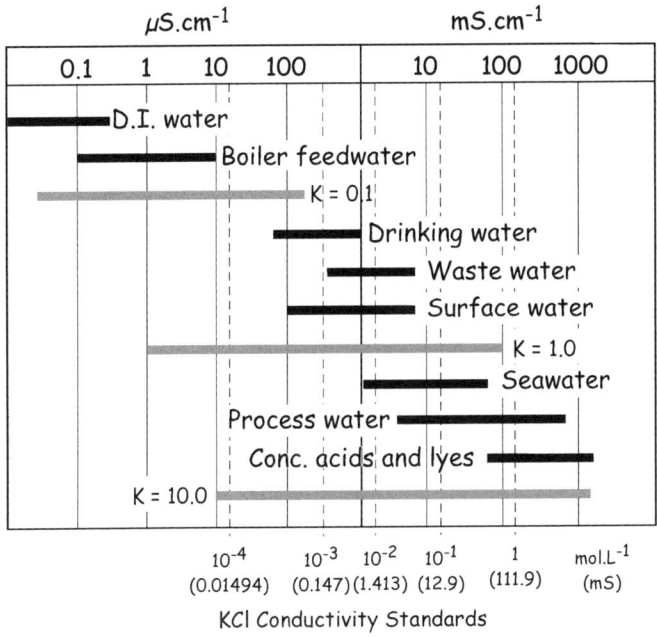

Figure 10.1. *Comparison of electrical conductivity values of waters (D.I. = "de-ionized water")*

Water quality is often characterized by the *total dissolved solids* (TDS), which is the total weight of cations, anions, and the undissociated dissolved species in 1 l of water. Water can be classified by the amount of TDS/l: fresh water <1,500 mg/l TDS < brackish water <5000 mg/l TDS < saline water. The standard method to determine TDS consists of evaporating a sample of water to dryness at 180°C. This method is time-consuming, tedious, and expensive. Consequently, conductivity readings can be converted to TDS using a conversion factor. Obviously, the TDS conversion factor depends on the standard solution used for the calibration. In dilute solution, the TDS can be estimated from the electrical conductivity by the equation:

$$\text{TDS (mg/l)} = 0.5\ \sigma(\mu\text{S/cm}) \quad\quad [10.2]$$

For concentrated solutions (TDS > 1g/l) the TDS factor has to be increased up to 0.9. Usually, the conductivity measurement provides an approximate value for the TDS concentration to within 10% accuracy.

Salinity is also a parameter used for evaluating water quality: it is a measure of the mass of dissolved salt in a given mass of solution. Salinity is generally reported in parts per thousand (ppt) or grams per liter. The salinity is calculated from an empirical relationship between conductivity and salinity of a seawater sample, taking into account the temperature dependence of conductivity. As an example, at 18°C, the salinity of a water exhibiting a conductivity of 20 mS/cm is 14 ppt.

10.3. Conductivity of liquid media

10.3.1. *Theoretical basis*

The electrical conductivity of a solution is given by:

$$\sigma = F^2 \sum_j z_j^2 \tilde{u}_j C_j \qquad [10.3]$$

where z_j and $\tilde{u}_j$ are the number of charge and the electrochemical mobility of ions j, with a molar concentration C_j. Obviously, the used units must be coherent, for example, if C is given in mol/l.

The conductivity is also expressed as a function of the electrical mobility u_j:

$$\sigma = F \sum_j z_j u_j C_j \qquad [10.4]$$

or as a function of the molar ionic conductivity λ_j:

$$\sigma = \sum_j \lambda_j C_j . \qquad [10.5]$$

The conductivity σ of a solution containing a single electrolyte depends on the concentration C of the electrolyte. The corresponding molar conductivity Λ_m is defined by:

$$\Lambda_m = \frac{\sigma}{C} . \qquad [10.6]$$

The molar conductivity of a strong electrolyte at low concentration obeys the empirical Kohlrausch's law:

$$\Lambda_m = \Lambda_m^o - k\sqrt{C} \qquad [10.7]$$

where Λ_m^o is the limiting molar conductivity (also referred to as conductivity at infinite dilution, i.e. no interactions between all the species present in the solution are considered) and k an empirical constant. Moreover, Kohlrausch found that the limiting conductivity of cations and anions are additive:

$$\Lambda_m^o = v_+ \lambda_+^o + v_- \lambda_-^o \qquad [10.8]$$

where v_+ and v_- are the molar quantities of cations and anions resulting from the dissociation of 1 mole of the dissolved electrolyte; λ_+^o and λ_-^o are the limiting molar conductivities of the individual ions. It should be pointed out that the limiting ion conductivity varies noticeably with the nature of the ions, as shown in Table 10.2.

Cations	λ_+^o (mS/m²/mol)	Anions	λ_-^o (mS/m²/mol)
H^+	34.96	OH^-	19.91
D^+	24.99	OD^-	11.9
Li^+	3.869	Cl^-	7.634
Na^+	5.011	Br^-	7.84
K^+	7.350	I^-	7.68
Mg^{2+}	10.612	SO_4^{2-}	15.96
Ca^{2+}	11.900	NO_3^-	7.14
Ba^{2+}	12.728	$CH_3CO_2^-$	4.09

Table 10.2. *Examples of limiting ion conductivities, at 25°C*

Onsager gave a theoretical explanation of the Kohlrausch' law by extending the Debye-Hückel theory. The following equation applies well at low concentrations:

$$\Lambda_m = \Lambda_m^o - (A + B\Lambda_m^o)\sqrt{C} \qquad [10.9]$$

where A and B are constants depending on temperature, charges of ions, and on the dielectric permittivity and viscosity of the solvent.

In the case of a weak electrolyte, the concentration of ions is less than the electrolyte concentration. As an example, for a monoprotic acid HA, with a dissociation constant K_a, the following equation is obtained, referred to as the Ostwald's dilution law:

$$\frac{1}{\Lambda_m} = \frac{1}{\Lambda_m^o} + \frac{\Lambda_m C}{K_a (\Lambda_m^o)^2} \qquad [10.10]$$

For most water solutions, the higher the concentration of dissolved salts, the higher the conductivity. However, the conductivity may actually decrease with increasing concentration. Various attempts have been made to extend the Onsager's treatment to more concentrated solutions; however, up to now, the interpretation of the conductivity maximum remains controversial.

10.3.2. Effect of temperature

Raising the temperature makes water less viscous, and the ions can move faster. Because the ions are of different sizes, and carry different amounts of water with them as they move, the temperature effect is different for each ion. Typically, the conductivity varies about 1–3%/°C, and this temperature coefficient may itself depend on concentration and temperature (Figure 10.2).

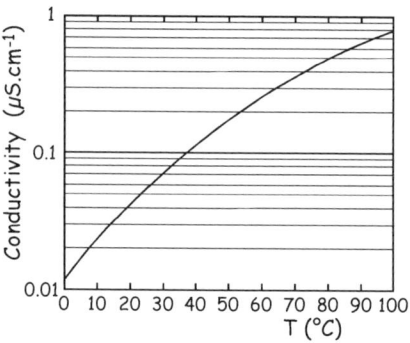

Figure 10.2. *Conductivity versus temperature of pure water*

Concentrated solutions have slopes of approximately 1.5%/°C, most industrial waters are in the 1.8–2%/°C range; however, ultrapure water has a slope of 5.2%/°C. A small difference in temperature can induce a noticeable change in conductivity; consequently, conductivity readings are normally referenced to at 25°C and temperature compensation, generally carried out automatically, is necessary for accurate measurements.

10.4. Impedance of first kind cell (direct measurement)

The conductivity measurement of an analyzed solution is generally carried out using alternating current (AC) impedance measurement. The two experimental parameters are the amplitude and the frequency of the signal. The recommended amplitude is generally small (in the order of 10–20 mV) in order to obtain a pseudolinear behavior of the cell. The electrical parameters of the analyzed solution can be deduced from the impedance versus the applied frequency curves. The interpretation of the diagrams is time-consuming, and impedancemetric sensors generally work at a constant frequency for a given conductivity range. The aim of the following section is to discuss the role of experimental parameters, i.e. conductivity range, cell configuration, etc, in the impedance diagram, and consequently, to estimate the optimal frequency required to obtain accurate measurements.

In the case of AC impedance measurements with a small amplitude signal, a two-electrode electrochemical cell involving an aqueous electrolyte can be schematically viewed as two electrical circuits connected in series, the first corresponding to the electrolyte contribution and the other to the electrode processes. The most simple, and most common model, is referred to as "Randles cell". In this R_1-(R_2/C_2) model, the electrolyte contribution is considered as a pure resistance (it is fair at low or moderate frequencies). The response of both electrodes to a small AC signal, can be modellized as a capacitor, schematizing the double layer capacitance, connected in parallel with a resistor, schematizing the charge transfer reaction or polarization resistance. The equivalent circuit for the Randles cell R_1-(R_2/C_2) is shown in Figure 10.3a. The corresponding variation of the impedance of this electrical circuit with applied frequency f is:

$$Z(\omega) = R_1 + \frac{R_2 - i\omega R_2^2 C_2}{1 + \omega^2 R_2^2 C_2^2} \qquad [10.11]$$

where ω is the angular frequency ($\omega = 2\pi f$), f is the signal frequency, R_1 is the electrolyte resistance, R_2 and C_2 are the charge transfer resistance and the electrode double layer capacitance, respectively. The impedance $Z(\omega)$ is the sum of a real $(Re(Z))$ and an imaginary $(Im(Z))$ parts. The absolute value of the impedance $/Z/$ (modulus) is equal to:

$$/Z/^2 = [Re(\omega)]^2 + [Im(\omega)]^2 \qquad [10.12]$$

Among the various representations proposed in the literature, we will consider only the Nyquist plot (- $Im(Z)$ vs. $Re(Z)$) and the Bode plot ($log/Z/$ vs. $log(f)$).

As shown in Figure 10.3b, the Nyquist plot for a Randles R_1-(R_2/C_2) cell is always a semicircle. The solution resistance, R_1, can be found by reading the real axis value at the high-frequency intercept. The real axis value at the low-frequency intercept is the sum of the polarization resistance and the solution resistance. The diameter of the semicircle is, therefore, equal to the polarization resistance R_2.

Figure 10.3c is the Bode plot for the same cell. In this diagram, the plateau (horizontal asymptote) reached at high frequencies gives the electrolyte resistance and the plateau observed at low frequencies corresponds to $R_1 + R_2$. It can be noted that the oblique asymptote corresponds to $log\ (1/C_2\omega)$ and the intersection between the latter and the horizontal asymptote $log\ (R_1 + R_2)$ is obtained at the angular frequency ω_2^o. This frequency corresponds to the point at the apex of the semicircle in the Nyquist diagram. It is equal to:

$$\omega_2^o = \frac{1}{R_2 C_2}.$$ [10.13]

The working frequency must be positioned on the plateau, $log\ R_1$. The value of ω_{lim}, which corresponds to the intersection between the oblique asymptote and the high-frequency horizontal asymptote $log\ R_1$ is then equal to:

$$\omega_{lim} = \frac{1}{R_1 C_2}.$$ [10.14]

It can be shown that for an angular frequency, ω_{meas}, higher than $10 \times \omega_{lim}$, the modulus of the impedance is equal to the electrolyte resistance within less than 1% [FAB 08].

More generally, if the electrolyte is considered a dielectric material, its capacitance C_1 is equal to:

$$C_1 = \varepsilon \frac{A}{d}$$ [10.15]

where ε is the associated permittivity, with $\varepsilon = \varepsilon_r\ 10^{-9}\ (36\ \pi)^{-1}$ (SI unit). ε_r is the relative permittivity, which is equal to 78 at 25°C for pure water and less than 20 for concentrated solutions. In these conditions, the electrical circuit (R_1/C_1)-(R_2/C_2) and the corresponding Nyquist diagram and Bode plot are given in Figure 10.4.

Impedancemetric Sensors 315

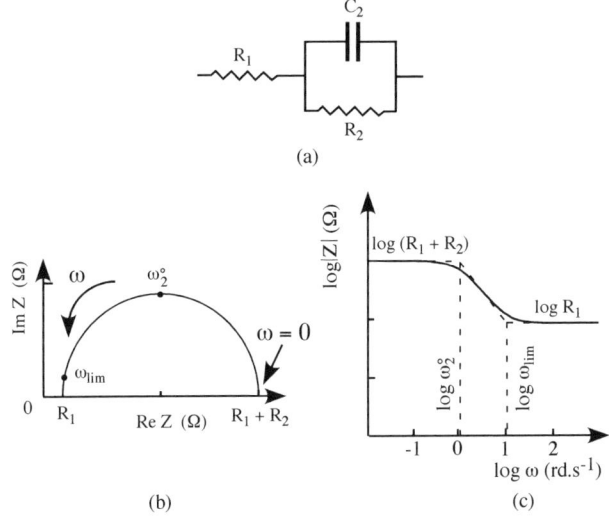

Figure 10.3. *a) Electrical circuit corresponding to Randles model R_1-(R_2/C_2); b) Nyquist diagram of the circuit (the arrow indicates increasing angular frequencies); c) Bode plot of the Randles circuit*

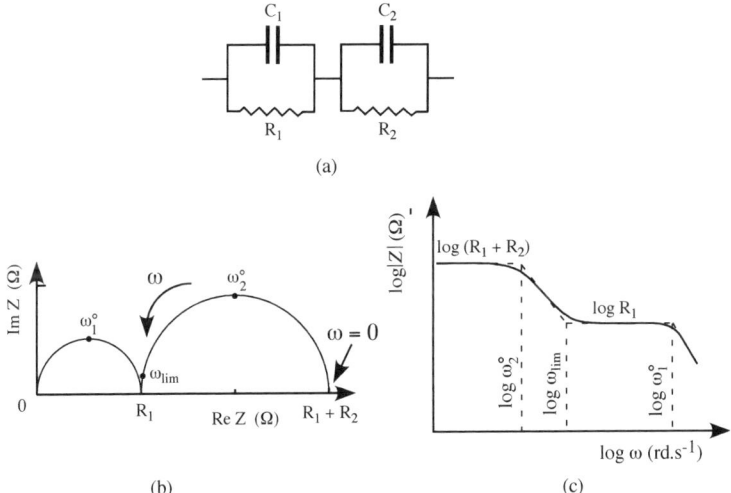

Figure 10.4. *(a) Electrical circuit corresponding to Randles model (R_1/C_1) -(R_2/C_2); (b) Nyquist diagram of the circuit (The arrow indicates increasing angular frequencies); c) Bode plot of the Randles circuit*

The characteristic frequency of the dielectric is then equal to:

$$\omega_1^o = \frac{1}{R_1 C_1} \quad [10.16]$$

In the case of measurements in aqueous solution, the double layer capacitance is generally noticeably higher than C_1. In this case, the impedance modulus equals the electrolyte resistance R_1 with an accuracy better than 1% for experimental angular frequencies between 10 ω_{lim} and 0.1 ω_1^o.

It should be pointed out that the Randles model gives an oversimplified interpretation of electrode behavior. As an example, the center of the semicircle obtained using the Nyquist representation is often not situated exactly on the real axis but below the x-axis. The true capacitance considered in the simple equivalent electrical circuit (Figure 10.4 a) has to be replaced by a "constant phase element" (CPE). The CPE's impedance is given by:

$$\frac{1}{Z} = Q^o (j\omega)^n \quad \text{with } 0 < n < 1 \quad [10.17]$$

A consequence is that the phase angle of the CPE impedance is independent of the frequency and has a value of $-90 \times n$ degrees. In these conditions, the Nyquist plot of a pure resistance in parallel with a CPE gives a semicircle with a centre depressed by an angle of $(1-n) \times 90°$. This behavior is ascribed to various phenomena, i.e. electrode roughness, inhomogenous reaction rates on the electrode surface, non-uniform current distribution, etc.

Obviously, when $n = 1$, the CPE resembles a pure capacitance, it is the case when there is no reaction at the electrodes (blocking electrodes). Then, the low frequency part of the Nyquist diagram is a vertical line, corresponding to the double layer capacitance at the electrodes. In this case, the oblique asymptote at low frequencies in the Bode representation is unlimited (it is equivalent to $R_2 \to \infty$). It is generally the case observed in water analysis. The double layer capacitance depends on the area of the electrodes. If they are very rough and porous (e.g. platinum black[1]), and if the average bump height is larger than the double layer thickness, the oblique asymptote is shifted to lower frequencies and the range of working frequency is improved. This point is discussed elsewhere [FAB 08].

---

1 Platinum black is a powder of very divided platinum metal, which can be formed by electrolysis (in chloroplatinic acid medium) on an electrode. The surface area is then much higher than the geometrical area. Its name is linked to its black color. The electrodes are named "platinized platinum electrodes".

Many other equivalent circuits have been proposed, but they are not considered in this chapter. For a more complete overview of modeling, the reader can refer to specific reviews, for example, [BAR 85, DIA 96].

Whatever the description of the electrochemical cell, the main objective is to determine the electrolyte resistance R_1. As shown in Figure 10.4c, this resistance is obtained for appropriate measurement frequency, i.e. for a frequency, f, in the range $0.1 \times f_o^1 < f_{meas} < 10 \times f_{lim}$, the ideal value being

$$f_{meas} = \sqrt{f_o^1 \times f_{lim}} \quad \text{or} \quad f_{meas} = \frac{1}{2\pi R_1 \sqrt{C_1 C_2}}. \qquad [10.18]$$

10.5. Cell configurations and sources of error

The most important considerations in an impedancemetric sensor are the following: i) the appropriate mounting of the cell, i.e. two-electrode cells, four-electrode cells, or non-contact cells, ii) the proper cell constant, iii) the measurement frequency, iv) the correct choice of materials.

10.5.1. *Types of conductivity cells*

10.5.1.1. *Two-electrode cell (Figure 10.5a)*

Two-electrode cells are the most commonly used design for analysis of liquids. Typically, the electrodes are coated with platinum black and are generally protected by a glass tube. Other body materials (such as polyvinyl chloride (PVC), Teflon, polyaryletheretherketone (PEEK), or even stainless steel) and measuring electrodes (graphite, stainless steel, titanium) are used [BAR 87]. The basic criteria are based on cost and performance requirements. However, according to the characteristics of the analyzed solution, various experimental errors can affect the signal: electrode polarization, field effects, etc.

The coaxial cylinder technique can be also used for liquid analysis. In this case, the cell constant is equal to:

$$K = \frac{\ln(r_1 / r_2)}{2\pi h} \qquad [10.19]$$

where h is the length of the radial part of the current path, r_1 and r_2 are the outer and inner electrodes' radii, respectively.

Planar interdigited cells are not often used for direct measurements in aqueous solutions because the cell constant is not easily determined, although it can be calculated [LVO 06]. There are used, for example, in non-aqueous fluids or in second kind transducer devices (see section 10.6).

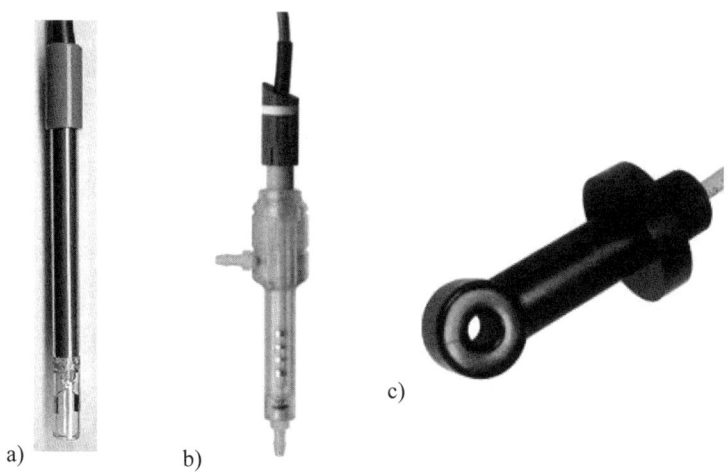

Figure 10.5. *Examples of cell designs: a) two-electrode cell; b) four-electrode design (flow-through type); c) toroidal cell*

10.5.1.2. *Four-electrode cell (Figure 10.5b)*

In a four-electrode cell, an AC is applied to the external electrodes, in such a way that a constant potential difference is maintained between the inner pair of electrodes, which are not polarized, because this voltage measurement takes place with a negligible current. The conductivity will be directly proportional to the applied current as no current flows in the measuring circuit. The electrodes can be of different forms: rings, pins, rectangular, and square-shaped, etc.

The four-electrode design reduces the problem of polarization error (due to the electrochemical reaction at the electrodes), helps to remove the field effect and reduces fouling of the electrodes. This type of probe is superior in operation in almost any solution.

A comparison of the respective advantages and disadvantages of both types of cells is given in Table 10.3.

Advantages	Disadvantages
Two-electrode cell	
Easier to maintain	Fields effects – cell must be positioned in the center of the measuring vessel
Usable with sample changer (no carryover)	Only cells with no bridge between the plates
Low cost	Polarization in high-conductivity samples
Recommended for viscous media or samples with suspension	Calibrate using a standard with a value close to measuring value
	Measurement accurate over two decades
Four-electrode cell	
Linear over a very large conductivity range (0.01 µS to 1000 mS/cm)	Unsuitable for microsamples; depth of immersion 3 to 4 cm
Calibration and measurement in different ranges	Unsuitable for use with a sample changer
Flow-through or immersion type cells	
Ideal for high-conductivity measurements	
Can be used for low-conductivity measurements if cell capacitance compensated	

Table 10.3. *Criteria for the selection of a conductivity cells [RAD]*

The Van der Pauw method is also based on a cell with four electrodes. It was essentially used for electrical characterization of solids (semiconductors, polymers etc), but it can also be used for liquids with low conductivity. In this method, four electrodes are positioned in a cell, with a symmetrical geometry as schematized in Figure 10.6 [DAV 05]. The current passes through electrodes A and B and the measurement is made between C and D. Other choices can be envisaged to obtain an average value allowing a compensation of error of symmetry. The conductivity is then given by the relation:

$$\sigma = \frac{\ln 2}{\pi h R} \qquad [10.20]$$

where h is the height of the electrodes (equal to that of the cell) and R the average value of the measured resistance.

320   Chemical and Biological Microsensors

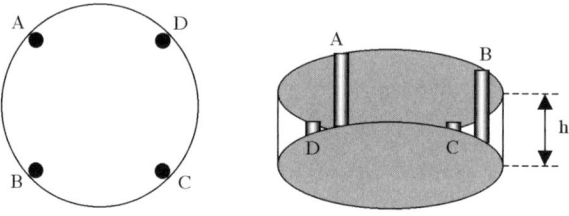

Figure 10.6. *Schema of a Van der Pauw cell (top view and perspective view)*

10.5.1.3. *Non-contact conductivity cell (Figures 10.5c and 10.7)*

Another type of technology is the non-contact (Toroidal) cell, which uses a magnetic field to sense conductivity [OEH 91]. By passing an AC current through the transmitting coil, a magnetic alternative field is generated in the analyzed solution which induces an alternating electric voltage in the liquid. The resulting current flow depends on the ionic concentration. The current in the liquid generates a magnetic alternating field in the receiving coil. The resulting current induced in the receiving coil is measured and used to determine the conductivity value of the solution.

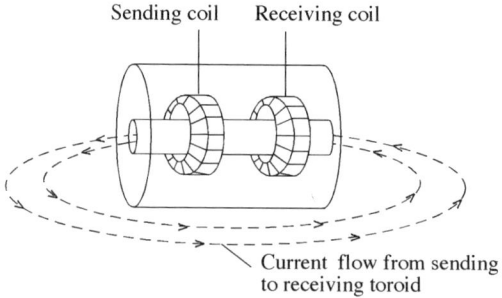

Figure 10.7. *Non-contacting conductivity cell*

The main advantages to this type of cell are: i) no polarization, ii) reduced maintenance and resistance to chemical attack, iii) complete galvanic separation of measurement from medium. Moreover, an anti-fouling coating can be envisaged in such a device, this point is crucial in sea applications. However, these devices exhibit less sensitivity and are larger than contact sensors.

For continuous conductivity monitoring, in a pipe line or an immersed tank, submersible conductivity probes are available. Flow-through conductivity cells can also be used in case of flow measurements or small sample volumes.

10.5.2. Characteristics – specifications

10.5.2.1. Cell constant

Cells of different physical configurations are characterized by their *cell constant*, K. As shown in section 10.1, in a simplified approach, the cell constant K (per cm) of a two-electrode device is defined as the ratio of the distance between the measuring electrodes to the electrode area. The cell formed by two 1 cm^2 surfaces spaced 1 cm apart has a cell constant $K = 1/\text{cm}$. For example, for an observed conductance reading of 200 µS using a cell with $K = 0.1/\text{cm}$, the conductivity value is 20 mS/cm. For low conductivity measurement, probes with low cell constant (0.1/cm or lower) are recommended, for high-conductivity measurement, devices with a high cell constant (1/cm) or even higher if possible, are preferred. Figure 10.1 and Table 10.4 give optimum conductivity ranges for three cell constants.

Cell constant	Optimum conductivity range (µS/cm)
0.1	0.5 to 400
1	10 to 2,000
10	1,000 to 200,000

Table 10.4. *Optimum conductivity range versus cell constant*

However, due to fringe effects at the edges of the electrodes (Figure 10.8), the electric field between both electrodes is not homogenous and the cell constant cannot be accurately calculated solely from the geometric dimensions, and so a cell calibration is necessary.

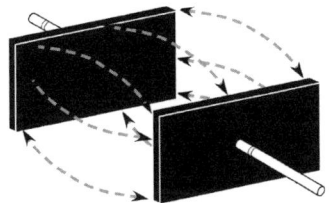

Figure 10.8. *Illustration of the fringe effects in a two-electrode conductivity cell*

Thus, it is recommended that the cell constant is recalculate periodically. This calibration is essential as the cell constant can vary as much as 10% from the nominal cell constant and the actual cell constant may change over time. Calibration frequency depends on the types of conductivity cell and of applications.

Two measurements of the cell resistance must be performed, either manually or automatically. First a calibration liquid with known electrical conductivity, σ_{cal}, is used in order to determine the cell constant:

$$K = R_{cal} \times \sigma_{cal} \qquad [10.21]$$

where R_{cal} is the measured resistance of the cell when it is filled with the calibration liquid. The most commonly used standard solution for calibration is 0.01 mol/l potassium chloride (KCl). This solution has a conductivity of 1412 µS/cm at 25°C. A further measurement of the resistance of the cell is then made when it is filled with the analyzed solution:

$$\sigma_{test} = \frac{1}{R_{test}} K(d) = \frac{R_{cal}}{R_{test}} \sigma_{cal} \qquad [10.22]$$

where R_{test} is the resistance of the cell when it is filled with the test liquid and σ_{test} is the electrical conductivity to be determined. This is a standard procedure for cell constant calibration.

This definition of the cell constant, however, neglects the existence of a *fringe-field effect*, which affects the apparent value of the area, i.e. the section of the current path. Schiefelbein et al. have proposed a calibration technique eliminating this phenomenon [SCH 98]. The measured resistance of the solution between the electrodes is given by the relation:

$$\frac{1}{R_{meas}} = \frac{1}{R_{radial}} + \frac{1}{R_{fringe}} \qquad [10.23]$$

where R_{radial} is the radial part of the measured resistance, R_{fringe} the part due to the fringe effect. For example, with a coaxial geometry the radial part can be written from relation [10.19]:

$$\frac{1}{R_{radial}} = \sigma \frac{2\pi h}{\ln(r_1/r_2)} \qquad [10.24]$$

The derivative vs. the parameter h is a constant and, consequently, the conductivity can be determined from several measurements of the resistance when h is modified with an accurate control:

$$\sigma = \frac{\ln(r_1/r_2)}{2\pi} \frac{d(1/R_{meas})}{dh} \qquad [10.25]$$

Such a determination needs only a relative knowledge of the position h, not an absolute value. Such a method was proposed when using a coaxial cell, but it could be extended to another geometry. The difficulty is the accurate knowledge of the h variation and the accuracy of the resistance measurement.

In summary, the calibration of a conductivity probe is necessary because K is not specifically known and K changes with the electrode ageing.

10.5.2.2. Temperature compensation

The conductivity measurements are temperature dependent. In moderately and highly conductive solutions, temperature correction can be based on a linear equation:

$$\sigma(T) = \sigma(T_{cal})[1 + \alpha(T - T_{cal})] \qquad [10.26]$$

where $\sigma(T)$ is the conductivity at any temperature T (in °C), $\sigma(T_{cal})$ is the conductivity at the calibration temperature, α is the temperature coefficient of solution at T_{cal}. Common α values are listed in Table 10.5.

All conductimeters have either fixed (generally 2%/°C) or adjustable automatic temperature compensation referenced to a standard temperature – usually 25°C. As shown previously, it should be pointed out that the temperature coefficients vary with concentration and temperature (see Figure 10.2), especially for highly resistive solutions.

Substance	Concentration (wt %)	α (%/°C)
HCl	10	1.56
KCl	10	1.88
H_2SO_4	50	1.93
NaCl	10	2.14
HF	1.5	7.20
HNO_3	31	31.0
Drinking water		2.0
Ultrapure water		5.2

Table 10.5. *Examples of temperature coefficients*

10.5.2.3. Measuring frequency

As previously discussed, the measuring frequency should be adapted to the conductivity of the sample. Ideally, the measuring frequency has to be at least 10 times lower than the relaxation frequency f_1^o of the electrolyte contribution and at least 10 times higher than the limit frequency f_{lim}. Generally speaking, low frequencies are applied for low conductivities, where polarization is negligible compared to the solution resistance, and high frequencies are applied for high conductivities, where the solution resistance is low.

The relaxation frequency ascribed to electrolyte contribution is given by:

$$f_1^o = \frac{1}{2\pi R_1 C_1} = \frac{\sigma}{2\pi \varepsilon^o \varepsilon_r} = \frac{1.8 \cdot 10^{12} \sigma}{\varepsilon_r} \quad (\sigma \text{ in S/cm}) \quad [10.27]$$

The upper measurement frequency limit for various solutions can be easily estimated, for example:

– the electrical conductivity of ultrapure water is in the order of 5.5×10^{-8} S/cm. Assuming a relative permittivity of 78 at 25°C, the relaxation frequency f_1^o is 1,270 kHz. According to the previous derivation, the measuring frequency has to be lower than 127 Hz. In this case the commercial analyzers use a frequency of 50 to 65 Hz;

– the conductivity of drinking water is in the range 100–1,000 μS/cm. Then, the limit of the measurement frequency is in the range 0.2 to 2 MHz;

– for a concentrated aqueous solution with a conductivity in the order of 500 mS/cm, the limit of the measuring frequency is 1 GHz.

The lowest frequency limit, i.e. $10 \times f_{lim}$, is not easy to calculate because the associated resistance and capacitance depend widely on the nature of the measuring electrode (essentially on the double layer capacitance C_2). This parameter depends on the ionic strength of the solution, but also on the actual area of the electrodes (see above). For instance, in the case of platinum black electrodes, the coating is highly pulverulent and, consequently, the real area is generally unknown.

The lowest frequency limit can be estimated from the error on the measured resistance of solutions of known conductivity. As shown in Figure 10.9a, the higher the conductivity of the analyzed solution, the higher the required measuring frequency. As an example, for low-conducting solutions, such as pure water, a measurement frequency of 50 Hz is adequate; however, for drinking water (conductivity in the order of 0.5–1 mS/cm) a higher frequency is required.

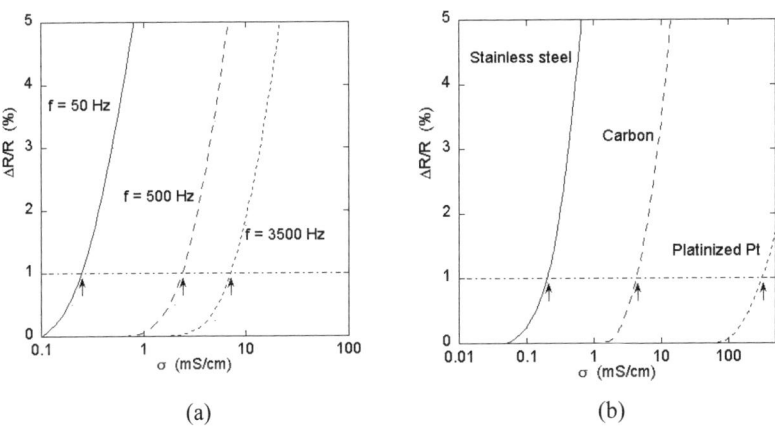

Figure 10.9. *Relative error on the measured resistance as function of the solution conductivity (two-electrode cell, stainless steel electrodes), values from [OEH 77] and [OEH 91]: a) for various measurement frequencies; b) for different electrode materials*

10.5.2.4. *Role of electrode materials*

The nature and surface state of the electrode materials have a noticeable effect upon the electrode polarization, and, consequently, upon the frequency range. Figure

10.9b compares the measurement errors as a function of the solution conductivity for three electrode materials: the usable linear conductivity range of a two-electrode cell, towards high conductivities, is larger by three decades with platinum black electrodes than with stainless steel electrodes. As mentioned previously, platinum black cells are covered with a pulverulent coating of platinum to create a more effective surface area for conductivity measurements. It should be noted that during storage of the cell, the platinum black layer of the plates dehydrates rapidly, and the cell should be dipped permanently in distilled water between two measurements. In many industrial applications, platinum black electrodes cannot be used, therefore, the measurement range in which the cell signal is linear is reduced.

10.5.2.5. *Role of the signal shape*

Most of the measurements are carried out using a sinusoidal signal. As mentioned previously, it must have a low amplitude (in the order of 10 mV). The choice of the frequency has been discussed above. To avoid any interference of the capacitance effect, essentially due to the dielectric properties of the analyzed medium, a triangular signal can be chosen, as proposed recently by Wu *et al.* [WU 07]. During the charge and discharge of the capacitance, using a linear signal ($U = a \times t + b$), the capacitive current part is constant:

$$I_c = C \frac{dU}{dt} = C \times a \qquad [10.28]$$

In these conditions, its first derivative is zero. Thus, the pure resistance of the cell can be simply obtained by the ratio of the first derivatives of voltage and current:

$$R = \frac{dU/dt}{dI/dt} \qquad [10.29]$$

This value is measured between a trough and a peak of the voltage signal, which can have a high amplitude (10 or 20 V). Such a technique (referred to as TWV, for triangular waveform voltage) allows a more accurate measurement of the conductivity.

10.6. Second kind cells

Devices with a second kind transducer are based on an indirect measurement of impedance. In this case, the changes of conductivity of a material according to its surrounding medium are exploited. It should be noted that these devices are not as widely used as the direct-type set-ups, at the present time they are essentially underdevelopment in the laboratory.

Several conductimetric biosensors are based on such a principle. Bio-organisms are deposited on the surface of a conductimetric cell, generally with an interdigitated planar geometry. In this case, the enzymatic reaction must generate ionic species that modify the conductivity at the surface of the sensor. For example, urease is an enzyme reacting with urea by an hydrolysis reaction [GAL 01]:

$$NH_2CONH_2 + 3H_2O \rightarrow 2\,NH_4^+ + HCO_3^- + OH^-$$

Such a reaction transforms a neutral species into several ions. The electrical conductivity is then strongly modified close to the enzyme. All these species diffuse through the layer, where the enzymes are immobilized by different manners, and a quasi-stationary regime is established during which the concentrations of ions are constant. Nitrate can be also analyzed in drinking water with similar devices, using nitrate reductase, an enzyme that reacts in the presence of methyl viologen as mediator [XUE 06]. These compounds are immobilized by a Nafion® membrane (see Chapter 5).

In this field, immunosensors (also named affinity biosensors) can also be mentioned. In these devices, a layer of bio-organism, or bio-molecule, is coated on a planar transducer. The reaction with the target modifies the impedance of the layer, for example, by a antigen–antibody reaction.

We can also mention here devices using a chemical species only, without a bio-organism. For example, Arnold *et al.* have proposed to use a thin layer of hydrogel, in which boronic acid reacts with glucose by a complexation equilibrium, through a bipolar electrodialysis membrane [ARN 00]. The use of a hydrogel matrix has also been proposed to measure the pH of aqueous solutions around the neutrality (i.e. a relative narrow range) [SHE 95]. Such a sensor is advantageous, compared to pH electrodes, because it does not need a reference electrode. Some oxides could be also promising for pH measurements, because the devices are simpler. Hydronium groups at the oxide surface and at the grain boundaries can be responsible for the sensitivity to H^+. For example, a layer of tin oxide (SnO_2) shows a sensitivity to pH in acid mediums, nevertheless the sensitivity was not repeatable for over long durations and such devices could be used only for short applications [ARS 07].

Similarly, the use of a sensitive membrane could be envisaged. For example Cammann *et al.* have proposed the use of membranes, based on PVC materials doped with sensitive molecules, coated on an interdigitated transducer, such as ISCOM (ion-sensitive conductimetric microsensors) [CAM 96]. The main advantage over ISE devices is that no reference electrode is required. Other devices using liquid membranes have also been proposed, for instance, to analyze potassium ions with valinomycin [SHV 01]. The working principle is based on co-extraction exchanges, which supposes that the membranes are not buffered in ionic targets.

Another possibility consists of associating a sensitive membrane with a semiconductor transducer. In this field, a Si:H (amorphous semiconductor based on silicium) has been coated on a NASICON membrane that is sensitive to Na^+ ions [BAR 99]. The transduction is due to a modification of the space charge in the semiconductor, close to the interface with the NASICON membrane, depending on the Galvani potential (see chapters on sensitivity of ISE). Nevertheless, such a device needs a reference electrode connected to the semiconductor. Its only advantage is the possibility of original shapes of devices, compared to those of ISE using solid ionic membranes.

10.7. Summary of practical precautions

Conductimetry covers a wide field of applications, and numerous sources of errors can affect the results if the impedancemetric cells are not used in appropriate conditions.

Highly conducting electrolytes

As indicated previously, with high concentrations, the conductance versus concentration curve may show a maximum. This can lead to ambiguity and major measurement errors.

Electrode polarization

Even though a high frequency AC signal is used in measurements, polarization effect can be reduced or prevented by optimizing the electrode areas, by an appropriate choice of the electrode material or of the measuring frequency, or by using a four-electrode conductivity cell. When blocking electrodes are used, the double layer capacitance plays also a major role at low frequencies. It can be improved with the roughness of the electrode surface (for example, with platinum black).

Dielectric interferences

The choice of the working frequency must also be made as a function of the dielectric properties of the tested material. The frequency must be lower than the characteristic frequency of the dielectric (at least a factor of 10). If a high frequency is needed, a triangular signal can improve the accuracy, as shown above [WU 07].

Improper calibration

As the response is not perfectly linear throughout the whole conductivity range, it is recommended that the probe is calibrated in the same range as that used for the samples being measured.

Cable correction

Cable correction takes into account the cable resistance and the cable capacitance:

$$\chi_{meas} = \frac{\chi_{sol}}{1+(R_{cab} \times \chi_{sol})} \quad [10.30]$$

where χ_{meas} is the measured conductance (S), χ_{sol} is the solution conductance, and R_{cab} is the cable resistance (Ω). The cable resistance influences measurements with two-electrode cells for high-conductivity measurements, but has no influence for four-electrode cells. The cable capacitance influences low-conductance measurements (below 4 µS).

Temperature variation

For automatic temperature compensation devices, it is important to verify that the thermistor reading is accurate. The linear temperature correction is not suitable for many aqueous liquids and the temperature dependency has to be described by non-linear functions.

Cleaning and maintenance

Conductivity measurements can become inaccurate when the measuring electrodes become coated with interfering substances. Cleaning should be undertaken with chemicals and a soft, non-abrasive cloth. The cell may occasionally need replatinization to refresh the cell plates.

10.8. Bilbliography

[ARN 00] ARNOLD F.H., ZHENG W., MICHAELS A.S., "A membrane-moderated, conductimetric sensor for the detection and measurement of specific organic solutes in aqueous solutions", *J. Memb. Science*, vol. 167, p. 227-239, 2000.

[ARS 07] ARSHAK K., GILL E., ARSHAK A., KOROSTYNSKA O., "Investigation of tin oxides as sensing layers in conductimetric interdigitated pH sensors", *Sens. Actuators B*, 127, p. 42-53, 2007.

[BAR 85] BARRAL G., DIARD J.P., LE GORREC B., DAC TRI L., MONTELLA C., "Impédances de cellules de conductivité. I. Détermination des plages de fréquence de mesure de la conductivité", *J. Appl. Electrochem.*, vol. 15, p. 913-924, 1985.

[BAR 87] BARRAL G., DIARD J.P., LE GORREC B., DAC TRI L., MONTELLA C., "Impédances de cellules de conductivité. II. Utilisation du dioxyde de ruthénium comme matériau d'électrodes", *J. Appl. Electrochem.* vol. 17, p. 695-701, 1987.

[BAR 99] BARROIL A., FABRY P., MURET P., "Optimization of an ion microsensor using a crystallized membrane and an amorphous semiconductor as impedancemetric transducer", *Sens. Actuators B*, vol. 59, p. 165-170, 1999.

[CAM 96] CAMMANN K. AHLERS B., HENN D., DUMSCHAT C., SHUL'GA A.A., "New sensing principles for ion detection", *Sens. Actuators B*, vol. 35, p. 26-31, 1996.

[DAV 05] DAVEAU S., PHILIPPE S., RIVIER C., CACHET H., KEDDAM M., TAKENOUTI H., 12th Int. Metrology Congress, Lyon, June 20–23, 2005.

[DIA 96] DIARD J.P., LE GORREC B., MONTELLA C., *Cinétique électrochimique*, Hermann, 1996.

[FAB 08] FABRY P., GONDRAN C., *Capteurs électrochimiques*, Ellipses, 2008.

[GAL 01] GALLARDO SOTO A.M., JAFFARI S.A., BONE S., "Characterisation and optimisation of AC conductimetric biosensors", *Biosens. Bioelectronics*, vol. 16, p. 23-29, 2001.

[LVO 06] LVOVICH V.F., LIU C.C., SMIECHOWSKI M.F., "Optimization and fabrication of planar interdigitated impedance sensors for highly resistive non-aqueous industrial fluids", *Sens. Actuators B*, vol. 119, p. 490-496, 2006.

[OEH 77] OEHME F., *GIT Fachz. Lab.*, vol. 21, p. 15-21, 1977.

[OEH 91] OEHME F., "Liquid electrolyte sensors: potentiometry, amperometry, and conductometry", Chapter 7 in GÖPEL W., HESSE J., ZEMEL J.N. (eds), *Sensors, a Comprehensive Survey*, vol. 2, VCH, p. 239-339, 1991.

[RAD 04] *Conductivity Theory and Practice*, Radiometer Analytical SAS, 2004 (www.radiometer-analytical.com)

[SCH 98] SCHIEFELBEIN S.L., FRIED N.A., BUHLMANN P., RHOADS K.G., SADOWAY D.R., "A high-accuracy, calibration-free technique for measuring the electrical conductivity of liquids", *Rev. Sci. Instrum.*, vol. 69, p. 3308-3313, 1998.

[SHE 95] SHEPPARD N.F., LESHO M., MCNALLY P., SHAUN FRANCOMACARO A., "Microfabricated conductimetric pH sensor", *Sens. Actuators B*, vol. 28 p. 95-102, 1995.

[SHV 01] SHVAREV A.E., RANTSAN D.A., MIKHELSON K.N., "Potassium-selective conductometric sensor", *Sens. Actuators B*, vol. 76, p. 500-505, 2001.

[WU 07] WU J., STARK J.P.W., "A high accuracy technique to measure the electrical conductivity of liquids using small test samples", *J. Appl. Phys.*, 101(5)-054520, p. 1-7, 2007.

[XUE 06] XUEJIANG W., DZYADEVYCH S.V., CHOVELON J.M., JAFFREZIC-RENAULT N., LING C., SIQING X., JIANFU Z., "Conductometric nitrate biosensor based on methyl viologen/Nafion®/nitrate reductase interdigitated electrodes", *Talanta*, vol. 69, p. 450-455, 2006.

List of Authors

Loïc BLUM
Institut de Chimie et Biochimie Moléculaires et Supramoléculaires (ICBMS)
Laboratoire de Génie Enzymatique et Biomoléculaire
Claude Bernard University – Lyon I
Villeurbanne
France

Philippe BREUIL
Laboratoire Sciences des Processus Industriels et Naturels
Ecole des Mines de Saint-Etienne
France

Jacques BUFFLE
Honorary Professor
CABE
Dept of Inorganic, Analytical and Applied Chemistry
University of Geneva
Switzerland

Serge COSNIER
Département de Chimie Moléculaire (DCM)
Joseph Fourier University
Saint-Martin-d'Hères
France

Pierre FABRY
Former Professor
Joseph Fourier University
L.E.P.M.I.
Saint Martin d'Hères
France

Jean-Jacques FOMBON
Former Research & Development Director
Radiometer Analytical SAS
Villeurbanne
France

Jacques FOULETIER
Joseph Fourier University
L.E.P.M.I.
Saint-Martin-d'Hères
France

Chantal GONDRAN
Département de Chimie Moléculaire (DCM)
Joseph Fourier University
Saint-Martin-d'Hères
France

Jean-Pierre GOURE
Emeritus Professor
Laboratoire Hubert Curien
Saint-Etienne
France

Nicole JAFFREZIC-RENAULT
LSA
Claude Bernard University
Villeurbanne
France

Claude MARTELET
Independent Biotechnology Professional
Former Professor
Claude Bernard University
Villeurbanne
France

Bernard MICHAUX
Former Scientific Director
EXERA
Paris
France

Jean-Claude MOUTET
Département de Chimie Moléculaire (DCM)
Joseph Fourier University
Saint-Martin-d'Hères
France

Annie PRADEL
Institut Charles Gerhardt
"Physicochimie de Matériaux Désordonnés et Poreux" Team
University of Montpellier II
France

Eric SAINT-AMAN
Département de Chimie Moléculaire (DCM)
Joseph Fourier University
Saint-Martin-d'Hères
France

Marie-Louise TERCIER-WAEBER
CABE
Dept of Inorganic, Analytical and Applied Chemistry
University of Geneva
Switzerland

Alain WALCARIUS
Laboratoire de Chimie Physique et Microbiologie pour l'Environnement (LCPME)
Henri Poincaré University – Nancy I
Villers-les-Nancy
France

Index

A

absorption, 214-215, 218
acid error, 17
activity, 2, 9-16, 19-20
  coefficient, 9, 20
adsorption, 119-120, 131-132, 141-142, 155-156
AdSV, 245-246, 249, 255
affinity biosensors, 193
agarose, 257, 260, 270
alkaline error, 17
amperometric, 64, 68, 70-71, 73, 76
  biosensors, 142
  detection, 155
  sensor, 36, 115
amperometry, 13
analysis of water, 235
angular frequency, 313, 314
antibodies, 5
antigen/antibodies, 193
antigens, 5, 7
aquatic media, 237, 240, 243, 258, 267, 273
artificial nose, 6
ASV, 245-246, 249, 255
asymmetry potential, 16, 84
avidine, 142

B

back-propagation, 305
band diagram, 174
BAW, 6
Beer-Lambert law, 289, 291
behavior model, 287-289, 303, 305
BIAcore™, 8
biochemical sensor, 5
BioFET, 140, 193, 195
bioluminescence, 221, 227-228
Bioluminescent biosensors, 228
biosensors, 36, 138, 146
blocking electrode, 316, 328
Bode plot, 313-315
buffer, 15
  solutions, 41

C

cable correction, 329
calibration, 18, 26, 28, 33, 36, 38, 40-43, 288, 309, 321-323, 327
  curves, 88
  procedure, 26
calomel electrode, 11, 13
cell constant, 308-309, 316-317, 321-322
cellulose acetate, 259
CHEMFETS, 12
chemical potential, 9, 51-52
chemically modified electrodes, 250
chemiluminescence, 221, 227, 229
chemiluminescent biosensors, 227
chemisorption, 5
chemometrics, 50

chiral electrode, 117
Clark, 119, 133, 149, 153
  electrode, 6, 14
CLS method, 293
CO_2-selective electrode, 91, 92
coaxial cylinder technique, 317
coefficient of selectivity, 87
coenzyme, 138, 155, 156
complexation, 49, 63, 69, 130
complexes, 234, 239, 241-243, 248
conductimetric, 47
  sensors, 36
conductimetry, 8
conducting polymers, 48-49, 68, 122, 129
conductivity, 36, 307-312, 317-327
conservation constraints, 29
constant phase element, 316
CPE, 316
cyclic voltammetry, 157, 158

D

Debye-Hückel theory, 311
degree of complexation, 242
dehydrogenase, 153, 156
desertion situation, 176
detection limit, 70-72, 75-76, 86-87, 95, 97, 108
dielectric, 311, 314-315, 326, 328
  interferences, 328
differential pulse, 245
diffusion, 55-57, 64-66
direct
  electrocatalytic detection, 118
  reduction, 245
disposable sensors, 122
dissolved oxygen, 257
DME, 248
DNA biosensors, 226
double layer capacitance, 313, 316, 325, 328
dropping mercury electrode, 248

E

Eadie-Hofstee, 144-145
Eisenman, 56, 63

electrical
  conductivity, 308-310, 321, 323, 326
  connection, 159-160
  mobility, 310
electroactive
  polymers, 121
  species, 238
electrocatalysis, 135-136
electrochemical
  chain, 81, 83, 95, 106
  mobility, 64, 310
  potential, 51
electrochemiluminescent sensors, 229
electrode
  first kind, 10
  materials, 325
  second kind, 11
  third kind, 12
electrodes for alkaline metals, 93
electroenzymatic
  amplification, 152
  biosensor, 146
electronic tongue, 200
ENFET, 189, 192
enzymatic biosensor, 142, 223
enzyme, 136, 138, 140-151, 157-158
EOS structure, 177
error matrix, 290
extraction work, 175, 178

F

factor analysis, 296-297, 305
feasibility, 26-27
Fermi level, 175, 177-178
FIA, 59, 227, 250, 261
Fick, 64
fidelity, 3
field effect, 317-318, 322
  transistors, 139-140
FIM, 74, 76
first generation, 148
flow-through conductivity, 321
fluorescence, 213, 216, 221, 223, 226
fluoride ions, 95, 104
fouling, 259, 265, 318, 320
four-electrode cell, 317, 318, 329

FPM, 74-75
free enthalpy, 9
fringe effects, 320-321

G-H

Galvani potential, 10, 15
gas-electrode, 91
GIME, 260
glass, 81, 89-91, 93, 99, 104-108
  electrode, 14, 81, 89-91, 104
glassy carbon, 249
glucose, 148-153, 158
  oxidase, 148, 151, 157
glutaraldehyde, 141, 158
grafting, 141-142, 158, 185-187
H_2O_2, 132, 147-151, 155, 158
hanging mercury drop electrode, 248
haptens, 5, 7
Henderson, 57
HMDE, 248
hydrogel, 15, 91
hysteresis, 17, 20

I

ILS, 291, 297, 299
IMFET, 193-194
immobilization, 131, 139-142, 149, 153, 156, 158
impedance, 307, 312-316, 325-326
impedancemetric, 140
indirect
  amperometric analysis, 118
  transduction, 308
industrialization, 26-27, 31
inert complexes, 241-242
influence parameters, 29
inhibition, 192
intensity modulation, 213, 215
interference, 49, 76, 84, 90, 118, 129-134, 140, 150, 152, 271
  coefficient, 28
interfering
  ion, 87, 88
  species, 128, 133-134, 150
interferometry, 216, 218

internal reference, 89, 96, 99, 103-104, 105, 106, 107
ion exchange, 128
  polymers, 121
ion selective electrode, 81
ionic
  electrode, 81-85, 91
  strength, 324
ionometry, 6, 14, 18
ionophores, 97
ion-sensitive membrane, 82, 87, 93, 96-99, 103-104, 107-108
ISE, 46, 50, 55-56, 59, 60, 72, 73, 81, 82, 83, 86, 87, 90, 92, 95, 97, 98, 99, 100, 105, 139, 173
ISFET, 12, 173, 179, 197
  fabrication, 180
ISFETs manufacturers, 197
IUPAC, 18, 28, 50, 71, 74, 86

J-L

junction potential, 83
knowledge model, 288-289
Kohlrausch's law, 311
labile complexes, 241-242
lab-on-a-cable, 125
lanthanum fluoride, 95
least squares, 290, 302, 305
lifetime, 28-29
limiting molar conductivity, 311
lineweaver-Burk, 144-145
luminescence, 7, 219

M

macro-electrodes, 239, 241242, 248, 251, 255-256, 260
maintenance, 27, 29, 33, 36-39
marketing, 29-30, 32
  department, 29
matched potential, 76
matrix effect, 2, 20
measuring
  chain, 39
  frequency, 323-324, 327
mediator, 128, 135-136, 148, 158, 160

membrane, 81, 87-89, 93, 95, 100, 103, 106, 108
  ionosensible, 97, 104
  potential, 84
mercury
  electrode, 236, 238, 244
  film electrode, 248
MFE, 248
Michaelis-Menten, 144-146, 188
micro total analysis system (µ-TAS), 5, 22
microchips, 127
microelectrodes, 241
microsensor, 105-109
miniaturizing, 139
mixed
  conduction membranes, 103
  potential, 54
MLR, 293-294, 297
modeling error, 294-295
modified electrodes, 115-124, 127-128, 134
modulus, 313, 314-316
molar
  conductivity, 310
  ionic conductivity, 310
molecular assembling, 142
monomolecular films, 119
MOS structure, 174
MOSFET, 178, 201
multidetection, 200
multimode fibers, 212
multimolecular films, 119-120, 122
multi-sensor, 255
multivariable function, 288
multivariate analysis, 288

N

NADH, 148, 150, 155-156, 159
NADPH, 148-150, 159
Nafion, 121, 128-151, 259, 327
NASICON, 89, 96
natural waters, 241, 245, 248, 250-251, 253, 259, 275
Nernst, 58
  equation, 10
  law, 6, 17-18, 184

nernstian, 56, 61, 63, 71, 74-75
  law, 85
neural networks, 303-305
NH_3 sensors, 223
Nikolskii-Eisenman equation, 4, 6, 19
noise, 71
  modeling, 294, 297
non-catalytic bioreceptors, 225
non-contact cell, 319
non-labile complexes, 241
Nyquist
  diagram, 313-316
  plot, 313, 316

O

Onsager, 311, 312
opt(r)ode, 7
optical fiber, 209, 211
  biosensors, 221
  sensor, 213-214, 220, 223
optical microsensors, 213
optode, 7
Ostwald's dilution law, 312
overfitting, 294, 296-297, 305
oxide-semiconductor, 174-175
oxidoreductases, 147
oxygen
  reduction, 238
  sensors, 222

P

partial least squares, 298
peak
  current, 237-238, 241, 274
  potential, 237-238, 241, 274
permittivity, 314
perm-selective membrane, 118
pH, 6, 14
  electrode, 31, 90-92, 99, 104
  measurements, 33, 39
  response, 183
  sensors, 33, 221
phase modulation, 216
pH-ISFET, 190
photodetectors, 214
physisorption, 5

planar guides, 212
platinum black, 324, 327
PLS calibration, 298
polarization modulation sensors, 217
polyaniline, 122, 133, 141
polymer membrane, 98, 105, 107, 109
polymeric membranes, 97
polypyrrole, 122, 129, 133, 141, 150
polysiloxanes, 97
polythiophene, 122
polyurethans, 97
polyvinyle chloride, 102
potentiometric, 46
   coefficient, 87
   sensor, 46, 52, 59, 76, 81-82, 85-89, 91, 99
potentiometry, 9
practical detection limit, 28
pre-concentrating, 118, 127, 131, 237, 245, 249, 250-251, 258, 265, 271, 274
prediction, 288
   error, 302
   matrix, 292-293, 297
principal component regression, 298
PSA, 265
PVC, 97, 100-102

Q-R

quality control, 31
quinhydrone electrode, 12
Raman spectrometry, 7
Randles cell, 314
recognition, 45, 63, 68
recycling, 152-153
re-dissolution, 245, 258, 271-272
redox
   mediator, 148
   polymers, 121
reference, 46, 53, 56-58, 60, 72
   electrode, 83, 177, 179, 235, 238, 252, 255-266
reflectance, 219
refractometry, 219
relaxation frequency, 324
reliability, 28, 32, 38-39

repeatability, 28
   in a batch, 3
reproducibility, 117, 120, 127, 197
residuals, 299, 300-301
resolution, 3, 71
response
   curves, 28
   law, 28
   range, 28-29
reticulation, 141, 160
rotogravure, 124

S

salinity, 310
saturation situation, 179
SAW, 6
scientific survey, 26
screen printing, 116
second generation, 148
second kind
   transducer, 325
   transducer devices, 318
   transduction, 308
selective, 4
   electrodes, 33
   permeability, 132
selectivity, 28, 45-51, 58, 59, 62-66, 68, 73-76
   coefficients, 19, 50, 59, 63, 87, 100, 129, 133-134
selenide, 95
sensitivity, 4, 45, 49, 51, 53, 55, 59-66, 69-70
   limit, 244
Severinghaus electrode, 6
signal shape, 326
silver
   chalcogenide, 89, 103
   halide, 85, 86
silver-silver chloride electrode, 11, 13
single mode fibers, 212
SMDE, 248
solid state electrode, 249
solubility product, 62
special fibers, 212

specific, 4
spectrometry, 7, 218
spherical electrode, 239
SSM, 73
stability, 28, 117, 142
standard, 29, 31, 32, 39-43
   hydrogen electrode, 11-13
   potential, 10
   solution, 20
static mercury drop electrode, 248
substrate, 123-124, 135, 143-147
sulphide, 95, 103
surface
   plasmons, 215
   potential, 175, 178
SWASV, 251-252, 258, 263, 265, 267, 272
synaptic coefficients, 304-305

T

TDS, 309
technological
   development monitoring, 26
   survey, 26-27
temperature
   compensation, 323
   effects, 197
third generation, 148
threshold voltage, 179-180
total dissolved solids, 309
trace elements, 233-234, 237, 244-245, 252, 255, 273

traceability, 29, 41-43
transducer, 115, 138-139, 140-142, 146
transducing function, 45
transduction, 45-49, 60
   modes, 46
trapping, 142
triangular waveform voltage, 326
two-electrode cell, 308, 317, 319, 325

U-Z

ultramicroelectrodes, 126
uncertainty, 29, 40-42
underfitting, 296
V cycle, 33
valinomycin, 98
Van der Pauw method, 319
VIP, 256, 269
VOC, 30
voice of the client, 30
voltammetric
   cell, 247
   curve, 236
   probes, 233-234, 247, 264, 267
voltammogram, 13, 236
water quality, 307, 310
waters, 307-309, 312
wavelength modification sensors, 217
working
   electrode, 235-236, 247, 255-256, 259, 266-267
   frequency, 314, 316, 328
zeolite, 137